Organic Farming in Plantation Crops

THE EDITORS

Dr. V. Krishnakumar is currently Principal Scientist (Agronomy) & Head, ICAR-Central Plantation Crops Research Institute, Regional Station, Kayamkulam, Kerala. He joined the Agricultural Research Service of the Indian Council of Agricultural Research as Senior Scientist during December 2001. After completion of his Ph.D programme in Agronomy from Tamil Nadu Agricultural University, Coimbatore, he has worked as Assistant Professor, Kerala Agricultural University; Assistant Agronomist, Rubber Research Institute of India, Kottayam (Rubber Board); and Senior Scientist, Indian Cardamom Research Institute (Spices Board) from 1982 onwards. He has served as Assistant Editor, Journal of Natural Rubber Research of RRII (1987-1989) and Editor, Journal of Plantation Crops (ISPC) during 2006-2011. He is a recipient of the ICAR Outstanding Team Research Award for Horticulture (2005-06), R L Narasimha Swamy Award (2010) and Fellow of the Indian Society for Plantation Crops. His main areas of research /specialization are cropping system for resource use efficiency, organic farming in plantation and spice crops. He has edited eight books and published more than 75 research and technical articles apart from contributing chapters in various books, technical bulletins *etc.*

Dr. Pallem Chowdappa received M.Sc. in 1980 from Sri Venkateswara University, Tirupathi, Ph.D in 1985 from Mangalore University, Mangalore, Karnataka and post doctoral research at CABI Bioscience, U.K. He joined as Scientist-SI in 1985 at ICAR-Central Plantation Crops Research Institute, Kasaragod, Kerala and was elevated to Principal Scientist in 2006 at Indian Institute of Horticultural Research, Bangalore. Dr. Chowdappa served as Scientist-in-Charge, Central Plantation Crops Research Institute Research Centre, Hirehalli and Head, Central Horticultural Experimental Station, Hirehalli from December, 2000 till April, 2006. He became Director, Central Plantation Crops Research Institute, Kasaragod in September, 2014. Dr. Chowdappa is specialized in molecular plant pathology and has over 30 years of research experience in molecular characterization and management of *Alternaria, Colletotrichum* and *Phytophthora* associated with diseases of horticultural crops. He attended international training program on 'Oomycetes bioinformatics' at Virginia Tech, USA in 2014. Dr. Chowdappa was awarded DFID fellowship for Post-Doctoral research at CABI Bioscience, UK in 1998 . Dr. Chowdappa has published more than 120 research papers in leading national and international journals, 12 books, 35 technical bulletins, 42 book chapters and 65 experimental manuals. He is a fellow of Scientific Academia and has won several awards of repute. He is also president of many scientific societies in India.

Organic Farming in Plantation Crops

— Editors —

V. Krishnakumar

P. Chowdappa

2017

Daya Publishing House®
A Division of

Astral International Pvt. Ltd.
New Delhi – 110 002

ISBN: 978-93-86071-70-5 (International Edition)

Publisher's Note:

Published by   : Daya Publishing House®
                        A Division of
                        Astral International Pvt. Ltd.
                        – ISO 9001:2015 Certified Company –
                        4736/23, Ansari Road, Darya Ganj
                        New Delhi-110 002
                        Ph. 011-43549197, 23278134
                        E-mail: info@astralint.com
                        Website: www.astralint.com

त्रिलोचन महापात्र, पीएच.डी.
एफ एन ए, एफ एन ए एस सी, एफ एन ए ए एस
सचिव एवं महानिदेशक

TRILOCHAN MOHAPATRA, Ph.D.
FNA, FNASc, FNAAS
SECRETARY & DIRECTOR GENERAL

भारत सरकार
कृषि अनुसंधान और शिक्षा विभाग एवं
भारतीय कृषि अनुसंधान परिषद
कृषि एवं किसान कल्याण मंत्रालय, कृषि भवन, नई दिल्ली 110 001

GOVERNMENT OF INDIA
DEPARTMENT OF AGRICULTURAL RESEARCH & EDUCATION
AND
INDIAN COUNCIL OF AGRICULTURAL RESEARCH
MINISTRY OF AGRICULTURE AND FARMERS WELFARE
KRISHI BHAVAN, NEW DELHI 110 001
Tel.: 23382629; 23386711 Fax: 91-11-23384773
E-mail: dg.icar@nic.in

Foreword

World over, agriculture has made many technological advances, increasing its level of productivity of various crops and other related enterprises, however, the successful growth of the sector has been accompanied by widespread concerns over food safety issues, damage to soil health, environmental issues and the loss of biodiversity from intensified agriculture. This has paved the way for thinking differently and adoption of farming practices, which promote and enhance agro-ecosystem health; enhance biodiversity, biological cycle and soil biological activity. Organic farming, being a unique production management system, is the best choice that we can make for our environment, animals and our own health. As organic farming is assuming greater significance both in developed and developing countries, and area is also increasing, one of the major challenges in the organic farming is to develop the package, which do not sacrifice the yield gains and at the same time utilizes natural resources fruitfully.

Plantation crops are high value commercial crops that play a vital role in improving economy, employment generation and poverty alleviation of millions of people, especially in the rural sector. These crops meet a wide variety of human needs such as food, fuel, oil, industrial raw materials, beverages and confectionary items. The growth habit and potential for supply of large quantities of recyclable biomass from the plantation crops makes it possible to adopt organic cultivation practices. Adoption of organic farming offers solutions, and to take advantage of them, it is necessary to keep the farmers, who adopt such practices, informed about the impact it can create on safer environments and foods that they produce. In this context the book on "Organic Farming in Plantation Crops" is being brought out. This book provides valuable information on the Status of Organic Farming in the world, besides chapters on Organic Plant Protection Technologies, Field Level Scenario and Future Strategies, Quality Control Standards and Organic Certification as well as Transition towards Organic Farming: Policies, Problems and Prospects. The other chapters cover various agro management practices to be adopted for organic farming, input management and biomass recycling, soil health management, plant health management, etc. of important plantation crops viz., coconut, arecanut, cocoa, cashew, coffee, tea, spices, and oil palm.

I compliment editors in compiling and editing this book. I am sure that this book will be of immense use to all those interested in organic farming of various plantation crops.

Dr. T. Mohapatra
Secretary, DARE and Director General,
ICAR, New Delhi

Preface

The increase in demand and consumption of organic foods has mainly been due to an increasing number of consumers, both in the developed and developing countries, associating significant environmental, biodiversity, ethical and food quality and safety benefits with organic and/or organic food production system. Organic farming has received wider acceptance among diverse categories of farmers operating in different parts of the world under varied agro-ecological and financial conditions. Currently, organic agriculture is being practiced in 172 countries in an area of 43.7 million hectares during 2014 with Oceania leading the list followed by Europe, Latin America, Asia, North America and Africa. The premium price for the organic foods in the international market enables the farmers to realize higher returns from organic production systems, which shows annual average growth rate of 20-25 per cent. In the past three decades, standards have been set by different countries for organic production, (a process certification and not the product certification), which needs careful understanding, and supervision for adoption of organic farming. However, the farmers had to face several problems while converting their farms from conventional to organic farming, some of which include non-availability of quality organic inputs, lack of storage and marketing facility.

Since, many of the plantation crops are consumed either directly (e.g. coconut, cocoa, coffee, tea, spices, cashew, oil palm) or used for industrial purposes (e.g. coconut oil and palm oil), organic production to meet consumer's needs is becoming imperative. In order to adopt organic farming, it is necessary that such technologies without the use of any inorganic fertilizers and plant protection chemicals are made available to the farmers. Many technologies have been generated but, there continues to be considerable gap between the needs and availability. In this context, the available information is being brought together for the benefit of farming community. This book on **"Organic Farming in Plantation Crops"** contains 13 chapters covering the organic farming practices of important plantation crops such as coconut, arecanut, cocoa, cashew, coffee, tea, oil palm and various spices.

Besides, chapters on Global and Indian Scenario of Organic Farming; Organic Plant Protection Technologies; Field Level Scenario and Future Strategies of Organic Farming Practices in Palms and Cocoa; Quality Control Standards and Organic Certification for Plantation Crops and Transition towards Organic Farming: Policies, Problems and Prospects are also included.

The editors are grateful to all the contributors to the book for providing latest information on diverse aspects of organic farming in various plantation crops. We hope that this book will prove to be a valuable source of information to all those involved in organic farming including scientists, developmental personnel, policy makers, NGOs and farmers. It is hoped that this comprehensive treatise will stimulate and motivate more intensified research, accelerate developmental efforts, favourable policy initiatives and spread of organic farming of plantation crops at the grass roots level for the production of safe food under healthy environmental conditions.

V. Krishnakumar

P. Chowdappa

Contents

List of Contributors

Alka Gupta
ICAR-Central Plantation Crops Research Institute, Kasaragod – 671 124, Kerala

Anithakumari, P.
ICAR-Central Plantation Crops Research Institute Regional Station,
Kayamkulam – 690 533, Kerala

Chandrika Mohan
ICAR-Central Plantation Crops Research Institute, Regional Station,
Kayamkulam – 690 533, Kerala

Chandran, K.P.
ICAR-Central Plantation Crops Research Institute, Kasaragod – 671 124, Kerala

Chowdappa, P.
ICAR-Central Plantation Crops Research Institute, Kasaragod – 671 124, Kerala

Durairaj, J.
UPASI Tea Research Institute, Regional Centre, Munnar – 685 612, Kerala

George V. Thomas
Council for Food Research and Development, Perinjottakkal-689 692, Kerala

Hamza, S.
ICAR-Indian Institute of Spices Research, Kozhikode – 673 012, Kerala

Jaganathan, D.
ICAR-Central Plantation Crops Research Institute, Kasaragod – 671 124, Kerala

Jayasekhar, S.
ICAR-Central Plantation Crops Research Institute, Kasaragod – 671 124, Kerala

Jose, C.T.
ICAR-Central Plantation Crops Research Institute Regional Station,
Vittal – 574 243, Karnataka

Josephrajkumar, A.
ICAR-Central Plantation Crops Research Institute Regional Station,
Kayamkulam – 690 533, Kerala

Kamala Bai, S.
Krishi Vigyan Kendra, Chandurayanahalli, Kalya – 562 120, Karnataka

Krishnakumar, V.
ICAR-Central Plantation Crops Research Institute, Regional Station,
Kayamkulam – 690 533, Kerala

Mathew Sebastian
Indocert, Thottumugham, Aluva – 683 105, Kerala

Murali Gopal
ICAR-Central Plantation Crops Research Institute, Kasaragod – 671 124, Kerala

Murugesan, P.
ICAR-Indian Institute of Oil palm Research, Research Centre, Palode,
Pacha-695 562,Kerala

Pushpalatha, P.B.
Cashew Research Station, Madakkathara, Kerala Agricultural University,
Thrissur – 680 651, Kerala

Ravi Bhat
ICAR-Central Plantation Crops Research Institute, Kasaragod – 671 124, Kerala

Radhakrishnan, B.
UPASI Tea Research Institute, Nirar Dam P.O., Valparai – 642 127, Tamil Nadu

Raghuramulu, Y.
Central Coffee Research Institute, Coffee Research Station P.O.,
Chikmagalur District – 577 117, Karnataka

Sajitha, M.P.
Indocert, Thottumugham, Aluva – 683 105, Kerala

Srinivasan, V.
ICAR-Indian Institute of Spices Research, Kozhikode – 673 012, Kerala

Subramanian, P.
ICAR-Central Plantation Crops Research Institute, Kasaragod – 671 124, Kerala

Sujatha, S.
ICAR-Indian Institute of Horticultural Research, Hessaraghatta Lake Post,
Bengaluru – 560 089, Karnataka

Thankamani, C.K.
ICAR-Indian Institute of Spices Research, Kozhikode – 673 012, Kerala

Thamban, C.
ICAR-Central Plantation Crops Research Institute, Kasaragod – 671 124, Kerala

Chapter 1

Organic Farming: Global Scenario

☆ *V. Krishnakumar and P. Chowdappa*

1. Introduction

The ever increasing demand for safe and healthy food, free from pesticide and other residues due to indiscriminate use of agrochemicals giving concerns on environmental pollution, are the major reasons responsible for the interest in alternate systems of farming in modern agriculture. The global market for organic food is about US $ 30 billion and projected to grow to 100 billion US $ in another five years. The demand for the organic food is increasing both in the developed and developing countries with annual average growth rate of 20-25 per cent. The premium price for the organic foods in international market enables the farmers to realize higher returns from organic production systems. Therefore, it is due to several advantages of organic farming over the modern agricultural practices, which is often high input demanding, that farmers across the globe are converting in to organic farming. This system of managing agricultural holdings restricts the use of chemical fertilizers, pesticides, growth regulators and livestock feed additives.

Varied crop farming practices such as cultural, mechanical and biological methods are practiced to achieve sustainable agricultural production by encouraging and enhancing biological cycles within farming system involving soil flora and fauna, plants and animals. The holistic approach also provides social and ecological advantages with the conservation of soil and water and enhancing long term fertility of soil. One of the basic principles of soil fertility management in organic systems is that plant nutrition depends on 'biologically-derived nutrients' instead of using readily soluble forms of nutrients supplied through fertilizers and therefore, the approach should be to 'feed the soil to make it living' rather than 'feeding the plants'. Often, organic farming has been criticized on the grounds that with the application of organic inputs alone, farm productivity and profitability might not be improved as the availability of organic sources is highly restricted. Though availability of organic resources is limited, it could be possible to raise various kinds of organic manure crops and prepare compost in the farm itself and apply to the crops.

Organic farming relies up on the basic concept of living soil. Therefore, the role of soil flora and fauna is given due recognition in this system of production as they are the prime drivers for various processes resulting in enhanced physical, chemical and biological health of soil. The emphasis in organic farming will be on the adoption of agro-management practices based on soil biodiversity aimed at enhancing the natural nutrient cycles and utilization of specific soil-plant-microbial associations in managing the soil in organic production. Organic farming is often considered as one that uses only organic inputs to meet the requirement of nutrients and management of pests and diseases. In fact, it is a specialized form of diversified agriculture, wherein farming is managed using local on farm resources to the extent possible.

In the larger context, organic farming is a production system which relies upon soil management techniques (*e.g.* mulching), crop rotations, various cropping systems (*e.g.* inter cropping), agro forestry (where woody perennials are grown in association with crop/livestock), recycling of crop residues, animal and green manures, legumes, recycling of on farm waste resources (*e.g.* fodder, organic wastes *etc*),mechanical cultivation, biofertilizers to maintain soil productivity, to supply nutrients, and to control weeds *etc*. The system also relies upon adoption of various other means of pest and disease management practices including cultural, mechanical and biological measures than use of any synthetic chemicals.

2. Organic Agriculture: Worldwide

Organic agriculture is being practiced in 172 countries in an area of 43.7 million hectares of organic agricultural land (including in-conversion areas) during 2014 with Oceania leading the list with 17.3 million hectares(40 per cent of the world's organic agricultural land) followed by Europe (11.6 million hectares, 27 per cent), and Latin America (6.8 million hectares, 15 per cent) followed by Asia (3.6 million hectares, 8 per cent), North America (3.1 million hectares, 7 per cent) and Africa (1.3 million hectares, 3 per cent). Among the countries with the most organic agricultural land, Australia (17.2 million hectares), Argentina (3.1 million hectares), and the United States (2.2 million hectares) are the top three leaders.

The organic share of total agricultural land is around one per cent in the countries surveyed by Helga and Lernoud (2016) and this is the highest in Oceania (4.1 per cent) followed by Europe with 2.4 per cent and Latin America with 1.1 per cent. In the other regions (Africa, Asia and North America), the share is less than one per cent. By region, the highest organic shares of the total agricultural land are in Oceania (40 per cent) and in Europe (27 per cent). In the European Union, 5.7 per cent of the farmland is organic. However, some countries reach far higher shares: Falkland Islands (36.3 per cent), Liechtenstein (30.9 per cent), and Austria (19.4 per cent). There were 2.3 million organic producers in the world during 2014 of which 40 per cent are in Asia, followed by Africa (26 per cent) and Latin America (17 per cent), Europe(15 per cent) and North America as well as Oceania (1 per cent each).

Table 1.1: Organic Agriculture 2016: Key Indicators and Top Countries

Indicator	World	Top Countries
Countries with organic activities	2014: 172 countries	New countries: Kiribati, Puerto Rico, Suriname, United States Virgin Island
Organic agricultural land	2014:43.7 million ha (1999: 11 million ha)	Australia (17.2 million ha-2013), Argentina (3.2 million ha), US (2.2 million ha-2011)
Organic share of total agricultural land	2014: 0.99 per cent	Falkland Islands (Malvinas)-36.3 per cent Liechtenstein -30.9 per cent Austria-19.4 per cent
Wild collection and further, non-agricultural areas	2014: 37.6 million ha (1994: 4.1 million ha)	Finland (9.1 million ha), Zambia(6.8 million ha), India (4.0 million ha)
Producers	2014: 2.3 million (1994:0.2 million)	India (650,000-2013), Uganda (190,552), Mexico (169,703-2013)
Organic market size	2014: 80 billion USD (1999: 15.2 billion US dollars)	US (35.9 billion, 27.1 billion euros), Germany (10.5 billion USD, 7.9 billion euros), France (6.8 billion USD, 4.8 billion euros)
Per capita consumption	2014: 11 USD (14 euros)	Switzerland (221 euros), Luxembourg (164 euros) Denmark (162 euros)
Number of countries with organic regulations	2015: 87 (2008: 73)	
Number of IFOAM affiliates	12015: 784 from 117 countries (2008: 734), (2000: 606)	Germany: 91 affiliates, China: 57 affiliates; India: 44 affiliates; United States: 40 affiliates

Source: FiBL Survey 2016.

Table 1.2: Region-wise Percentage Distribution of Organic Agricultural Land and Organic Producers (2014)

Region	Organic Agricultural Land	Organic Producers
Africa	3	26
Asia	8	40
Europe	27	15
Latin America	15	17
North America	7	1
Oceania	40	1

Source: FiBL Survey 2016.

African countries had around 1.3 million hectares of certified organic agricultural land during 2014 (three per cent of the world's share) with almost 0.6 million producers. Among the various countries, Uganda had the largest organic area (> 0.24 million hectares) and with the largest number of organic producers

(0.191 million). The main crops organically cultivated are coffee, olives, nuts, cocoa, oilseeds, as well as cotton and the major share of certified organic production in Africa is destined for export markets. There is a growing recognition among policymakers of African countries that organic agriculture has a significant role to play in addressing food insecurity, land degradation, poverty, and climate change in Africa. Asian countries maintained 3.6 million hectares of total organic agricultural area during 2014 (eight per cent of the world's share) with more than 0.9 million producers; most of these being in India (0.6 million). The leading countries with organic cultivation by area are China (1.9 million hectares) and India (0.7 million hectares). Organic production and domestic markets have established themselves throughout the region, and Asia has the third-largest market for organic products.

As of 2014, European countries had 11.6 million hectares of agricultural land (27 per cent of the world's share) which were managed organically by almost 0.34 million producers. The countries with the largest organic agricultural areas are Spain (1.7 million hectares), Italy (1.4 million hectares), and France (1.1 million hectares). Eight countries have more than 10 per cent organic agricultural land: Liechtenstein has the lead (30.9 per cent), followed by Austria (19.4 per cent) and Sweeden (16.3 per cent). Retail sales of organic products totaled approximately 26.2 billion euros in 2014. The largest market for organic products in 2014 was Germany, with retail sales of 7.9 billion euros, followed by France (4.8 billion euros), and the UK (2.3 billion euros). In Latin America and the Caribbean, almost 0.39 million producers managed 6.8 million hectares of agricultural land organically in 2014. This constituted 15 per cent of the world's organic land and 1.1 per cent of the region's agricultural land. The leading countries are Argentina (3 million hectares), Uruguay (1.3 million hectares), and Brazil (0.7 million hectares, 2012). The highest shares of organic agricultural land are in the Falkland Islands/Malvinas (36.3 per cent), French Guiana (8.9 per cent), and Uruguay (8.8 per cent). Many Latin American countries remain important exporters of organic products such as bananas, cocoa and coffee.More than 3 million hectares of farmland were managed organically by 0.17 million producers in North America during 2014, which is seven per cent of the world's organic agricultural land. United States and Canada are the leading countries in the list.

In Oceania region, more than 0.22 million producers managed 17.3 million hectares, which is 40 per cent of the world's' organic land. More than 98 per cent of the organic land in the region is in Australia (17.2 million hectares) followed by New Zealand (0.11million hectares). The rapidly growing overseas as well as domestic demand has strongly influenced the growth in the organic industry in Australia, New Zealand, and the Pacific Islands.

3. Global Market for Organic Produce

According to Organic Monitor, the global retail sales of organic food and drink during 2014 were 80 billion US dollars and North America and Europe together constituted around 90 per cent of the same. The countries with the largest organic markets were the United States (27.1 billion euros). Germany (7.9 billion euros), and France (4.8 billion euros). The largest single market is the United States (approximately 43 per cent of the global market), followed by the European Union

(23.9 billion euros, 38 per cent) and China (3.7 billion euros, 6 per cent). By region, North America has the lead (29.6 billion euros), followed by Europe (26.2 billion euros) and Asia.

4. Plantation Crops

Out of the 43.7 million hectares of organic agricultural land, organically cultivated permanent crops account for more than 3.4 million hectares, which is 2 per cent of the world's permanent cropland. With around eight per cent share of the organic agricultural land, permanent cropland has a higher share in organic agriculture than in total agriculture. Most of the permanent cropland is in Europe (1.4 million hectares), followed by Latin America (0.8 million hectares), and Africa (0.6 million hectares). The most important crop is coffee, with more than 0.76 million hectares reported and constituting almost one quarter of the organic permanent cropland. Cocoa and coconut have 0.16 and 0.25 million hectares, respectively.

Table 1.3: Region-wise Land Area (ha) under Organic Management in the World (2014) (Permanent crops)*

Region	Cocoa	Coffee	Coconut	Tea/Mate etc.
Africa	38,609	2,23,351	8,501	5,140
Asia	3,282	1,13,061	1,21,781	58,084
Europe	–	–	–	3,897
Latin America	2,06,242	4,07,776	13,689	1,903
North America	–	–	–	–
Oceania	1,060	18,728	12,401	–
Total	2,49,193	7,62,916	1,56,372	69,024

* Data available for a few plantation crops.

Source: FiBL survey 2016.

As per FAOSTAT about 0.25 million hectares of cocoa are grown organically during 2014, which constitutes 2.5 per cent of the world's harvested cocoa beans from an area of 10 million hectares (2013). The organic cocoa bean area has grown almost fivefold since 2004 (approximately 0.05 million hectares) and thus faster than most other crops/crop groups. The available data on the conversion status indicate that at least 3 per cent of the organic cocoa area was in conversion in 2014 (6'200 hectares). Thus, a slight increase in the supply of organic cocoa in the near future may be expected. Though the world's leading cocoa producers are Cote d'Ivoire (2.5 million hectares), Indonesia (1.8 million hectares), Ghana (1.6 million hectares), and Nigeria (almost 1.2 million hectares), the largest organic cocoa areas are in the Dominican Republic (0.12 million hectares), Peru (over 0.03 million hectares) and Mexico (0.02 million hectares). Over 83 per cent of the world's organic cocoa area is in Latin America.

The world's leading coffee producers are Brazil (2.1.million hectares), Indonesia (1.2 million hectares), Colombia (0.8 million hectares), Mexico (0.7 million hectares), and Vietnam (almost 0.6 million hectares). According to FAOSTAT, almost 0.76 million hectares of coffee are grown organically during 2014, which is 7.7 per cent of the world's harvested coffee area from 9.9 million hectares (2013). The organic coffee area has more than quadrupled since 2004. More than 50 per cent of the world's organic coffee area is in Latin America and almost 30 per cent in Africa. In organic farming, the largest areas are in Mexico (0.24 million hectares), Ethiopia (0.15 million hectares), and Peru (0.09 million hectares). Nepal had the highest share, with almost 46 per cent of organic coffee, followed by Timor-Leste (45 per cent), Bolivia (37 per cent), and Mexico (almost 35 per cent). Some of these high percentages must be attributed to the fact that coffee is grown more extensively in organic agriculture, and often in association with other crops.

5. Definitions and Principles of Organic Agriculture

The word "organic" is legally protected in some countries. In the EU, for example, this word has been protected since the early 1990s in English-speaking countries. The equivalent in French, Italian, Portuguese and Dutch-speaking countries is "biological" and "ecological" in Danish, German and Spanish-speaking countries.

5.1. IFOAM Definition

The International Federation for Organic Agricultural Movements (IFOAM), established in the early 1970s, represents over 600 members and associate institutions in over 100 countries. IFOAM (1996) defines the "organic" term as referring to the particular farming system described in its Basic Standards.

5.2. US Definition

In 1980 the US Department of Agriculture defined the concept of organic agriculture as follows: ".a production system which avoids or largely excludes the use of synthetically compounded fertilizers, pesticides, growth regulators, and livestock feed additives. To the maximum extent feasible, organic agriculture systems rely upon crop rotations, crop residues, animal manure, legumes, green manure, off-farm organic wastes, mechanical cultivation, mineral bearing rocks, and aspects of biological pest control to maintain soil productivity and tilth, to supply plant nutrients, and to control insects, weeds, and other pests'. The report also included the following observation: "The concept of the soil as a living system which must be "fed" in a way that does not restrict the activities of beneficial organisms necessary for recycling nutrients and producing humus is central to this definition."

5.3. CODEX Definition

Most recently, the Codex Committee on Food Labeling has debated "Draft Guidelines for the Production, Processing, Labeling and Marketing of Organically Produced Foods"; for adoption of a single definition for organic agriculture by the Codex Alimentarius Commission. According to the proposed Codex definition, "organic agriculture is a holistic production management system which promotes and enhances agro-ecosystem health, including biodiversity, biological cycles, and soil biological activity. It emphasizes the use of management practices in preference to the use of off-farm inputs, taking into account that regional conditions require locally adapted systems. This is accomplished by using, where possible, agronomic, biological, and mechanical methods, as opposed to using synthetic materials, to fulfill any specific function within the system."

6. Principles of Organic Farming (IFOAM)

1. To produce food of high nutritional quality and sufficient quantity;
2. To interact in a constructive and life enhancing way with all natural systems and cycles;
3. To encourage and enhance biological cycles within the farming system, involving micro organisms, soil flora and fauna, plants and animals;
4. To maintain and increase long-term fertility of soils;
5. To promote the healthy use and proper care of water, water resources and all life therein;
6. To help in the conservation of soil and water;
7. To use, as far as is possible, renewable resources in locally organized agricultural systems;
8. To work, as far as possible, within a closed system with regard to organic matter and nutrient elements;
9. To work, as far as possible, with materials and substances which can be reused or recycled, either on the farm or elsewhere;
10. To give all livestock conditions of life which allow them to perform the basic aspects of their innate behaviour;
11. To minimize all forms of pollution that may result from agricultural practices;
12. To maintain the genetic diversity of the agricultural system and its surroundings, including the protection of plant and wildlife habitats;
13. To allow everyone involved in organic production and processing a quality of life conforming to the UN Human Rights Charter, to cover their basic needs and obtain an adequate return and satisfaction from their work, including a safe working environment;
14. To consider the wider social and ecological impact of the farming system;
15. To produce non-food products from renewable resources, which are fully biodegradable;
16. To encourage organic agriculture associations to function along democratic lines and the principle of division of powers;
17. To progress towards an entire organic production chain, which is both socially just and ecologically responsible.

The Key Characteristics are that Organic Farming

1. Relies primarily on local, renewable resources;
2. Makes efficient use of solar energy and the production potential of biological systems;
3. Maintains the fertility of the soil;
4. Maximises recycling of plant nutrients and organic matter;
5. Does not use organisms or substances foreign to nature (*e.g.* GMOs, chemical fertilisers or pesticides);
6. Maintains diversity in the production system as well as the agricultural landscape;
7. Gives farm animals life conditions that correspond to their ecological role and allow them a natural behaviour.

Organic farming is also a sustainable and environmentally friendly production method, which has particular advantages for small-scale farmers. Organic farming contributes to poverty alleviation and food security by a combination of many features, such as:

1. Increasing yields in low-input areas;
2. Conserving bio-diversity and natural resources on the farm and in the surrounding area;
3. Increasing income and/or reducing costs;
4. Producing safe and varied food;
5. Being sustainable in the long term.

The International Federation of Organic Agriculture Movements (IFOAM) has formulated four broad principles of organic farming, which are the basic roots for organic agriculture growth and development in a global context. These principles of organic agriculture serve to inspire the organic movement in its full diversity. The principles are to be used as a whole, which are composed as ethical principles to inspire action. They are:

6.1. Principle of Health

Organic agriculture should sustain and enhance the health of soil, plant, animal, human and planet as one and indivisible. It is the maintenance of physical, mental, social and ecological well-being. The role of organic agriculture, whether in farming, processing, distribution, or consumption, is to sustain and enhance the health of ecosystems and organisms from the smallest in the soil to human beings. In particular, organic agriculture is intended to produce high quality, nutritious food that contributes to preventive health care and well-being. It should avoid the use of fertilizers, pesticides, animal drugs and food additives that may have adverse health effects.

6.2. Principle of Ecology

Organic agriculture should be based on living ecological systems and cycles, work with them, emulate them and help to sustain them. Organic farming, pastoral and wild harvest systems should fit the cycles and ecological balances in nature. These cycles are universal but their operation is site-specific. Inputs should be reduced by reuse, recycling and efficient management of materials and energy in order to maintain and improve environmental quality and conserve resources. Organic agriculture should attain ecological balance through the design of farming systems, establishment of habitats and maintenance of genetic and agricultural diversity. Those who produce, process, trade, or consume organic products should protect and benefit the common environment including landscapes, climate, habitats, biodiversity, air and water.

6.3. Principle of Fairness

Organic agriculture should build on relationships that ensure fairness with regard to the common environment and life opportunities. Fairness is characterized by equity, respect, justice and stewardship of the shared world, both among people and in their relations to other living beings. Fairness requires systems of production, distribution and trade that are open and equitable and account for real environmental and social costs.

6.4. Principle of Care

Organic agriculture should be managed in a precautionary and responsible manner to protect the health and well-being of current and future generations and the environment. It should prevent significant risks by adopting appropriate technologies and rejecting unpredictable ones, such as genetic engineering. Decisions should reflect the values and needs of all who might be affected, through transparent and participatory processes.

7. Conversion to Organic Farming

Organic farming is a process of developing a viable and sustainable agro-ecosystem, and the establishment of an organic management system and building of soil fertility requires an interim period. The time between the start of organic management and certification of crops and/or animal husbandry is known as the conversion period. Though the conversion period may not always be of sufficient duration to improve soil fertility and re-establish the balance of the ecosystem, but it is the period in which all the actions required to reach these goals are started. The whole farm, including livestock, should be converted according to the standards over a period of time. A farm may be converted step by step but both crop production and all animal husbandry should be converted into organic. If the whole farm is not converted, the certification programme shall ensure that the organic and conventional parts of the farm are separate and inspectable. All the standards requirements shall be fulfilled during the conversion period itself. Before products from a farm/project can be certified as organic, inspection is to be carried out during the conversion period.

8. Length of the Conversion Period

The length of conversion period depends largely on the past land use and the ecological situation. It shall be at least 12 months before sowing or planting in the case of annual production, 12 months before grazing or harvest for pastures and meadows and 18 months before harvest for other perennials. Plant products may be used or sold as "in-conversion" provided that they have undergone a 12 month conversion period. Products under conversion are to be sold as "Produce of organic agriculture in process of conversion", or a similar description, when the National Standards stipulations have been met for at least 12 months. Animal products may be sold as "product of organic agriculture" only after the farm or relevant part of it has been under conversion for at least 12 months and provided the organic animal production standards have been met for the appropriate time. With regard to dairy and egg production, this period shall not be less than 30 days. Animals present on the farm at the time of conversion may be sold for organic meat if the organic standards have been followed for 12 months.

9. Issues in Organic Farming

Although, world over, the commercial organic farming with its rigorous quality assurance system is a new market controlled, consumer-centric agriculture system, it has grown almost 25-30 per cent per year during the last decade. Though the movement initiated in the developed world is gradually picking up in developing countries as well, the demand is still concentrated in developed and economically advanced countries. With the increasing awareness about the safety and quality of foods, long term sustainability of the system and experiences of being equally productive, the organic farming has emerged as an alternative system of farming which not only addresses the quality and sustainability concerns, but also ensures a profitable livelihood option to the farmers. The main bottlenecks experienced by many farmers to switch over to organic farming are non-availability of sufficient amount of organic supplements, bio-fertilizers *etc.* for use in fertilization and soil amendments, as well as for plant pest and disease control, the lack of unrestricted veterinary medicines and the lack of experience in marketing organic products. Absence of recognized/established marketing channels leads to poor quality as well as adulteration of organic inputs. Application of poor and adulterated organic inputs looses the confidence of the farmers on organic farming due to their poor performance. Besides, lack of access to guidelines, lack of market information and vocational training, risk of low yield, high cost of certification and inputs coupled with capital-driven regulation by contracting firms strongly discourage farmers of small holdings to switch over to organic farming. The domestic market for organic products is not yet well developed as that of export market. The products available in the domestic organic market are mostly cereals, pulses, fruits, vegetables and a few plantation crops (tea, coffee). The small farmers, spread across the country, often can offer only an incomplete product range that is mostly available as local brand. On the other hand, in countries like the US and Europe, every supermarket houses a complete range of certified organic products. Therefore, organized retailing and marketing from the prevalent unorganized pattern is very much needed.

The growing demand for organic products in industrialized countries, particularly, the EU, United States and Japan, has lead to a growing international trade over the past 15 years. Apart from the producer-driven approach to organic farming, a market-driven approach is taking shape in developing countries. However, the typical character of Indian organic food market, which is in the nascent stage, compared to the developed countries, is that it is buyers/consumers driven rather than producers/supply driven. This calls for creating awareness about organic food and its benefits when compared to conventional food. The development of organic farming will not occur at fast pace as expected, but will be responsive to technological advancements, which can take care of unforeseen factors that will challenge agricultural development as a whole. Within Europe, the development of organic agriculture took 30 years to occupy 1 per cent of agricultural lands and food markets. Success in organic farming depends greatly on local conditions. One of the main characteristics of organic agriculture is the use of local resources to optimize present and future output. Therefore, deciding up on the suitability of organic farming must include agro-ecological, economic, and social and institutional considerations.

The agro-ecological considerations include availability of natural resources (land, soil quality, vegetation, access to material which can be used in compost and mulch *etc.*); suitability of enterprises (crops to be grown or livestock to be raised, based on the availability of natural and other resources); and problems likely to occur (which pests are common, what is the cause, what can be done to avoid them within available resources *etc.*). Suitability of organic farming depends on its profitability and, therefore, the main economic considerations include the availability of labour (quantity and timing of labour); total net return (income from main and other crops as well as livestock); long-term productivity of the system (effect of present production on the soil and implications for future yields); and marketing possibilities. The major social and institutional constraint is the belief among the farmers, scientists, researchers, and extension officers that organic farming is not a feasible option to improve food security. Unless this attitude is changed, no positive consideration towards this farming system can be expected. The performance of organic farming on production depends on the previous management practices adopted by the farmer. In developed countries, where the intensity of use of external inputs before conversion is high, the organic systems is found to decrease yields, whereas, conversion to organic farming usually leads to almost identical yields in irrigated lands and in traditional rain fed agriculture (with low-input external inputs), organic agriculture has the potential to increase yields.

10. Benefits of Organic Farming

Although, there are several benefits of changing over to organic farming from traditional/conventional farming practices, yet all these advantages, in general, may not be feasible considering the rural economy of many countries. However, some are really feasible enough to be considered as benefits for organic farming conditions. The following are some of the advantages that are relevant.

10.1. High Premium for the Produce

Very often, the organic food is priced 20 - 30 per cent higher than conventional food, thus, there is ample scope for organic farmer to get a high premium so that the farm family income could be increased to higher level.

10.2. Low Investment

The capital investment needed for organic farming is not that high as compared to the traditional chemical farming techniques. Moreover, no sophisticated techniques are required for the production of organic manures, and pesticides which could mostly be produced locally in the farm itself. The farm family labour also is put to use in the organic farms, which could turn out to be highly productive.

10.3. Synergy with Life Forms

Organic farming involves synergy with various plant and animal life forms inside the organic farms. Small farmers can understand this synergy very easily and hence find them easy for adoption.

10.4. Traditional Knowledge

The rich heritage of traditional knowledge in farming practices, especially that for management of pests and diseases, the farmers of developing countries have, and traditional land races, can be utilized very successfully to reap rich benefits in organic farming.

11. Biodiversity and Organic Farming

Organic farming utilizes locally available resources to minimize competition for food and space between different plant and animal species. Thus, the manipulation of the temporal and spatial distribution of biodiversity is the main productive "input" of organic farmers. Organic farmers are both custodians and users of biodiversity at all levels as locally adapted seeds and breeds are preferred for their greater resistance to diseases and resilience to climatic stress; diverse combinations of plants and animals optimize nutrient and energy cycling for agricultural production; and the maintenance of natural areas within and around organic farms and non utilization of chemical inputs create suitable habitats for wildlife. Reliance on natural pest and disease control measures maintains rich species diversity, especially that of beneficial natural enemies and avoid the development of pest species that are resistant to chemical control methods. Ultimately, the diversity of landscape and wildlife brings people in the form of eco-tourism, providing an important source of off-farm income.

12. Plantation Crops and Organic Farming

Plantation crops are, those crops, cultivated on an extensive scale in contiguous area, owned and managed by an individual farmer or a company. They are high value commercial crops having greater economic importance and play a vital role in improving economy, mainly because of their export potential, employment generation and poverty alleviation of millions of people, especially in the rural

Table 1.4: Summary of SWOT Analysis on Organic Farming

Parameter	Potential Benefits
Agriculture	Increased diversity, long-term soil fertility, high food quality, reduced pest/disease, self-reliant production system, stable production. Well-suited for smallholder farmers, who comprise the majority of the world's poor. Resource poor farmers are less dependent on external resources, experience higher yields on their farms and enjoy enhanced food security.
Environment	Reduced pollution, reduced dependence on non-renewable resources, builds soil fertility and enhances biodiversity on and around the farm, negligible soil erosion, wildlife protection, resilient agro-ecosystem, compatibility of production with environment, more resilient to climatic stress, including drought and floods, more energy efficient than conventional agriculture and holds carbon in the soil.
Social conditions	Farmers and other members of the families and labourers are no longer exposed to hazardous agro chemicals, which is one of the leading causes of occupational injury and death in the world. Improved health, better education, stronger community, reduced rural migration, gender equality, increase employment, good quality work.
Economic conditions	Stronger local economy, self-reliant economy, income security, increase returns, reduced cash investment, low risk.
Organizational/institutional	Cohesiveness, stability, democratic organizations, enhanced capacity building.

Strengths	Weaknesses
Safety food	Productivity gaps
Comparative advantage in organic food production	Lack of established markets
Low cost of production	Poor quality management in production and processing
High quality and improved nutrition	Less incentives from Government
Improved soil health	Low R and D investments on Organic farming research
Premium prices	Organic market buyers/consumers driven market
Environmental sustainability	Lack of strategy for development of organic market
High water-use efficiency	Disjointed producers, processors and traders
Government policies (like NPOP)	Adulteration and poor quality of organic inputs
Preserves traditional varieties/species and high shelf-life	Large number of small farms with weak organizational building
	Intensive in nature and high labor costs

Opportunities	Threats
Big and growing market potential	High cost of organic food
Growing purchasing power of consumers	Costly and complex organic certification process
Growing health awareness	Lack of infrastructure facilities (like labs) and certification bodies
70 per cent of Gross Cropped Area is under rainfed agriculture	Only export regulated organic market
Reduce heavy subsidies on food and fertilizers	Low awareness about organic inputs
Control the nitrate losses and CO_2 emissions	Most of the fields are contiguous and problem of contamination
Earn high export earnings	Introduction of GM crops

sector. A range of plantation crops including arecanut, cashew, cocoa, coconut, coffee, oil palm, tea, rubber, and spices are cultivated in the humid tropics and tropical belts in different parts of the world. Plantation crops meet a wide variety of human needs such as food, fuel, oil, industrial raw materials, beverages and confectionery items. Arecanut, cashew, cocoa and coconut are the major small holder's plantation crops cultivated in India. The major socio-economic features in which these crops are cultivated include predominance of fragmented, small and marginal holdings, medium to resource poor farm environment, often with less marketable and marketed surplus.

The low level of adoption of management practices particularly chemical fertilizers by the farming community is the major factor responsible for low productivity in farmers' gardens especially in small holders' plantation crop like coconut. Lack of adequate resources is also another reason for low level of adoption of technologies.

Wherever continuous use of chemical fertilizers without application of adequate organic inputs and farming practices which leads to over exploitation of natural resources take place, degradation of soil takes place and results in poor productivity. Pollution of water bodies is also happening due to the leaching of chemicals from farm lands. Reduced crop and soil health due to poor organic matter content and micronutrient deficiencies are also being experienced by the farmers. All these factors point towards the urgent need for soil and plant health management to achieve sustainable production, protecting the environment and safeguarding the natural resources. It is also necessary to reduce the cost of cultivation and other inputs to enhance competitiveness in the international market under the changing global scenario.

Selected References

Anonymous (2012). *The IFOAM norms for organic production and processing.* Version 2012. Published in Germany by IFOAM. p.132.

Bhattacharyva, P. and Chakraborty, G. (2005). Current Status of Organic Farming in India and other Countries. *Indian Journal of Fertilizers.* **1** (9):111-123.

Elisa Morgera, Carmen Bullón Caro and Gracia Marín Durán (2012). *Organic agriculture and the law.* FAO Legislative Study 107,FAO, Rome. p.302.

Willer Helga and Julia Lernoud (2016). *The world of organic agriculture: Statistics, and Emerging trends-2016.* Research Institute of Organic Agriculture (FiBL Frick and IFAOM-Organics International, Bonn, p. 334.

Chapter 2

Organic Farming in Coconut

☆ *P. Subramanian, Murali Gopal, Alka Gupta,*
George V. Thomas, V. Krishnakumar
and P. Chowdappa

1. Introduction

There is an increasing demand among the consumers all over the world for organically grown agricultural commodities with increasing health consciousness. The production to meet this demand could be made possible only through preserving soil health, plant health and environmental health. Organic farming has gained importance globally as holistic production system so as to produce safe food for consumption aimed at minimizing all forms of pollution within the agro-ecosystem as well as protecting the environment. This system of managing agricultural holdings restricts the use of chemical inputs such fertilizers, pesticides, growth regulators and livestock feed additives *etc*. The focus on soil fertility management shall be on development and application of innovative agro technologies based on soil biodiversity to enhance the natural nutrient cycles and utilization of specific soil-plant-microbial associations in managing the soil nutrient status. The holistic approach in farming provides social and ecological advantages with the conservation of precious inputs such as soil and water and ensures long term fertility of soil for sustainability in cropping.

The coconut palm is one of the five legendary Devavrikshas and is eulogized as Kalpavriksha – the all giving tree – in Indian classics. Also called as *"Tree of Heaven"*, *"Tree of life"* or *"Nature's supermarket"*, coconut has considerable global significance as a versatile tree crop providing essential needs of human life *viz*. income, livelihood opportunity, food supplement, energy requirement, environmental stability and raw material for a number of enterprises. It stabilizes farming systems, especially in fragile environments such as small island states, and in coastal zones. No religious function is complete without coconut, which is an essential component of several rituals in Hindu tradition throughout India. It is considered a symbol of auspiciousness, blessing and prosperity. The coconut palm has significant influence on the rural economy particularly in the coastal regions of the world, where it provides livelihood security to over 12 million people. Coconut is an eco-friendly plant which helps to conserve soil, provides aesthetic beauty to the nature, perennial and less exhaustive and ideal tree for bio-hedging on the coastal ecosystem.

The demand for organic coconut products and spices (which are usually grown in coconut plantations as mixed crops) is increasing over the years and this trend is likely to continue. It provides opportunity for farmers who produce coconut (both mature and tender nuts) as well as various other value added products by following organic methods of farming to realize better returns through the premium price available for organic products. Though there is declining viability in some of the coconut producing countries, they continue to produce because of the importance given to coconut as a social crop. These countries have realized the potentials coconuts hold in economic development and poverty alleviation particularly, among the rural population. For most of these countries, coconut is still the backbone of their economy and it could be the base on which their rural economies are based.

The coconut is a 'no-waste tree' because even its waste products provide ample opportunity to augment farmers' income. Through R and D, the once considered wastes have been put into use and even turned up to be an income generating industry. Coir dust, the major by-product of coir production and considered a pollutant, is now being sought to conserve the environment. It is now used as substitute for peat as a potting medium for plants. Coir peat or dust, is now being exported and is becoming a significant foreign exchange earner in Sri Lanka. Coconut sap can be processed into a number of products, one of which is sugar. The midribs, twigs, spathe, leaves, shell *etc.*, are also now being utilized in the manufacture of handicrafts - another non-traditional export product from the Philippines.

2. Production Scenario

2.1. Global Scenario

Coconut is grown in more than 93 countries worldwide in 12.5 million hectares of land (2012-13), which constitutes about 0.7 per cent of the net crop of the world (Table 2.1). Close to ten million hectares is contributed by four countries, namely Indonesia, Philippines, India and Sri Lanka and they contribute 79.09 per cent of the total area under coconut and its production in the world. The Asian Pacific Coconut Community has 18 coconut producing member countries accounting for about 95 per cent of world area under coconut cultivation and production as well as exports of coconut products. The palm exerts a profound influence on the rural economy of many countries where it is grown extensively and provides livelihood security to several millions of farm families across the globe. The crop is grown in the coastal lands of continental south Asia and spread along the Indian and Pacific Ocean. The coconut oil ranks sixth among the eight major vegetable oils of the world.

**Table 2.1. Area and Production of Ten Major Coconut Growing Countries
in the World (2012-2013)**

Country	Area ('000ha) and Per cent Share	Production (million nuts) and Per cent Share	Productivity (nuts/ha)
Indonesia	3,787.00 (30.35)	16,463.00 (22.84)	4,347
Philippines	3,550.00 (28.45)	15,353.00 (21.30)	4,325
India	2,137.00 (17.12)	22,680.03 (31.46)	10,615
Sri Lanka	395.00 (3.17)	2,513.32 (3.49)	6,363
Tanzania	310.00 (2.48)	427.51 (0.59)	1,379
Brazil	279.00 (2.24)	3,326.57 (4.61)	11,923
Papua New Guinea	221.00 (1.77)	1,482.59 (2.06)	6,709
Thailand	209.00 (1.67)	838.00 (1.16)	4,010
Mexico	176.00 (1.41)	1,463.74 (2.03)	8,317
Vietnam	158.00 (1.27)	1,235.45 (1.71)	7,819
Others	1,257.3 (10.07)	6,311.38 (8.75)	
Total	12,479.00	72,094.58	5,777

Source: APCC.

2.2. Indian Scenario

The coconut is not only significant in socio cultural needs of Indian society, but also has gained considerable importance in the national economy as a potential source of rural employment and income generation among the plantation crops. India, with an area of about 2.14 million ha under cultivation and production of 23 billion nuts (15.61 million tonnes) (2012-2013), stands third and first among the coconut growing countries in area and production, respectively. India shares 17.12 per cent of area under coconut cultivation and 31.46 per cent of coconut production in the world. Considering the area under cultivation and total production, India stands first in productivity of coconut among the major coconut producing countries of the world at 10615 nuts/ha during 2012-13. Unlike other world countries, India has the comparative advantage of having the crop grown under varied agro-climatic zones and hence, there is distinct difference in the pattern of distribution of coconut in the country. As a result of this unique distribution, a steady supply of coconut is ensured in the country throughout the year. Most of coconut production in India comes from the Western plains comprising of the states of Kerala, Karnataka, Maharashtra, Goa and Gujarat followed by the East Coast plains comprising of Tamil Nadu, Puducherry, Andhra Pradesh, Telengana and Odisha. Andaman and Nicobar and Lakshadweep are the two major coconut growing islands. Four southern states of India, *viz.*, Andhra Pradesh, Karnataka, Kerala and Tamil Nadu together account for 91.30 per cent of the total production in the country. In terms of area, they contribute to 89.11 per cent (Table 2.2).

In India most of the production comes from small and marginal farms and more than 90 per cent of the holdings are below one hectare with the average size being 0.22 ha. In the west coast of India, the palm is an essential component in the homestead system of farming where it is mostly grown as a rain fed crop.The countrywide demand for coconuts both for edible and non-edible purpose, and the adaptability of coconut palm to grow under varying soil and climatic conditions has generated keen interest among the farmers of even non-traditional zones in the country to plant a few seedlings in their homestead gardens. West Bengal, North Bihar, Chattisgarh, Assam, Tripura and Nagaland are the other non- traditional areas where coconut cultivation is being taken up.

Table 2.2: Area, Production and Productivity of Coconut in different States of India (Area with more than 2.0 per cent share) (2012-2013)

State	Area ('000 ha) and Per cent Share	Production ('000 mt) and Per cent Share	Production (million nuts)	Productivity (nuts/ha)
Kerala	796.16 (37.36)	3990.39 (25.60)	5798.04	7264
Karnataka	513.10 (24.01)	4169.90 (26.70)	6058.86	11808
Tamil Nadu	465.11 (21.71)	4760.67 (30.50)	6917.25	14872
Andhra Pradesh	128.90 (6.03)	1330.40 (8.50)	1933.07	14997
Odisha	54.29 (2.54)	262.17 (1.70)	380.93	7017
Other states	179.11 (8.35)	1095.57 (7.00)	1591.88	–
All India	2136.67	15609.10	22680.03	10615

Source: Coconut Development Board, 2013.

In most of the producing countries, about 50 per cent of the production is consumed domestically and balance is made available for global trade, whereas, the bulk of coconut produced in India goes for the domestic consumption. The decelerating growth in production of coconut in the major exporting countries in the world like; Philippines, Indonesia and Sri Lanka, on account of displacement of coconut area for more profitable enterprises like; oil palm cultivation, urbanization followed by real estate business *etc.*, have caused deficit in global supply of coconut and its products. Indian coconut and coconut products are, therefore, now gaining competitive advantage in occupying world market. However, in many of the EEC countries, Indian coconut products are yet to make the presence felt due to the fact that the products do not satisfy the quality standards prescribed in the European market as well as for want of organic certifications. In view of the emerging economic scenario and also due to the tremendous growth in the production and productivity of coconut in the country, it is necessary to explore new markets for these products abroad for the stability of domestic market. Promotion of organic cultivation in coconut gardens is, therefore, considered very much essential.

3. Tall and Dwarf Varieties

Two main categories of coconut palms *viz.*, Talls and Dwarf are cultivated globally. The tall plants of the Typica group are generally cross-pollinated, whereas, in dwarf types of the Nana group, self-pollination is the norm. Tall varieties should always be chosen for agroforestry systems, because they only can reach up to the upper levels intended for them, and thus fully develop. Dwarf palms get easily overshadowed in the system, hindering their full development. In addition, the Nana varieties are more sensitive to drought as well as some pests and diseases than Typica varieties.

The tall cultivars are predominantly grown for fresh, oil yielding kernel, whereas, the dwarf cultivars for their attractive bright coloured tender fruits having sweet tender nut water. Tall varieties are the common type that occurs throughout the world. They are widely planted both for household and commercial production in all the coconut cultivating regions of the world. Generally, they are slow maturing, flower 6-10 years after planting and grow to a height of about 20-30 m. They are long-lived with an economic life of about 60-70 years, although much older palms are known to exist and yield well. They produce copra of good quantity and quality, and have fairly high oil content as compared to dwarf cultivars. Many *talls* are grown for the production of copra for oil extraction and coir for fiber. The different cultivars of the *tall* are generally named after the place where they are largely cultivated. The *tall* cultivars are most commonly grown in India are the West Coast Tall (WCT), Tiptur Tall (TPT) and East Coast Tall (ECT).

'Dwarfs' represent about 5 per cent of coconut palms and are also cultivated worldwide. They are more commonly found near human habitation and show traits closely associated with human selection. They are predominantly self-pollinated and with slow trunk growth in nature. These are believed to be mutants from tall types with short stature, 8-10 m when 20 years old. They begin bearing at an early age of around three to four years, however, with short productive life of 30-40 years. The nuts are smaller and the copra soft, leathery and low in oil content. The dwarf cultivars are generally grown for tender nuts and also used for hybrid production. The common dwarfs available in India are Chowghat Orange Dwarf (COD), Chowghat Green Dwarf (CGD), Kenthalli (KTOD) and Gangabondam (GBGD). Among the dwarfs, Chowghat Orange Dwarf has very good quality of tender nut water and has been released as a tender nut variety suitable for cultivation throughout India. Some of the common local varieties in India and their features, improved coconut varieties, coconut hybrids released in India are given in Tables 2.3–2.5.

4. Amenability of Coconut to Organic Farming

Being a perennial horticulture crop and the economic life span extending to more than 50-60 years after planting, the cultivation of the palm involves heavy investments during the juvenile phase and continuous recurring expenditure for its subsequent maintenance. The coconut palm, therefore, is characterized by high rate of investments in the initial years and for realizing the output in succeeding periods. Due to its unique nature of continuing growth and yielding phases, any change in the cultivation practices in the middle may upset the growth and physiology of the crop. A vast stretch of area under coconut in the country is rain fed, and hence, the success of coconut growing depends on the uniform distribution of an annual rainfall also.

Table 2.3: Local Coconut Varieties grown in India

Name of Variety	Salient Features	Recommended States
West Coast Tall (WCT)	The palms are tall, robust and bear large green nuts but have wide range of variation in size, shape and colour of nuts. Suitable for production of copra and tender nut. It comes to bearing in 5-7 years after planting. The average yield 15000 nuts/ha, copra yield 3.6 t/ha and oil 2.5 t/ha.	Kerala, Karnataka, Maharashtra, Goa, Tamil Nadu, Andhra Pradesh, Bihar, Andaman and Nicobar Islands, Gujarat, Lakshadweep and Puducherry
East Coast Tall (ECT)	The palms are tall, compact and bear medium sized oval shaped green nuts. Suitable for production of copra and tender nut. The average yield 14500 nuts/ha, copra yield 3.4 t/ha and oil 2.2 t/ha.	Odisha, Assam, West Bengal, Bihar, Andhra Pradesh, Meghalaya and Tripura
Tiptur Tall (TPT)	It comes to bearing in 5-7 years after planting. The palms are tall, robust. Suitable for production of copra and tender nut. The average yield 15050 nuts/ha, copra yield 3.7 t/ha and oil 2.6 t/ha.	Karnataka
Andaman Ordinary (ADOT)	The palms are tall, robust and bear large green nuts but have wide range of variation in size, shape and colour of nuts. Suitable for production of copra and tender nut. It comes to bearing in 5-7 years after planting. The average yield 16450 nuts/ha, copra yield 3.6 t/ha and oil 2.4 t/ha.	Kerala and Andaman and Nicobar Islands
Gangabondam (GBGD)	The palms are semi tall, robust and bear large green nuts. It comes to bearing in 5-7 years after planting. Suitable for tender nut. The average yield 16000 nuts/ha.	Andhra Pradesh
Chowghat Green Dwarf (CGD)	It is an early bearing cultivar and takes about 3-4 years for initial flowering. The average yield 14000 nuts/ha.	Kerala, Karnataka, Andhra Pradesh, Tamil Nadu, Odisha, Andaman and Nicobar Islands, West Bengal, Goa, Gujarat, Assam, Bihar and Maharashtra
Malayan Orange Dwarf (MOD) and Malayan Green Dwarf (MGD)	These are two introductions from Malaysia and are suitable as tender nut varieties with 400 ml and 370 ml tender nut water/nut, respectively. Both these are early bearers and start flowering during 3rd or 4th year of planting.	Kerala, Tamil Nadu, Karnataka, Andhra Pradesh, Odisha and West Bengal.

Table 2.4: Improved Coconut Varieties

Variety	Important Traits	Nut Yield (ha/year)[@]	Copra Yield (t/ha/year)[@]	Recommended states/regions	Agency Responsible for Release
		Tall			
Chandra Kalpa	Drought tolerant, high copra oil content, suitable for neera tapping	17,700	3.12	Kerala, Karnataka, Tamil Nadu, Andhra Pradesh, Maharashtra	Central Plantation Crops Research Institute (CPCRI)
Kera Chandra	High yield, dual purpose variety for copra and tender nut; suitable for soap industry	19,470	3.86	Kerala, Karnataka, Konkan region, Andhra Pradesh, West Bengal	CPCRI
Kalpa Pratibha	High nut, oil yield, dual purpose variety for copra and tender nut	16,107	4.12	Kerala, Andhra Pradesh, Tamil Nadu, Maharashtra	CPCRI
Kalpa Mitra	High nut, oil yield, drought tolerant	15,222	3.68	Kerala, West Bengal	CPCRI
Kalpa Dhenu	High nut, oil yield, drought tolerant	14,160	3.41	Kerala, Tamil Nadu, Andaman and Nicobar Islands	CPCRI
Kalpa Haritha	Dual purpose variety for copra and tender nut; less eriophyid mite damage	20,886	3.70	Kerala, Karnataka	CPCRI
Kalpatharu	Drought tolerant, ball copra, high yield, coir fibre amenable for dyeing	20,709	3.64	Kerala, Karnataka, Tamil Nadu	University of Hort. Sciences (UHS), Bagalkot, Karnataka
Pratap	High yield	26,727	4.01	Konkan region of Maharashtra	Dr. Balasaheb Sawant Konkan Krishi Vidyapeeth (Dr. BSKKV), Maharashtra
Kamrupa	High yield	17,877	2.90	Assam	Assam Agricultural University (AAU), Assam
ALR (CN) 1	High yield	22,302	3.50	Tamil Nadu	Tamil Nadu Agricultural University (TNAU), Tamil Nadu
Kera Bastar	High yield	19,470	3.18	Bastar region of Chhattisgarh, Konkan region of Maharashtra, Coastal zone of Tamil Nadu, Andhra Pradesh	Indira Gandhi Agricultural University (IGAU), Chhattisgarh
Kalyani Coconut 1	High yield	14,160	3.84	West Bengal	Bidhan Chandra Krishi Viswavidyalaya (BCKV), West Bengal

Contd...

Table 2.4–Contd...

Variety	Important Traits	Nut Yield (ha/year)@	Copra Yield (t/year)@	Recommended states/regions	Agency Responsible for Release
Kera Keralam	High yield, drought tolerant, suitable for neera tapping; soap industry	26,019	3.53	Tamil Nadu, West Bengal, Kerala	TNAU
ALR (CN) 2	High yield	21,240	2.89	Tamil Nadu	TNAU
VPM-3	High yield, drought tolerant	14,868	3.41	Tamil Nadu	TNAU
Kera Sagara	High yield	17,523	3.64	Kerala	Kerala Agricultural University (KAU), Kerala
Dwarf/Semi Tall					
Chowghat Orange Dwarf	Tender nut purpose, orange colour fruit, coarse fibre	11,505	1.80	All coconut growing regions for tender nut	CPCRI
Kalparaksha	Semi-tall, high nut and oil yield in RWD prevalent areas; tender nut purpose	13,260 (17,748)#	2.85 (3.34)#	Kerala, Root (wilt) disease prevalent tracts	CPCRI
Gouthami Ganga	Tender nut purpose, green colour fruit	15,930	1.54	Coastal zone of Andhra Pradesh	Acharya N. G. Ranga Agril. University (ANGRAU), A.P.
Kalpasree	Early flowering, green colour fruit; superior oil – rich in linoleic acid, recommended for root (wilt) diseased areas	20,178	2.86	Root (wilt) disease prevalent tracts	CPCRI
Kalpa Jyothi	Tender nut purpose, yellow colour fruit	21,771	4.07	Kerala, Karnataka, Assam	CPCRI
Kalpa Surya	Tender nut purpose, orange colour fruit	9,133	2.20	Kerala, Karnataka, Tamil Nadu	CPCRI
Kera Madhura	Semi-tall, dual purpose variety for copra and tender nut	24,480	4.80	Kerala	KAU
CARI-C1 (Annapurna)	High copra content, tender nut purpose, green colour fruit	20,231	1.41	Andaman and Nicobar Islands	Central Agricultural Research Institute (CARI), Andamans
CARI-C2 (Surya)	Ornamental purpose, orange colour fruit	24,072	1.77	Andaman and Nicobar Islands	CARI
CARI-C3 (Omkar)	Ornamental purpose, yellow colour fruit	16,373	1.67	Andaman and Nicobar Islands	CARI
CARI-C4 (Chandan)	Ornamental purpose, orange colour fruit	11,505	1.80	Andaman and Nicobar Islands	CARI

@ : Yield estimated at 7.5 m x 7.5 m spacing, population of 177 palms per ha under experimental conditions. It will vary according to organic farming practices being adopted. #: Figures in parenthesis indicate yield in root (wilt) disease free tracts.

Table 2.5: Coconut Hybrids Released in India

Hybrid Variety	Source Population of Parents	Important Traits	Nut Yield[@] (/ha/year)	Copra Yield[@] (t/ha/year)	Area Recommended	Agency Responsible for Release
Chandra Sankara	COD x WCT	High yield	20,532	4.27	Kerala, Karnataka, Tamil Nadu	CPCRI
Kera Sankara	WCT x COD	High yield, drought tolerant	19,116	3.78	Kerala, Karnataka, Maharashtra, Andhra Pradesh	CPCRI
Chandra Laksha	LCT x COD	High yield, drought tolerant	19,293	3.76	Kerala, Karnataka	CPCRI
Kalpa Samrudhi	MYD x WCT	Dual purpose variety, Drought tolerant, higher nutrient use efficiency	20,744	4.35	Kerala, Assam	CPCRI
Kalpa Sankara	CGD x WCT	Tolerant to root (wilt) disease, high yield	14,868	3.20	Root (wilt) disease prevalent tracts	CPCRI
Kalpa Sreshta	MYD x TPT	Dual purpose variety, High yield	29,227	6.28	Kerala, Karnataka	CPCRI
Laksha Ganga	LCT x GBGD	High yield	19,116	3.73	Kerala	KAU
Ananda Ganga	ADOT x GBGD	High yield	16,815	3.63	Kerala	KAU
Kera Ganga	WCT x GBGD	High yield	17,700	3.56	Kerala	KAU
Kera Sree	WCT x MYD	High yield	23,364	5.05	Kerala	KAU
Kera Sowbhagya	WCT x SSAT	High yield	23,010	4.49	Kerala	KAU
VHC-1	ECT x MGD	High yield	21,240	2.87	Tamil Nadu	TNAU
VHC-2	ECT x MYD	High yield	25,134	3.74	Tamil Nadu	TNAU
VHC-3	ECT x MOD	High yield	27,612	4.47	Tamil Nadu	TNAU
Godavari Ganga	ECT x GBGD	High yield	18,585	2.79	Andhra Pradesh	ANGRAU, Andhra Pradesh
Konkan Bhatye coconut hybrid 1	GBGD x ECT	High yield	20,532	3.47	Maharashtra	Dr. BSKKV, Maharashtra

Contd...

Table 2.5—Contd...

Hybrid Variety	Source Population of Parents	Important Traits	Nut Yield[@] (ha/year)	Copra Yield[@] (t/ha/year)	Area Recommended	Agency Responsible for Release
Kalpa Ganga	GBGD x FJT	High yield, suitable for ball copra production	21,417	3.38	Karnataka	UHS, Bagalkot, Karnataka
Vasista Ganga	GBGD x PHOT	High yield	22,125	3.88	Andhra Pradesh, Karnataka	Dr YSR Horticultural University (Dr.YSRHU), Andhra Pradesh
Ananta Ganga	GBGD x LCT	High yield	22,656	3.84	Andhra Pradesh, Karnataka	YSRHU
VPM 5	LCT x CCNT	High yield			Tamil Nadu	TNAU

@ : Yield estimated at 7.5 m x 7.5 m spacing, population of 177 palms per ha under experimental conditions. It will vary according to organic farming practices being adopted.

SSAT: Straits Settlement Apricot Tall; ADOT: Andaman Ordinary Tall; FJT: Fiji Tall; PHOT: Philippines Ordinary Tall; CCNT: Cochin China Tall; LCT: Laccadive Ordinary Tall; MYD: Malayan Yellow Dwarf; SSAT: Straits Settlement Apricot Tall.

The growth habit and planting methods of coconut make it highly suitable for managing through organic farming. About 74 per cent of the roots produced do not go beyond 2 m from the bole and most of the roots also confine to the 30 to 120 cm depth, thus, utilizing only limited extent of land area for growth of palms and leaving considerable area for inclusion of other crops to maintain crop diversity, an important requirement for organic farming. The orientation of leaves in the coconut crown helps penetration of sunlight into the soil and provides opportunities for exploitation of land and solar energy for inter/mixed cropping. Such a cropping system approach will also add large quantities of organic wastes to the system and their recycling within the system help to increase organic content of soil and improve microbial activity and make the entire production system more productive even in the absence of external inputs.

Coconut plantations are highly amenable to organic farming as they produce large quantities of waste biomass which, if recycled, can meet the nutrient demand of the crops to a great extent. The total availability of waste biomass from 2.14 million hectare of coconut plantation in the country has been estimated as 15.92 million tonnes annually. The natural decomposition of these wastes and the nutrient release are very slow due to the high lignin content and the nature of lignocellulose complex of the coconut waste materials. Substantial saving in terms of fertilizer input is possible through effective recycling of the waste biomass. Coconut is socially, culturally and religiously associated with millions of people around the world. Apart from healthy food and drink, it provides shelter, health, wealth and aesthetic values. Coconut not only provides livelihood to people who are directly or indirectly depending on the crop, but also ensures food security, nutritional security and alleviates poverty.

5. Climatic Requirements

Coconut palm is a unique plantation crop and stands apart from all other palms because of its high degree of consistency and continuity in flowering and fruit production, month after month (once it starts flowering), year after year, for decades. Depending on the site, coconut palms can be suited to cultivation on agroforestry systems. As a plant of the upper storey, with requisite light requirements, the coconut palms grow above such crops as citrus plants, cocoa, nutmeg, banana and many other crops.

5.1. Geographical Position

Coconut is essentially a tropical plant, growing mostly between 20°N and 20°S latitudes. The ideal elevation is up to 600 m above MSL and because of its temperature requirements, the palms cannot grow above 750 m, even near to the equator. Relative humidity plays a very important role in pollination and fertilization, thus directly influences yield. High humidity provides congenial conditions for the rapid spread of fatal diseases such as bud rot, leaf rot *etc*.

5.2. Temperature

The ideal mean annual temperature is 27°C with 5-7°C diurnal variation and humidity > 60 per cent. Very high humid conditions right through the growth of palms is not considered good from two aspects. One is that it reduces transpiration and thereby reduces the uptake of nutrients. High temperature might cause the young developing inflorescence to dry up, and limit production to those months in the year when the temperature remains at favourable level. In shallow and well-drained soils, effect of dry spell is more pronounced. Under such conditions and also where annual rainfall is less than 100 cm, economic production is possible only under irrigation. In places where the climate is characterized by long dry spells of hot day weather during summer and severe cold associated with cold wave and frost during winter; cyclone prone areas *etc.* are not suitable for coconut cultivation. Increase in temperature is also expected to have a detrimental effect on coconut oil quality, aroma and flavour. Increase in temperature is reported to reduce the oil content in coconut endosperm, resulting in increased content of starch, carbohydrates and reducing sugars in copra.

5.3. Rainfall

Rainfall is the one of the important factors that affects successful growth of coconut palm under natural conditions. Though the palms can tolerate wide range in intensity and distribution, a well distributed rainfall of about 200 cm per year is the best for proper growth and maximum yield. Since the tree does not store moisture and has no tap roots, it is not suited for regions with long and prolonged dry spells during which the water table goes considerably low. Very heavy rainfall may affect pollination due to washing of pollen grains. Since coconut is a perennial crop, delay in monsoon by two to three weeks would not adversely affect the yield and so no special cultural practices is followed. Flooding for shorter duration will not cause much damage to the palms. However, coconut palms do not tolerate high water table and stagnant water for longer periods. Prolonged water stagnation causes palms to suffer from physiological drought, where palms will not be able to uptake water and nutrients due to hampered root function because of lack of O_2 for respiration. Water logging in coconut basins is found to cause yellowing in leaves of palms. Microbial activity is also adversely affected under such situations.

5.4. Impact of Drought

In coconut, spikelets on the inflorescence are formed about 15 months before the opening of spathe and the pistillate flowers before 12 months. Even after the spathe is opened, female flowers remain for about 11 to 12 months to develop into full mature nuts. Drought affects coconut palms and the impact can be seen from the year of drought till four years. Severe drought during early formative phase of the inflorescence may kill the growing points due to desiccation resulting in the abortion of spadix. Any coincidence of drought or dry spell with critical sensitive stages such as inflorescence primordial initiation, ovary development, button size nut stage adversely affects the nut yield. In severe drought situations, drooping of leaves due to low leaf water potentials, shedding of buttons and immature nut fall, bending and breaking of leaves *etc.* may happen.

6. Soil Requirement

Coconut is grown in different soil types such as lateritic, coastal sandy, alluvial and also in reclaimed soils of the marshy lowlands and it does best in relatively coarser textured soils likes sandy loams, sandy coastal alluviums and sandy river valleys. It is found to grow well on littoral (coastal) sand which is generally unsuitable for many other crops provided it is managed carefully in the early stages with organic manuring and watering. It tolerates salinity and a wide range of pH (from 5.0-8.0). Soil with a minimum depth of 1.2 m and fairly good water holding capacity is preferred for coconut cultivation. Shallow soils with underlying hard rock, low lying areas subject to water stagnation and clayey soils are to be avoided as it will be difficult to raise successful coconut plantations under such conditions. Proper supply of moisture either through well distributed rainfall or irrigation and sufficient drainage in waterlogged soil are essential for coconut.

7. Environmental Services of Coconut

Coconut is a perennial plantation crop and committed to the land for decades. At a time when there is considerable impact of climate change on growth and production of various crops, the above and below ground portion of coconut provides an opportunity to C-sequestration of 20 to 35 t ha^{-1} year^{-1}, which is the highest among many of the plantation crops. Coconut, cultivated along with compatible companion crops, provides around 20 t of biomass for recycling and effective utilization of such biomass can improve the soil fertility status as well as the soil microbial community.

8. Production of Planting Materials

Seedlings will be required when planting is to be taken up for new plantation or replanting the existing old and senile plantations. In such cases, selection of seed nuts, raising seedlings and selection of good quality seedlings are very important in coconut, because evaluation of the performance of the new progeny is possible only several years after planting. Location specific varieties which are adapted to local soil and climatic conditions are to be considered, since they are time tested. Many varieties and hybrids have been released for cultivation in different states.

8.1. Mother Palm Selection

For raising nursery, seed nuts are to be collected from selected mother palms. Such mother palms should be of 20 years old or more. Wherever possible, it is advisable to select middle-aged trees as they will be in their prime of life and it is easier to distinguish good yielder from low/poor yielder.

The important characteristics to be considered are:

1. Straight stout trunk with even growth and closely spaced leaf scars
2. Spherical or semispherical crown with short fronds
3. Short and stout bunch stalks without tendency to drooping
4. More than 30 leaves and 12 inflorescences carried evenly on the crown
5. Inflorescence with 25 or more female flowers

6. Consistent yield of about 80 nuts under rainfed conditions and 120 nuts under irrigated conditions

7. 150 g copra per nut

8. Free from any disease and pest incidence

8.2. Collection of Seed Nuts and Sowing

Only fully matured nuts, which are about 12 months old, should be harvested. The nuts should not be allowed to fall, but should be cut down, and carefully lowered, using a rope to avoid possible damage to the nuts. The ideal time for collecting seed nuts and their sowing is April-May and June in west coast region of India, while in the east coast region, sow nuts during October-November to coincide with the monsoon. However, when irrigation facilities are available, seed nuts can be collected and sown at any period of time depending of the requirement. Nuts which are too big or too small in the bunch and also the nuts of irregular shape and size are to be discarded. Following the harvest, the nuts should be stored in a covered, well-ventilated place. In case of tall varieties, sowing is to be done one or two months of storage after collection, whereas, seed nuts should be sown for dwarf varieties either immediately or within 10-15 days after harvest. Before sowing, the nuts are to be sorted again and only those nuts containing water are to be used.

8.3. Raising Nursery

The site selected for raising nursery should be well-drained with coarse-textured soil and close to irrigation water source. The seed nuts can be sown in flat beds if there is no drainage problem. Take raised beds for sowing wherever water stagnation is a problem. Seedlings can be raised in nurseries either in the open area provided with artificial shade or in coconut gardens where the palms are tall and the ground is not completely shaded. The seed nuts should be planted in long and narrow beds at a spacing of 40 cm x 30 cm during May-June or September-October, either vertically or horizontally in 20-25 cm deep trenches. The advantage of vertical planting is that it causes less damage only during transit. However, if sowing is delayed due to some reason, the nut water goes down considerably, and in such cases, adopt horizontal sowing for better germination.

8.4. Selection of Seedlings

Seed nuts, which do not germinate within 6 months after sowing as well as those germinated but have dead sprouts, are to be removed to ensure only good quality seedlings are produced. Seedlings, which are 10 to 12 months old, are to be selected for planting through rigorous selection based on the following characteristics:

1. Early germination, rapid growth and seedling vigour

2. Six to eight leaves for 10-12 month old seedlings

3. Collar girth of 10 cm and above

4. Early splitting of leaves

Figure 2.1: Coconut Nursery Seedlings.

8.5. Poly Bag Nursery

Good quality seedlings can be raised using poly bags (500 gauge thickness, 45 cm x 60 cm with 8-10 holes at the bottom) by transplanting germinated seeds in to the bags. The common potting media are top fertile soil mixed with sand (3:1), and top fertile soil, sand or coir dust and well rotten and powdered cattle manure (3:1:1). Potting mixture using sand + vermicompost (3:1) is also ideal for raising poly bag seedlings. As the entire ball of earth with the root system, after removal of poly bag, can be placed in the pits, no transplanting shock will be experienced by the seedlings. The poly bags should be removed from the coconut plantation

Figure 2.2: Coconut Poly Bag Seedlings.

after the planting is completed.Application of 25 g each of biofertilizers such as *Azospirillum* spp. and Phosphobacterium *Bacillus* sp. results in production of more vigorous and robust seedlings.

8.6. Bio-Priming of Seedlings

Bio-priming of seedlings with bio-inoculants such as *Pseudomonas fluorescens* imparts tolerance to diseases as well as promote their better growth. Initial establishment of such seedlings will be superior in the main field with enhanced growth and field tolerance to diseases. Application of talc based preparation of *P. fluorescens* @ 50 g (10^8cfu) per seedling during four, seven and ten months after sowing in the nursery will be beneficial. At the time of planting seedlings in the main field, coconut seedlings are to be dipped in 100 g (10^8cfu) of talc based preparation of *P. fluorescens* in slurry mode. Use of Plant Growth Promoting Rhizobacteria (PGPR) based bioinoculants, '*Kera Probio*', (talc formulation of *Bacillus megaterium*) @25 g/seedling and '*Ker*AM' (Arbuscular Mycorrhizal bioinoculant) @50 g/seedling also helps in producing robust coconut seedlings. The seedlings in the nursery, in general, do not require any additional nutrition, as the endosperm provides them with sufficient nutrients.

9. Planting and After Care

9.1. Preparation of Land and Planting

In loamy soils with low water table, planting in pits (1.0 m x 1.0 m x 1.0 m) filled up to 50 cm depth is generally recommended. However, in places where the water table is high, planting at the surface or even on mounds becomes necessary. Even while planting at the surface or mounds, digging pits and filling has to be done. If planting is taken up in the littoral sandy soil, application of 0.15 m^3 of red earth is recommended to improve the physical characteristics of soil. The type of soil decides the depth of pits to be taken for planting. In areas with laterite soil and rocky substratum, deeper and wider pits (1.2 m x 1.2 m x 1.2 m) are to be dug and filled up with top soil, powdered cow dung and ash up to a depth of 60 cm before planting. Addition of 2 kg of common salt will help in loosening the soil in such areas. In order to conserve soil moisture in the pits, arrange two layers of coconut husk with concave surface facing up at the bottom of the pit before filling up the soil. Plant the seedling at the center of the pit by removing the soil mixture and press the soil well around the seedling and shade using coconut leaves or palmyrah leaves or any other suitable shading materials.

9.2. Spacing

For tall varieties of coconut, a spacing of 7.5 m x 7.5 m is generally recommended. If triangular system is adopted, an additional 20 to 25 palms can be planted. If wider spacing is adopted, it provides ample opportunity for intercropping a number of annual and perennial crops in the interspaces, which ensures crop diversity in the plantation. In case dwarf varieties are used for cultivation, there is possibility of reducing the spacing to be adopted.

9.3. Time of Planting

Planting the seedlings during May/October, with the onset of pre-monsoon rains is ideal. In places where assured irrigation is available, planting can be done at least a month before the monsoon sets in to allow seedlings to establish well before the onset of heavy rains. In low-lying areas, planting is to be done in September once the heavy rains ceases.

9.4. Care of Young Palms

The seedlings are to be protected from heavy sunlight by proper shading and irrigated during summer months. In sandy soils, irrigate once in four days with 45 l of water. In areas subject to water logging, proper drainage is to be provided by making drainage channels. Care should be taken to avoid soils covering the collar region of coconut seedlings due to heavy rain. The soil washed down by the rains and covering the collar of the seedlings is to be removed. Widen the pits every year before the application of organic manures and gradually fill up the pits as the young plants grow. Inspect the plants at regular intervals for any insect or fungus attack and necessary remedial measures should be taken up as and when required.

9.5. Irrigation

The coconut palm responds well to summer irrigation. Methods such as drip irrigation or basin irrigation (hose irrigation) can be adopted depending on the situation. Drip irrigation is a micro irrigation system in which the water is applied to the root zone at the rate at which the palm can take up. It is ideal considering the advantage of water saving. Another advantage is that vermiwash and other organic solution also can be applied through drip irrigation *i.e.* fertigation. Make small pits of 30 cm^3 one metre away from the bole of the palm at equidistant and fill the pits with coir pith or locally available mulch. Place a conduit tube of 40 cm length diagonally in the pit and allow water to drip through the tube. Apply water at 66 per cent of open pan evaporation through drip irrigation based on the evaporation. In order to supply 32 l/day, four drippers at a discharge rate of 4 l/ hour will be required so that daily two hours of irrigation is sufficient. Where basin irrigation is practiced, 200 l/palm once in 4 days will be beneficial. Irrigation is to be commenced during November when the soil moisture depletes to 50 per cent ASM. In other parts of India especially Tamil Nadu and Andhra Pradesh, irrigation is to be practiced throughout the year except rainy period. Care should be taken to use only uncontaminated water for irrigation. Under west coast conditions of India, irrigation through perfo-sprays 2 cm once in 5 days during December-February and once in 4 days during March-May has been found to be beneficial in sandy loam soil in improving growth of palms.

9.6. Weeding

Generally, weeding is to be done when the monsoon recedes preferably during September- October. Application of any kind of chemical herbicide is not allowed under organic system of cultivation. Therefore, depending on the intensity of weed growth, they are to be removed by hand around the palm base and the weeds in the inter space need only be slashed with sickle. Clean weeding is to be avoided. While weeding, dried shoots and other thrashed materials can be used as mulch around the base of palms, which will help to conserve moisture in the ensuing dry months and help in vermicomposting process in the basin as well as in the interspaces.

9.7. Nutrient Management for Palms in the early Growth Phase

Seeds of any one of the green manure crops such as *Pueraria phaseoloides*, *Vigna unguiculata* (cowpea), *Crotalaria juncea* (sunhemp), *Calopogonium mucunoides* or *Sesbania aculeata* (daincha) may be sown in the palm basin with the receipt of first monsoon rains (May-June and September-October for the areas benefited by South-West monsoon and North-East monsoon, respectively) according to the soil and climatic conditions. The plants are to be uprooted and incorporated *in situ* at the time of flowering one or two plants. If no intercrops are grown in the interspaces, green manures crops can be grown in the interspaces which helps to check the weed growth also.

9.8. Soil and Water Conservation under Rainfed Situations

Conservation of natural resources like soil and water for sustainable crop production under organic farming is very much essential. This is very important when coconut is grown under rainfed condition with undulating terrain and sloppy conditions. Coconut cultivation provides various kinds of organic materials in the plantation. Coconut leaves, husk and coir pith could be utilized as mulches to reduce the loss of soil moisture. Decomposition of such mulch materials after a period of time will result in enrichment of soil organic matter pool and create conditions for proper root growth and proliferation of soil flora and fauna. Coconut husk is an important organic material and a good source of plant nutrients. On dry weight basis, the average composition of material is 0.23 per cent N, 0.04 per cent P_2O_5, 0.78 per cent K_2O, 0.08 per cent Ca and 0.05 per cent MgO. On an average, husk constitutes 45 per cent of the weight of nut and, on this basis, a nut weighing 1,000 g will have 450 g of husk with 20 per cent moisture.

9.8.1. Mulching

Most of the organic wastes from coconut have high moisture holding capacity and can be very profitably used as moisture regulators and conservators rather than nutrient sources. This gains more practical significance in the light of the fact that soils cannot be rejuvenated with organics in the absence of sufficient moisture. Similarly the full benefits from irrigation can be obtained only if there is sufficient quantity of soil organic matter. Keeping in mind the complementary roles of soil organic matter and moisture conservation, coconut leaves, husks and coir pith can be utilized directly for mulching. Spreading of these materials in basin areas will protect the soil from direct sunlight and heavy rains. Mulches can reduce the loss of soil moisture and create good microclimate in soil for the proper growth of plant roots and soil flora and fauna. Over a period of time, mulches will decompose and add to the soil organic matter reserves. The best time for mulching is before the end of the monsoon rains and before the top soil dries up. For mulching, coconut leaves cut into two or three pieces can be used. In order to cover 2 m radius of coconut

Figure 2.3: Coconut Leaves as Mulch.

basin, 15 to 25 fallen coconut leaves are required to spread in 2- 3 layers. Mulching with composted coir pith to 10 cm thickness (approximately 50 kg/palm) around coconut basin is also ideal method to conserve moisture as it can hold moisture five times of its weight. Because of its fibrous and loose texture, incorporation of coir pith considerably improves the physical properties and water holding capacity of soil and thereby increases the coconut productivity. The applied material may last

Figure 2.4: Mulching Coconut Basin using Husk.

Figure 2.5: Mulching Coconut Basin using Coir Pith.

for about 4 to 5 years. The weeded materials also can be used for mulching and should be properly dried before applying in the basins.

9.8.2. Burying Coconut Husk

Burying coconut husk in trenches in between the rows of palms is also effective for moisture conservation in coconut gardens. This cultural operation is to be done at the beginning of the monsoon in linear trenches of 1.5 m to 2 m wide and about 0.3 m to 0.5 m deep between rows of palms with concave side of husk facing upwards. Each layer is to be covered with soil. Husks can also be used as surface mulch around the base of the palms by placing them in a single layer and keeping the convex side upwards up to a radius of 2 m from the base. Coconut husks are also important sources of potash, which becomes available to the palms over a period of time.

9.8.3. Catch Pit filled with Coconut Husk

Construction of catch pits of 1.5 m length x 0.5 m width x 0.5 m depth in sloppy gardens helps in conserving soil and water. A bund is to be made at the downside using the excavated soil and pineapple suckers planted on it. This pit also is to be filled with coconut husk while taking up planting of pineapple.

9.8.4. Contour Trench filled with Coconut Husk

In plantations, where the land slope is high, this approach is to be adopted. Trenches of 50 cm width x 50 cm depth and convenient length are to be made in between two rows of coconut palms. These trenches are to be filled with coconut husk in layers by keeping the bottom layer facing up and top layer facing down. A bund of 20 cm height and suitable width (>50 cm) is to be prepared at the downstream using the excavated soil and two rows of pineapple plants are to be planted on the bund with a spacing of 20 cm x 20 cm. The pineapple plants are found to stabilize the bund and provide additional income to the farmer. The runoff water from the

Figure 2.6: Catch Pit with Pineapple on Bund.

upper side along with soil particles will be collected in the trenches. Coconut husk kept in the trenches will retain the moisture and makes it available for plants during summer months.

10. Nutritional Management

In an organic production system, the fertility program should be based on materials of microbial, plant or animal origin, such as green manure, compost or mulch, which are organically produced on the farm; or of organic quality, obtained from the surrounding farms or natural environment; or other inputs which are allowed. One of the basic principles of soil fertility management in organic systems is that plant nutrition depends on 'biologically-derived nutrients' instead of using readily soluble forms of nutrients supplied through fertilizers and the idea should be to feed the soil to make it living rather than feeding the plants. Hence, sufficient quantities of biodegradable material of microbial, plant or animal origin should be returned to soil to increase or at least maintain its fertility and the biological activity within it.

Once start flowering, coconut palms produce a spadix in the axil of each leaf, and the yield depends on the number of leaves produced per year. Vegetative as well as reproductive growth goes on simultaneously and hence, nutrition is important all the time. Coconut palms export nutrients to the above ground parts continuously from a limited volume of soil throughout its existence. It is, therefore, essential that a nutritionally rich environment is provided in the root zone of coconut all round the year to realise adequate yields. The organic farm is to be maintained as a closed ecosystem to the extent possible. Soil and nutrient loss through soil wash, run off and percolation water should be minimized through proper agronomic practices. All the crop residues and farm wastes should be recycled through suitable composting techniques and applied to fields so that soil fertility is maintained and nutrient requirement of coconut palms and other crops met.

Regular nutrition from the first year of planting is essential for good vegetative growth, early flowering and bearing and high yields of coconut palms. The inflorescence primordium can be detected about 4 months after the first leaf primordium is differentiated; the male and female flowers, 22 months thereafter. The opening of the fully grown spathe occurs one year later. Any hindrance to successful growth of the tree during these productive phase results in reduction in yield and the cumulative effect of these adverse growths appears only after two-three years. Therefore, the crop production practices require the need for uninterrupted application of inputs over longer periods to ensure continuous flow of returns. A deficiency in potash will result in a large reduction of yield for coconut palms. The vast majority of the potassium is thereby contained in the husk and fruit water of the coconuts. In cultivation systems which include cocoa, returning the cocoa shells to the site will supply sufficient potassium to balance out the extraction. The continual pruning of crops on diversified agroforestry systems also will provide an important source of nutrients (*e.g.* of potassium).

10.1. Cultivation of Legumes (Green manures or deep-rooted plants)

The coconut palm is cultivated in humid tropical soils characterised by low organic matter content due to the higher pace of degradation of organic materials caused by heavy rain fall, optimum temperature and porous soil texture. The major portion of inorganic N, wherever applied, is lost through leaching. Organic manures are important in sustaining soil fertility and productivity especially with a perennial crop like coconut. Application of organic materials like coconut coir pith, miscellaneous tree leaves, cattle manure *etc.* are recommended for coconut. However, due to the non-availability of land for exclusive cultivation of green manure crops and also the short supply of cattle manure for use as organic manure, organic manuring is practiced only in limited scale in coconut cultivation.

The Soil Science Society of America in 1971 described green manure as "plant material incorporated with the soil while green or soon after maturity for improving the soil". In practice, it includes *in situ* growing and incorporation of biomass or collection and ploughing of the biomass grown outside the main field. The best known and popular green manure crops are those, which fix nitrogen in association with *Rhizobium*. Biomass production of green manure crops and their nitrogen and phosphorus contents vary widely according to the species, their growth stage, environmental conditions, soil fertility and crop management practices. The basin area of 1.8 to 2.0 m radius around the bole of coconut palm generally is left unutilized by most of the farmers for any other purpose. Wherever the inter space between the palms is cultivated with intercrops such as tuber crops, cocoa, pepper and banana, green manure crops can be raised in the coconut basins. If vermicomposting is practiced in trenches taken in between coconut rows, green manure crops can be grown in the basins. Cultivation and incorporation of green manure result in significant increase in the population of specific groups of microorganisms as well

Figure 2.7: Green Manure Crop in Coconut Basin.

as the enzymatic activities, which suggest a modification in the soil environment to the benefit of plant growth.

Nitrogen fixed symbiotically by legume-rhizobium association can form an important source of nutrients and organic manure for coconut palms. Leguminous green manure crops such as *Pueraria phaseoloides, Mimosa invisa, Calopogonium mucunoides, Crotalaria juncea* (sunhemp) and *Vigna unguiculata* (cowpea) can be successfully raised in coconut basins. The biomass production and nitrogen contribution by green manure legumes and their influence on soil fertility parameters vary with soil type, climatic factors and type of green manure raised. Sow 100 to 150 g seed of any of the green manure cover crop mentioned above in coconut basins with the onset of the pre-monsoon in May. Sowing during heavy rainfall should be avoided. Among these leguminous crops, the former two species are preferable because of shade tolerance, self seeding nature and higher biomass productivity. Stir the soil around the basin in 1.8 to 2.0 m radius area lightly and broadcast the seeds uniformly and later on cover them by slight raking. Allow the plants to grow in the basin and harvest the biomass as and when one or two plants start flowering and put back into the basin and cover with soil. While doing so, the soil should be disturbed to the bare minimum.

On an average about 15-20 kg green biomass could be generated in the basin of coconut palm and their incorporation can contribute around 100-150g nitrogen/ basin and other major nutrients as well as enhance the population of specific groups of beneficial micro organisms (bacteria and nitrogen-fixers) in the basin thereby improving the soil fertility. The significant increase in the population of micro organisms and the enzymatic activity modifies the soil environment for the benefit of palm growth. When irrigation is practiced, sowing of green manures may be undertaken twice in a year. The method of cultivation of green manure in coconut basin is simple, inexpensive and can be adopted even by small farmers. With continuous cultivation of legumes, it is possible to improve soil organic matter resources for sustaining soil fertility and enhance coconut yield. Besides legumes, several other plants may be used as green manure. For example, water hyacinth (*Eichhornia crassipes*) which grows wildly in stagnant and backwaters in Kerala contains 2.0 per cent N on dry weight basis and some sea-weeds containing 1-2 per cent N, may be used for manuring crop in coastal areas like Kerala.

10.2. *Glyricidia sepium* as Green Manure

Generation of large quantities of nitrogen rich biomass is also possible through the cultivation of the fast growing perennial leguminous green leaf manure tree crops. A fast growing multipurpose tropical leguminous tree, *Glyricidia sepium* with high nitrogen fixing potential is well adapted in coconut growing soils. This can be very well grown along the borders of coconut plantation also and can generate adequate amount of nitrogen rich green leaves. The boundary of one hectare coconut garden can accommodate 450 to 500 cuttings. It can also be raised in littoral sandy soils where no other green manure can establish. *Glyricidia sepium* is propagated either through vegetative cuttings or seeds. One metre long stem cuttings or 3 to 4 months old seedlings raised in poly bags or raised beds can be used for planting. It is preferable that the planting season coincides with the monsoon (South West/ North East monsoon). For better establishment, a basal dose of 50 g of rock phosphate per pit (30 cm^3) may be applied. Height of the plants should always be maintained at 1 m by pruning.

Figure 2.8: Growing Glyricidia as Green Manure Crop.

The best growth and biomass of Glyricidia leaves could be obtained with planting of three rows of Glyricidia (at 1 m x 1 m spacing between two rows of coconut) and three pruning of leaves (February, June and October). Pruning can commence one year after planting. This could produce around 8 t of biomass in one hectare of coconut garden and about 10 kg green manure/palm/year can be made available to the palm. The loppings can be chopped and incorporated into the soil as green manure. Application of Glyricidia prunings from the interspaces of one hectare of coconut garden to palms could supply around 90 per cent, 25 per cent and 15 per cent of the requirement of N, P and K, respectively.

10.3. Biofertilizers in Organic Cultivation

The traditional additives for improving soil health and fertility in organic farming comprises of farm yard manure, composts and vermicomposts, sewage and sludge, night soil, green manure, oil cakes, meat, blood and fish meal as well as crop residues. With increasing awareness of contribution of soil microorganisms towards the soil health and fertility, use of microorganisms as biofertilizers has gained importance for improving the soil health and fertility. Bio-fertilizers are microbial inoculates containing active strain of selective microorganisms like bacteria, fungi, and algae or in combination. Biofertilizers are important components of organic farming, which help to nourish the crops through required nutrients. These microbes help to fix atmospheric nitrogen, solubilize and mobilize phosphorus, translocate minor elements like zinc, copper, *etc.*, to the plants, produce plant growth promoting hormones, vitamins and amino acids and control plant pathogenic fungi, thus helping to improve the soil health and increase crop production.

The group of micro-organisms responsible for biological nitrogen fixation, phosphorus mobilization and production of plant growth promoting substances have been found to be closely associated with the coconut palms and the palms can benefit from the use of beneficial micro-organisms as biofertilizers. The diazotrophic bacteria isolated from coconut roots include different species of *Azospirillum*, *Herbaspirillum* sp. *Azoarcus* sp., *Burkholderia* sp., *Arthrobacter* sp.,*Pseudomonas* sp. and *Bacillus* sp. Phosphate solubilizing microbes of coconut soil include *Pseudomonas* sp., *Bacillus* sp. and *Micrococcus* sp. Efficient strains of nitrogen fixing bacteria and phosphate solubilisers can be used for preparation of biofertilizers employing locally available materials such as vermicompost and coir pith as carrier materials. Enhancing the soil fertility by using *Rhizobium* sp. in conjunction with green manure crops like *Pueraria phaseoloides, Mimosa invisa* and *Calopogonium mucunoides* supplements 187 to 196 g N per coconut basin in laterite soils and 102 to 153 g N per coconut basin in sandy soils. Stem nodulating legumes such as *Sesbania rostrata*, *Aeschynomene* sp. and *Neptunia oleracea* have also become popular in improving the soil fertility. The N-fixing bacteria associated with such stem nodulating legumes belong to *Azorhizobium* and fast growing species of *Rhizobium*.

Bacteria such as *Pseudomonas* and *Bacillus* excrete acids into growth medium and hence solubilize bound phosphates. These organisms are quite useful in utilization of rock phosphates with low content of P_2O_5.

The population of phosphate solubilizing bacteria and fungi will be higher in coconut based high density multiple species cropping, multi-storied cropping with cocoa and mixed farming with napier grass as compared to that of coconut monocrop. Rhizosphere bacteria that exerts beneficial effect on plant growth, referred to as plant growth promoting rhizobacteria (PGPR),belonging to several genera *e.g. Actinoplanes, Agrobacterium, Alcaligenes, Amorphosporangium, Arthrobacter, Azospirillum, Azotobacter, Bacillus, Bradyrhizobium, Cellulomonas, Enterobacter, Erwinia, Flavobacterium, Pseudomonas, Rhizobium, Streptomyces* and *Xanthomonas. Bacillus* spp. are suitable due of their endospore producing ability. More recently *Pseudomonas* spp. are also receiving much attention as PGPR because of their multiple effects on plant growth promotion. Species of *Pseudomonas* and *Bacillus* can produce phytohormones or growth regulators that cause crops to have greater amounts of fine roots, which have the effect of increasing the absorptive surface for uptake of water and nutrients. These PGPR are referred to as biostimulants and the phytohormones they produce include IAA, cytokinins, GA and inhibitors of ethylene production. Bacteria in the genera *Bacillus, Streptomyces, Pseudomonas, Burkholderia* and *Agrobacterium* are the biological control agents. They suppress plant disease through induction of systemic resistance, and production of siderophores or antibiotics.

Utilization of vermicompost produced from coconut leaves results in the production of high quality biofertilizers with more than 10^8 cfu bacteria per gram of the carrier material. The recommended dose of biofertilizer for coconut is 100 g of carrier based inoculants per palm. The biofertilizer is to be applied in the coconut basin, twice in a year (pre monsoon and post monsoon), by mixing with top soil followed by application of organic amendments. Organic amendments such as vermicompost, coir pith compost, farm yard manure, neem cake, green manures *etc.* can be combined with biofertilizers. While applying biofertilizers, organic amendments such as vermicompost are added @ 20 kg/palm. Use only certified biofertilizer inoculants. The biofertilizer should be mixed with one kg vermicompost and applied to soil and incorporated. Care should be taken to use only biofertilizer containing adequate number of living micro organism and before the expiry period mentioned in the packet.

Optimum soil moisture is essential after biofertilizer application to ensure the survival of the introduced microbial inoculum in the soil. Hence, biofertilizer application should coincide with the onset of monsoon especially when the palms are maintained under rainfed condition. However, under irrigated conditions, it can be applied at any time, since maintaining optimum moisture is not a problem. Use of biofertilizers in coconut gardens can reduce dependency on chemical fertilizers and thus bring both economic and ecological benefits. Among the microorganisms that are promising enough to fit well into the coconut based cropping systems are the mycorrhizal fungi and certain other free-living microorganisms with specific functions. The group of microorganisms responsible for nitrogen fixation, phosphorus mobilization and production of plant growth promoting substances are used as biofertilizers in coconut based cropping systems.

The inoculation of associative diazotrophs such as *Azospirillum, Arthrobacter, Azoarcus, Herbaspirillum, Bacillus, Burkholderia* and *Pseudomonas* enhances growth and vigour of polybag raised coconut seedlings. These bioinoculants are highly effective in enhancing root biomass and branching of the secondary roots of the coconut seedlings. Organic amendment along with microbial inoculation brings

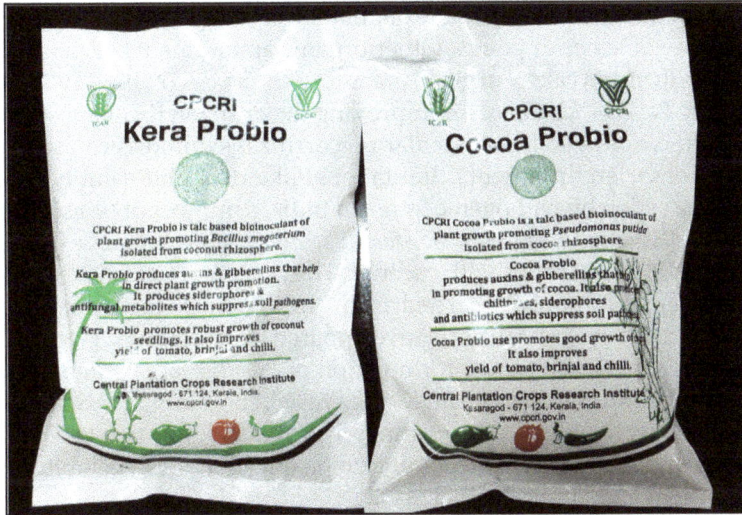

Figure 2.9: Kera Probio and Cocoa Probio.

about a greater level of plant response.The beneficial nature of biofertilizers becomes even more important in coconut based mixed cropping/farming systems as the component crops continually add plant residues to the soil which undergo organic recycling. This leads to alterations in the composition of the rhizosphere, which promotes the growth and population of beneficial microorganisms. Inclusion of livestock enterprises also creates a favourable environment for proliferation of beneficial microflora. In mixed cropping, dominated nitrogen-fixing microbial group is the bacterium *Beijerinckia* and phosphate-solubilizers such as *Pseudomonas* sp., *Bacillus* sp., *Aspergillus* sp. and *Penicillium* sp. are present in higher numbers. Not only this, higher inhibition potential of resident soil bacteria to phytopathogens is seen in coconut based cropping systems when compared to coconut monocropping systems.

When coconut is grown with cocoa, rhizosphere activity increases and a better mobilization of phosphates take place coupled with fixation of nitrogen and production of growth substances such as auxins and gibberellins in rhizosphere, which is observed to enhance yield. Inoculation of *Azospirillum brasilense* in polybags helps in enhancing the vigour of coconut seedlings by inducing profuse growth of root biomass of coconut seedlings.

Soil amendments as well as farming practices also bring about a protracted change in rhizosphere microflora, which favour the growth of specific microorganisms, thus leading to better plant growth and crop yield. For example, organic amendments like cow dung increase VA-

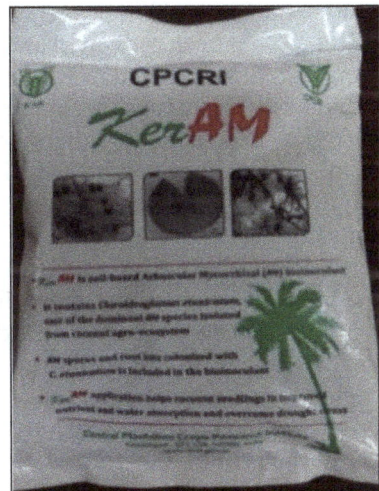

Figure 2.10: CPCRI 'Ker AM'.

mycorrhizal colonization as well as the population of phosphate solubilizing bacteria in the root zone of coconut palms. Other organic amendments such as farm yard manure, coir pith, neem cake and green manures, *etc.* can be combined with microbial inoculants like *Beijerinckia indica* for improving the nitrogen fixation by indigenous diazotrophs in coconut soils.Arbuscular mycorrhizal symbiosis can improve host responses to other environmental limitations, like drought, salinity, pollutants, erosion, and infection by pathogenic fungi. In India, a number of fungi belonging to four genera *viz., Glomus, Gigaspora, Sclerocystis,* and *Acaulospora* have been found to form mycorrhizal associations with coconut. The occurrence of a mixed population of AM has been commonly recorded from the coconut rhizosphere soils. The colonization rate is higher in tall varieties compared to dwarf ones. Two PGPR based bioinoculants, *'Kera Probio'*, a talc formulation of *Bacillus megaterium*, effective for raising robust coconut seedlings, and *'Cocoa Probio'*, containing *Pseudomonas putida*, effective for raising healthy cocoa seedlings, have been developed at ICAR-CPCRI. Both these bioinoculants were also found to be effective for vegetable crops such as tomato, brinjal and chilli.

Similarly an Arbuscular Mycorrhizal bioinoculant, *'Ker*AM', has been developed at ICAR-CPCRI, which is a soil based AMF bioinoculant for coconut seedlings. The bioinoculant contains *Claroideoglomus etunicatum*, one of the dominant AM species isolated from coconut agro-ecosystem with high potential to increase the growth parameters of coconut seedlings.

10.4. On Farm Biomass Recycling

Organic materials influence the physical, chemical and biological characteristics of soil, which in turn affect growth, development and yield of crops grown in a particular soil. The level of nutrient extraction in a coconut palm/mixed cropping system can be balanced by encouraging the decomposition of organic materials made available through mulching material, green manure crops and application of composts.The availability of waste biomass in the form of coconut leaves, stipules, spathe, bunch waste and husk from a well-managed coconut plantation with 175 palms/ha has been estimated as 14 to 16 tonnes per hectare per year. Coconut based cropping/farming system with inter/mixed crops generate still higher quantities of biomass. The natural decomposition of these wastes and the nutrient release are very slow due to the high lignin content and the nature of lignocellulose complex of the coconut waste materials. By effective recycling of this waste-biomass, the requirement of a major portion of nitrogen and a part of other nutrients of crops could be met. It also helps to replenish the nutrients exhausted by the palms internally without depending on the external sources. The estimates indicate that the biomass production by coconut-pineapple cropping system will be 4.3 times that of pure coconut stands. Studies at Kasaragod have revealed that cocoa as a mixed crop in coconut garden added to the soil 818 and 1985 kg (oven dry wt.) per hectare per year of biomass through shed leaves and pruning under single and double hedge systems, respectively. As much as 50 kg N, 4.85 kg P and 29.1 kg K per hectare could be returned to the soil every year under the double hedge cocoa system (Varghese *et al.*, 1978). The addition of large amount of organic refuses by the cocoa was found to improve the soil physico-chemical characteristics and thereby positively influence the yield of coconut. The availability of biomass for recycling in the high density multispecies cropping system was around 12.7 t to 18.2 t per hectare in terms of fronds, bunch waste and other organic materials in case of coconut, fallen leaves in clove, above ground biomass for banana, crown and leaves in case of pineapple *etc*. This biomass, if recycled, and applied into the land can enhance the productivity and sustenance of the system in terms of nutrient need besides economic benefits.

Integration of crops and livestock in mixed farming is also widely practiced in majority of the coconut growing areas and they generate additional income, provide relief against the fluctuating prices of nuts, generate more employment as well as provide large quantity of biomass for recycling. Mixed farming involves integrating animal enterprises such as dairy, poultry, duck rearing and aquaculture and cultivation of shade tolerant fodder crops in the interspaces of coconut as well as effectively recycling all the organic residues. The animals not only enhance the nutritional status of the household members but also help to augment the farm income by the sale of milk, eggs and other products. While the crop residues and fodder provide animal feed, the manure and litter of the livestock provide renewable sources of organic matter and plant nutrients. They help reduce dependence on fertilizers and maintain soil health, resulting in a high degree of organic recycling. Here the system is kept productive by maximizing the complementary and synergistic effects of the components involved. Such recycling of waste will improve the soil health and, thus, provide ecological sustenance to the system. Microbial biomass and activities of soil enzymes (phosphatase and dehydrogenase) will also be more in the farming system when compared to the mono culturing of coconut. Fodder grasses such as hybrid Napier and Guinea grass can yield about 50 to 60 tonnes of fodder per hectare in a year under coconut shade. This will be sufficient to maintain five crossbred milch cows and provide enough farm yard manure that can be used as a component for meeting the on farm organic manure requirement of the system. This will also increase the labour opportunities in the farm. Biogas plant of suitable capacity can also be installed in the farm for production of biogas for use in the farm house and slurry for manuring coconut and other component crops.

10.5. Vermicomposting for on farm Waste Recycling

Vermicomposting involves conversion of biomass into useful compost using very efficient strain of epigeic earthworm belonging to *Eudrilus euginae*, or other earthworms which can be easily done *in situ* in coconut plantations using coconut leaves and other biomass including wastes from intercrops especially from banana. When the organic wastes available from a plantation are recycled through vermicomposting, it can supply the major portion of nitrogen and a part of other nutrients required by the palms. The leaf dry matter production by tall coconut palms is around 32 kg/palm/year and hence the availability of leaf from one hectare of coconut plantation can be estimated as 5.6 t per hectare per year. In this manner all the leaves produced from one coconut palm can be converted into very good organic manure. Apart from coconut leaves, other agro-wastes like pineapple waste, banana pseudo stem and leaves and glyricidia green manure can also be effectively used along with coconut leaves for vermicomposting. Hence, the agro-wastes generated from coconut based cropping system can also be recycled efficiently in the production system. Various methods such as cement tanks, trenches as well as composting in the coconut basin itself can be adopted for vermicomposting wherein composting in the basins itself reduces the cost of transportation of leaves and application of vermicompost. Vermicomposting can also ideally be done in permanent cement or brick tanks constructed under shaded conditions to maintain appropriate quantity of substrate, optimum moisture, temperature which are necessary for efficient vermicomposting.

The weathered wastes obtained during rainy seasons are preferred. These organic wastes are to be treated with cow dung @ 10 per cent by weight in the form of slurry and must be allowed to undergo a preliminary decomposition for about 2 –3 weeks. The earthworms @ 1000 worms per tonne of coconut leaves are to be introduced. The compost bed should be mulched properly using any locally available plant material or gunny bags and has to be protected from direct sun light. Regular watering is to be done to maintain enough moisture.The low cost vermicomposting technology enables production of high quality organic manure from coconut leaves in a period of 60-75 days, leaving behind only a portion of un-decomposed material.The indigenous earthworm *Eudrilus* sp. also has affinity for wastes other than coconut leaf wastes. A coconut garden, where other intercrops/ mixed crops are grown, generates leaf wastes from these intercrops also. All these mixture of wastes can be successfully composted using *Eudrilus* sp. earthworm. It has been found that coconut leaves can be mixed with pineapple waste, banana pseudo stem or glyricidia leaves in 3:1 ratio for effective utilization of other wastes commonly produced in coconut based cropping system.

The average nutrient composition of the vermicompost recovered will be around 1.2-1.8 per cent (N), 0.1-0.2 per cent (P) and 0.2-0.4 per cent (K), organic carbon (17.84 per cent), and C/N (9.95:1). The composted palm wastes also contain higher levels of micro nutrients such as Fe, Zn, Cu and Mn when compared to that of the untreated substrate.

The C/N ratio of the organic matter ingested by the earthworm decreases and bound nutrients are converted into easily available forms. The passage of organic matter through the gut of earthworms results in enhanced availability of nutrients, increased counts of microbes and enrichment with a number of bio-active compounds. Thus, vermicompost increases soil fertility through addition of plant nutrients, growth hormones, increased level of soil enzymes and important micro-organisms as they are rich in microbial diversity, population and activity. Total microbial counts and beneficial microbial population will also be more in the coconut leaf compost compared to the base material. This enables disposal of coconut wastes in a less expensive way and eco-friendly manner with the benefit of producing high quality organic manure in the coconut plantation itself.

10.6. Vermicomposting in the Coconut Basin

The same technology for vermicomposting could be taken up in large pits taken in the inter spaces of four coconut palms in sandy loam and coastal sandy soils. If vermicomposting is done in the field itself, lot of labour required for transportation of the biomass and compost can be saved. This technology is also suitable for plantations having limited irrigation facilities, as only a few pits or trenches are to be watered. With minimum soil disturbance, open a basin with a radius of 2 m from the trunk. Cut the fallen dried coconut leaves from individual palm into two or three pieces and spread them in the basin in such way that the leaflets can hold water to hasten pre composting process. As a first step, begin the initial vermicomposting with the addition of five to six coconut leaves. Later, add more leaves in respective basins as and when available. In areas, where palms are grown under rainfed condition, vermicomposting is feasible only in the rainy period and hence, it may be started with the beginning of monsoon season. In fields where irrigation is adopted, maintenance of optimum moisture is easy and hence, year round vermicomposting is possible. After completion of the process of pre curing for three weeks, apply cow-dung @ 10 per cent of leaf weight and release around fifty earthworms of *Eudrilus* sp. to each basin. Wherever possible, the cow dung should be obtained from the organically maintained dairy unit. Cover the basin with suitable mulch material such as dried weed material *etc.* In case of continuous rainfall, care should be taken to drain excess water in the basin to avoid water stagnation.

The vermicomposting system should be kept adequately moist at 40 to 50 per cent as deficit in soil moisture will result in death of worms and it necessitates further application of earthworms. Vermicomposting directly in coconut basins not only provides major, secondary and micro nutrients, but also acts as mulch, improves soil physico chemical properties, suppresses weed growth and ensures higher microbial population build up thereby enhancing soil health. Continuous supply of biomass materials in the basins is to be ensured for effective functioning of earthworms. Advantage of this system is that all the coconut leaves and weeds added into the basin are completely composted. By this method, around 25 to 30 kg of vermicompost is added per palm per year.

10.7. Vermicomposting in Trenches

Vermicomposting in trenches can also be done by opening trenches of 1.5 m width and 1.0 m depth in interspaces of coconut garden. Apply fallen dried leaves in such trenches, apply cow dung @ 1:10 (cow dung: coconut leaves) and release earth worms @ 1:1 (1 earthworm per kg of dried leaves).Other organic materials *i.e.*, dry weeded materials can be also applied in the trenches. Proper turning of the added materials should be done for adequate aeration and to hasten vermicomposting process. Newly fallen dried leaves can also be placed in the trenches as and when available. Once the vermicomposting is completed, remove the vermicompost and apply to coconut palms. This method can be practiced even under limited irrigation or when palms are maintained under rainfed conditions. By this method, it could be possible to produce vermicompost for application @25 to 30 kg per palm per year. *In situ* recycling of coconut wastes by vermicomposting in trenches dug in interspaces of four coconut palms yield on an average recovery of 70 per cent in a composting period of 90 days.

A long term study (>10 years from 2003-04 to 2013-14) conducted at ICAR-CPCRI, Kasaragod clearly indicated the possibility of organic farming in coconut under coastal ecosystem. The increase in yield for West Coast Tall variety of coconut was 75 per cent (from pre-experimental yield of 55 nuts/palm/year to 96 nuts/palm/year), while that of Chandrasankara (COD x WCT) was 59 per cent (from pre-experimental yield of 68 nuts/palm/year to 108 nuts/palm/year). The practice of vermicomposting the recyclable biomass in trenches made in the interspaces, application of biofertilizers (Phosphobacterium and Azospirillum @100 g/palm) and raising leguminous cover crop in basins and its incorporation was found to be the best for improvement in soil nutrient status, enhancement of soil microbial population, as well as nut yield and copra content.

10.8. Vermicomposting in Tanks

Vermicomposting of coconut wastes can also be done in cement tanks. The weathered coconut leaves collected from the garden should be kept for two weeks

Figure 2.11: Vermicomposting in Tanks.

after sprinkling with cow dung slurry. Apply cow dung @ one tenth of the weight of dried leaves and release earth worms @ one kg for one tonne of the material. Sufficient moisture is to be provided for the decomposing material by frequent sprinkling of water. Direct sunlight is to be avoided. Vermicomposting will be completed in about 75-90 days. Stop providing water one week before collecting the compost. On an average, 70 per cent recovery of vermicompost could be obtained.

The vermicompost produced from coconut leaves using the technology developed at ICAR-CPCRI is now available by the trade name 'Kalpa Organic gold'.

Figure 2.12: Kalpa Organic Gold.

10.9. Multiplication of *Eudrilus eugeniae* CPCRI Strain

The earthworms can be multiplied in a 1:1 mixture of cow dung and decaying coconut leaves taken in a cement tank or wooden box or plastic bucket with proper drainage facilities. The nucleus culture of earthworms is to be introduced into the above mixture @ 50 per 10 kg of organic wastes and properly mulched with grass, straw or wet gunny bag. Keep the unit in shade and maintain sufficient moisture level by occasional sprinkling of water. The earthworms will multiply to 300 times within 30 to 60 days, which can be used for large scale composting.

Figure 2.13: Multiplication of Earthworm.

10.10. Production of Vermiwash

Vermiwash is a by-product obtained from vermicompost production technology that can be used as a liquid organic manure for improving crop growth and yield. The water-soluble components from vermicomposting tanks may be collected as leachate by passing water slowly through the composting beds or by simple suspension of vermicompost in water. This vermiwash is honey-brown in colour with a pH of 8.5 and contains both major and minor nutrients in appreciable quantity (Table 2.6). Growth promoting hormones like IAA and GA are also present in vermiwash. Vegetables and ornamental plants respond very well to its application. Vermiwash has been found to be effective as foliar spray for growth promotion and bio suppression of pathogens in crop plants. All the physiologically active water soluble components of vermicompost such as humic acids, plant growth regulators, amino acids, vitamins, micro nutrients and microbial cells are extracted in water and is known as vermiwash. Vermiwash produced from actively vermicomposting substrates of coconut leaf + cow dung by *Eudrilus* sp has an alkaline pH, contains

Figure 2.14: Collection of Vermiwash.

major and minor nutrients, growth hormones, humic acid and plant beneficial
bacteria.

Table 2.6: Characteristics of Vermiwash from Coconut Leaf Vermicompost

Nutrient	Concentration (ppm)	Nutrient	Concentration (ppm)
N	2.8	P	10.2
K	205	Ca	37.9
Mg	6.5	Fe	Traces
Cu	Traces	Zn	0.07
Mn	0.17		

Field trials conducted in farmers' plots with bitter gourd, cowpea, amaranthus,
and chillies indicated that application of vermiwash results in yield of crops on par or
slightly lower than the plots that received regular fertilizer inputs. Such application
results in healthy plant with lesser pest and disease damage, larger and deep leaf
colours, longer ability of plants to stay without wilting in field as well as longer time
of remaining fresh in case of amaranthus after harvesting, more freshness of fruits
etc. There will also be increase in soil microbial populations, soil dehydrogenase,
phosphatase and urease enzyme activities and organic carbon content of soil. The
population of the nematode, *Meloidogyne incognita* and gall formation could also be
checked with application of vermiwash.Vermiwash can be applied either through
foliar spray or soil drenching. It should be applied @ 1: 5 dilution. It has been
found to be effective as foliar spray for growth promotion and bio suppression of
pathogens in crop plants.

10.11. Vermicomposting of Coir Pith

Coir pith, a by-product obtained after extraction of coir fibers from husk, often accumulates as waste around the coir fiber extraction units and cause environmental hazard. Extraction of one kilo gram of coir fibre generates two kilo grams of coir pith. This spongy cork-like material left as such is normally resistant to biodegradation. Approximately 180 g of coir pith is obtained from the husk of one coconut. It is acidic in nature and has low bulk density and porosity. Though coir pith has a number of beneficial properties like improving soil physical properties and moisture holding capacity to a great extent, its direct utilization as manure is not advisable as it contains large amounts of lignin (75 per cent) and phytotoxic polyphenols and less of nitrogen. The C:N ratio is more than 100:1. Hence, it is to be applied to soil only after composting. This coir pith can be converted into compost either through the use of local species of *Eudrilus* or through microbial degradation, and the final product can be used to improve soil physical properties and moisture holding capacity to a great extent. The composted coir pith has near-neutral pH, a C: N ratio in the range nearly 20–27, per cent N and per cent K ranging above 1 and stable CO_2 evolution after 60–65 days of composting. *Pleurotus* spp. also has the capacity to degrade part of the cellulose and lignin present in coir pith by production of enzymes *viz.*, cellulases and lactases. The lignin content also reduces considerably.

Preparation of vermicompost from coir pith on a large scale would bring back both economic and environmental benefits. Composting of coir pith can also be achieved by inoculation of biopolymer degrading microorganisms at 0.2 per cent level. Studies conducted at ICAR-CPCRI have resulted in the isolation of efficient strains of fungi with high lignocellulose degradation capabilities, from naturally decomposing coconut wastes. *Marasmiellus troyanus* and *Trichoderma* inoculations were found to be effective in production of quality compost from coir pith. Co-composting of coir pith can be done using poultry manure, lime and rock phosphate @ 10 kg, 0.5 kg and 0.5 kg, respectively for every

Figure 2.15: Kalpa Soil Care.

100 kg of coir pith. This brings about bioconversion of coir pith to a final product in 45 days. The composted coir pith can be used as manure in coconut plantations and can increase the capability of soils to store moisture and nutrients. The coir-pith compost, thus, produced using the co-composting technology developed at ICAR-CPCRI has been released by the trade name 'Kalpa Soil Care'.

11. Biochar

Biochar is a charred solid material obtained from thermochemical conversion of plant derived biomass in an oxygen limited environment. Biochar could be produced from coir pith and tender nut husks, a waste produced from coir industries and tender nut parlours, respectively, using a charring kiln. Biochar is alkaline in nature

Figure 2.16: Biochar from Coir Pith.

and its application improves C sequestration in soil, promotes microbial activity and soil nutrients and physical properties.

Figure 2.17: Biochar from Tender Nut Waste.

12. Coconut-based Cropping System

In many coconut growing countries, coconut as a mono crop is only marginally productive and profitable and hence, a cropping system involving inclusion of compatible crops is necessary. The interplay of various factors like limited size of holding, number of trees, needs of the family, labour requirement for crop, year round returns, easiness of marketing are some of the considerations for the farmer to diversify his farm operations for higher returns by adopting intercropping, mixed cropping or introducing other enterprises like dairy, poultry *etc.* in the system. Under coconut based cropping system, the same land can be put to use to produce other crops so that the productivity of the land can be increased.

The rooting pattern of coconut palm, like other monocots, has a typical adventitious root system and about 74 per cent of the roots produced by a palm under good management do not go beyond 2m lateral distance and 82 per cent of the roots are confined to 30 to 120 cm depth of soil. In coconut gardens with a square system of planting with 7.5 m x 7.5 m spacing, the active root zone of coconut is confined to 25 per cent of the available land area and the remaining 75 per cent of the planted area left unutilized, which could be profitably exploited for raising subsidiary crops. The venation structure of the coconut crown and orientation of leaves allow considerable portion of incident solar radiation to pass through the canopy and fall on the ground. The space utilization of coconut is very low, and therefore, plenty of sunlight falling on the ground remains unutilized. As much as 56 per cent of the sunlight is transmitted through the canopy during peak hours (10.00-16.00 hrs) in palms aged around 25 years. This diffused sunlight facilitates growing a number of crops in the interspaces.

Based on the growth habit of the palm and the amount of light transmitted through its canopy, the life span of coconut palm is divided into three distinct phases. Good light transmission will be there in the initial growth phase of the palms (from planting till full development of canopy *i.e.* up to 8 years) and this period is suitable for growing intercrops with minimal competition. There will be maximum ground coverage (80 per cent) for young palms (9-25 years) with low canopy due to shorter trunk and therefore, the poor light availability in the plantation makes it not suitable for growing of many crops in the interspaces. However, shade tolerant crops could be accommodated as intercrops. Grown up palms above 25 years facilitates gradual increase in the amount of light penetration to the ground and the decrease in ground coverage of canopy makes the condition suitable for raising annual/perennial crops. Such cropping system with cultivation of a variety of biotopes will provide congenial habitats for useful insects and special bees – which both contribute to the fertilising of coconut palms and improve the productivity.The crops are to be selected based on their shade tolerance and amount of solar radiation available. It should not grow as tall as coconut and should not have an economic life longer than the main crop. Availability of resources like rainfall, irrigation facilities, soil characteristics, labour, farmers needs and market demands are some of the factors to be considered while selecting the crop combinations in a coconut based cropping system.

By applying the basic principles of organic farming for designing and developing low cost technologies based on the local resources, soil productivity as well as coconut yield could be considerably improved. These agrotechnologies integrate ecological principles into intensification process and ensure that plant nutrients are in constant and close cycling within soil and plant compartments with minimal losses from the system. Technologies have been developed for production of sufficient quantities of organic matter for recycling, production of high quality composts, conservation of soil organic matter, prevention of loss of nutrients, recovery of lost nutrients *etc.* Enhancement of biodiversity is the primary principle used to evoke self-regulation and sustainability in agro-ecosystems and when biodiversity is restored, a number of complex interactions between soil, soil organisms, plants and animals are established, giving stability to the organic farming system. The integrated farming system in coconut holdings is an interactive practice in which integration of coconut farming with suitable inter/mixed crops, livestock and other allied enterprises are undertaken with the aim of increasing income. The production alternatives can be a single intercrop, a mixture of crops, or a crop/livestock combination which are compatible with each other and other environmental factors. One of the most common farming systems practiced by coconut growing traditional farmers is the coconut-based farming system (CBFS). This is a multiple cropping or crop/livestock production system aimed at maximizing or complementing the benefits that can be derived from coconut. Different crop combinations are recommended by research to suit the availability of resources, sunlight, rainfall, irrigation and soil characteristics. It ensures optimal utilization and conservation of available resources and effective recycling of farm residues within the system. Coconut, being a voracious feeder, removes large quantity of nutrients from the soil for its growth and production of numerous energy giving materials. A bearing palm producing 50 nuts per year removes 390g nitrogen, 100 g phosphorus and 1 kg potash in each year.

Since almost all parts of the palms are used, chances of recycling biomass from coconut are less. Hence, regular application of manures is essential, especially in the traditional areas to compensate this and to maintain the soil fertility at optimum level for sustainable coconut production. Coconut based integrated cropping systems enables better utilization of natural resources and improves the soil fertility due to the continuous biomass addition by the subsidiary crops. Hence, it is recommended as one of the management practices to increase the productivity of coconut by enriching the soil fertility and also for generating higher income from unit holdings.

The farming/cropping systems designed based on local resources and needs consider the whole farm as a single unit and all components are given importance in the functioning of the system. These systems create conditions and microclimate suitable for the multiplication and activity of a variety of beneficial organisms. They protect soils from direct sunlight and rainfall and thus preserve soil organic matter reserves. As many component crops are involved, soil resources are utilized to the maximum extent, thus preventing the loss of nutrients from the system. As the biomass production per unit area will be very high, when the available organic wastes are recycled, soil health and coconut yields can be sustained even in the absence of external inputs. These systems can be adopted even by small scale

Figure 2.18: Cocoa as Mixed Crop with Coconut.

farmers and coconut based homestead farming. Coconut based farming system (CBFS) can be adopted by many small scale farmers as a self sustaining and risk minimizing strategy. The rationale is that the productivity of the coconut land can be increased. In large scale farm operations, CBFS is also adopted because it provides an efficient resource allocation strategy and minimizes input costs. The adoption of CBFS encourages improved husbandry practices, increases the productivity of coconut land, and enhances the viability of coconut ventures.

13. Coconut-based Multi-storey Cropping System

A multi-storey cropping system is a more complex CBFS, developed to accommodate two or more intercrops of different heights, canopy patterns and rooting systems, to maximize the use of available sunlight, nutrients, moisture and land area under coconut, with the fundamental objective of increasing the productivity of coconut land. The high density multispecies cropping system (HDMSCS) involves growing a large number of crops at very high plant population per unit area to meet the diverse needs of the farmer such as food, fuel, timber, fodder and cash. They are ideally suited for smaller units of land and aim at maximum production per unit area of land, time and inputs with minimum or no deterioration of land. The salient features of the system are as follows:

- ☆ HDMSCS models consist of a large number of crop species with at very high plant density.
- ☆ It includes annuals, biennials and perennials.
- ☆ The crops selected include cash crops, food crops and fodder crops.
- ☆ It includes large, medium, and small canopy crops arranged in a systematic way.
- ☆ The soil disturbance should be kept minimum - only slash weeding is done.

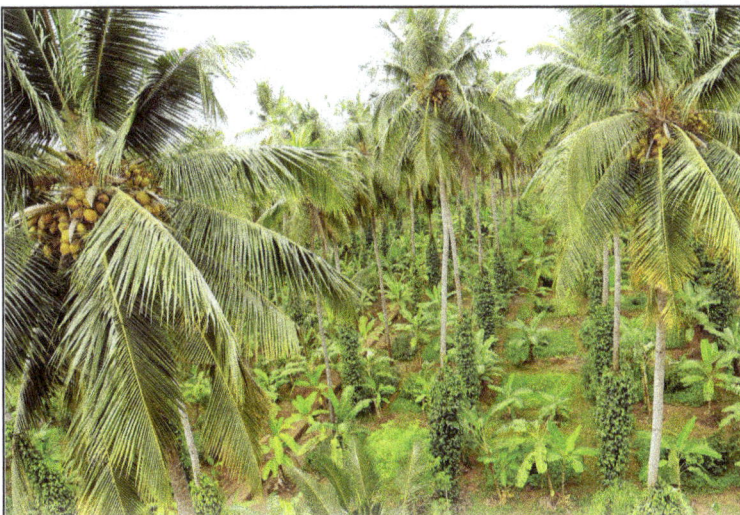

Figure 2.19: Coconut-based High Density Multispecies Cropping System.

☆ The biomass (other than the economic part) is recycled within the system.

☆ The annual crops are removed as the canopy size of perennial crops increases.

The coconut palm serves as the 'top floor', whereas, perennials such as cocoa, bananas, papaya *etc.*, form the mid-storey crops, and short-growing crops such as spices, vegetables, pineapple, fodder *etc.*, form the ground floor. As coconut palms do not have deep root system, the nutrients that are leached down are lost to the palms. When plants possessing deeper roots and greater root volume are included in the coconut based cropping system, the nutrients available below the root zone of the palms are captured and deposited on the soil surface via shed leaves, fallen twigs and other plant parts. These materials on decomposition release nutrients for the uptake by the palms.The micro climate inside the multi-storeyed cropping system is characterised by lower maximum temperature, smaller diurnal variation and less evaporative demand compared to mono cropping system. Cultivation of different crops in a particular field results in the continuous addition of bio mass and higher level of nutrient supply which have a positive impact on the physico-chemical and biological properties of soil. The beneficial effects of such a system are evidenced by the enhancement of microbial population, improvement of soil fertility status and better utilization of natural resources for the benefit of plant growth and sustainable crop yields.

14. Coconut-based Mixed Farming System

Mixed farming system integrating livestock with crop husbandry is an integral part of the organic production system. It involves cultivation of shade tolerant fodder crops in the interspaces of coconut and integrating animal enterprises like dairy, poultry, fisheries *etc.* and recycling the organic residues and by-products.

Figure 2.20: Mixed Farming in Coconut Garden with Dairy Cows.

Maintenance of livestock and other components as well as production of fodder (feed) needed for them are to be based on organic standards. The animals not only enhance the nutritional status of the household members but also help to augment the farm income by the sale of milk, eggs and other products. While the crop residues and fodder provide animal feed, the manure and litter of the livestock provide renewable sources of organic matter and plant nutrients.They help reduce dependence on inorganic chemical fertilizers and maintain soil health, resulting in a high degree of organic recycling. Such integration will also maximize the beneficial impact of species diversity on soil fertility. Here the system is kept productive by maximizing the complementary and synergistic effects of the components involved.

The suitable grasses for rainfed condition are guinea grass, Congo signal and for irrigated condition, hybrid bajra Napier. The rainfed crop yields about 35-45 t/ha of green fodder, whereas, the yield could be increased to about 75-100 t/ha under irrigated conditions. Intercropping of hybrid bajra Napier Co 3 in coconut and maintained organically produces more than 100 t fresh fodder/ha/year, which would be sufficient to manage 10 milch animals.

15. Plant Protection

15.1. Pests and their Management

The organic production should aim at minimizing losses from pests, diseases *etc*. As organic system of cultivation does not permit use of chemical pesticides for the management of pests, other measures such as use of cultural, mechanical, biological and use of botanical and bio pesticide are to be adopted. Being a perennial crop, coconut is subjected to attack by an array of pests round the year. Even though there are over 750 insect species (including the ones that directly feed and those which are only associated) recorded on coconut palm, only a few are considered to be of economic importance (Table 2.7). All parts of the palm *viz*., leaves, stem, root, inflorescence and the nuts are subjected to attack by pests. Damage when caused to the leaves, leads to reduction in photosynthetic efficiency and decrease in value for thatching purpose, but when done to inflorescence and nuts leads to direct economic loss. The major insect pests of coconut are rhinoceros beetle (*Oryctes rhinoceros*), red palm weevil (*Rhynchophorus ferrugineus*), leaf eating caterpillar (*Opisina arenosella*), root eating white grub (*Leucopholis coneophora*), Coreid bug (*Paradasynus rostratus* Dist.) and Coconut eriophyid mite (*Aceria guerreronis* Keifer). The important pests of coconut and their management are listed in Table 2.7.

15.2. Diseases and their Management

The coconut palm is affected by a number of diseases, some of which are lethal while others gradually reduce the vigour of palms causing severe loss in yield. As in the case of management of pests, no chemicals are allowed for disease control in organic cultivation. The important diseases of coconut and their management strategies are given in Table 2.8.

Organic Farming in Plantation Crops

Table 2.7: Important Pests of Coconut and their Management

Name of Pest	Symptoms	Management
a) Rhinoceros beetle (*Oryctes rhinoceros*)	The adult beetle bores through into the unopened fronds and spathes. The affected frond when fully opened shows the characteristic geometric cuts. Infestation on spathes often destroys the inflorescence and thus prevents production of nuts. The beetle breeds in a variety of materials such as decaying organic debris, farmyard manure, dead coconut logs, compost pit etc.	Field sanitation by proper disposal of decaying organic debris. Mechanically extract beetles with hooks without causing any further injury to the growing point of the palm. Apply powdered neem cake or "Marotti cake" (*Hydnocarpus wightiana* Blume) @ 250 g mixed with equal quantity of sand in topmost three leaf axils three times a year or fill the innermost two leaf axils with 12 g naphthalene balls covered with sand at 45 days interval. Treat manure pits and other possible breeding sites with leaves and tender stems of *Clerodendron infortunatum* or with the culture of *Metarhizium anisopliae* (green muscardine fungus). Spray 250 mg fungal culture diluted with 750 ml water/m³ of breeding site. The fungus can be mass multiplied on local materials such as coconut water and cassava chips. Release *Oryctes* rhinoceros virus (ORV) infected adult beetles @ 10-15/ ha of coconut plantation.
b) Red palm weevil (*Rhynchophorus ferrugineus*)	Young palms below 20 years succumb to severe damage by this pest. Bud rot and leaf rot disease and rhinoceros beetle attack are predisposing factors for red palm weevil infestation. Being an internal feeder, it is very difficult to detect the damage caused by the pest at an early stage. Wilting of the central spindle, presence of chewed fibers and cocoons in the trunk, presence of holes in the trunk with brown fluid oozing out are the important symptoms. The symptom of infestation becomes clear in advanced stages when the crown of the affected palm topples. The weevil multiplies enormously in young coconut plantations causing loss to an extent of 5 to 10 per cent.	Avoid injury to the palms, as they would attract the weevil to lay eggs. Injuries caused by rhinoceros beetle, mechanical injury during cutting of leaves or steps cut on the trunk for climbing give a favourable condition for egg laying. Mechanical injury, if any, caused should be treated with coal tar. While cutting of fronds, leave petiole to a length of 120 cm from the trunk to prevent the entry of weevils through the cut end. Periodically clean the crown to avoid decaying of debris in leaf axils. Remove palm in the advanced stage of infestation, split open the stem and burn. Adopt prophylactic leaf axil filling as suggested for rhinoceros beetle. Set longitudinally split coconut log traps (50 cm length) after smearing the cut surfaces with fermenting toddy or pineapple or sugarcane activated with yeast or molasses to attract weevil. Coconut petiole pieces smeared with fermented toddy kept in pots @ 10 pots/ha also serve as weevil traps. The traps should be placed at dusk and the weevils trapped are destroyed next morning. Install traps with aggregation pheromone to mass trap and destroy the weevils. This technology should be taken up on community basis.

Contd...

Table 2.7–*Contd...*

Name of Pest	Symptoms	Management
c) Leaf eating caterpillar (*Opisina arenosella*)	Leaf eating caterpillar commonly occurs in the coastal and backwater tracts. In recent years, it has assumed severe proportions in interior tracts as well. The caterpillars live on the under surface of leaflets inside silken galleries and feed voraciously on the chlorophyll containing functional tissues. This affects the health of the palm adversely by reducing the photosynthetic area and results in reduction of yield. The severity of infestation by this pest will be marked during the summer months from February to June. With the onset of southwest monsoon, the pest population begins to decline. In severe outbreaks of leaf eating caterpillar, the older leaves of the palms are reduced to dead brown tissue and only three or four youngest leaves at the centre of the crown remain green. In case of severe infestation, the whole plantations present a scorched appearance.	Cut and burn the heavily affected and dried outer most 2 to 3 leaves. Adopt biological control by periodical release of larval/pupal parasitoids such as *Goniozus nephantidis, Elasmus nephantidis* and *Brachymeria nosatoi*. Combined release of the parasitoids is required in multistage condition of the pest.
d) White grub (*Leucopholis coneophora*)	The soil inhabiting white grubs cause damage to the roots of coconut. Besides coconut, it infests tuber crops like tapioca, colocasia, and sweet potato etc., grown as inter-crops in coconut gardens. In coconut nursery, the grubs feed on the tender roots and tunnel into the bole of the collar region resulting in drying up of the spindle followed by yellowing of the outer leaves and gradual death of the seedling. In older coconut plantations continuous infestation by the grub results in yellowing of leaves, premature nut fall, delayed flowering, retardation in growth and reduction in yield.	Collect and destroy adult beetles during peak period of emergence in May–June to reduce the population.

Contd...

Table 2.7–Contd...

Name of Pest	Symptoms	Management
e) Coreid bug (*Paradasynus rostratus* Dist.)	Coried bug occurs in coastal areas and in high ranges of Kerala. Apart from coconut it feeds on tamarind, cashew, cocoa and guava. The peak population occurs during post monsoon period. The adults and nymphs feed by desapping the contents on buttons and developing nuts below the perianth region. The feeding points develop to brownish necrotic lesions, which later turn to furrows or cracks. The symptoms are easily identified by cracks and gummosis. Severe damage leads to nut fall and malformation of mature nuts.	Apply neem based bio pesticide on the newly opened inflorescence.
f) Eriophyid mite (*Aceria guerreronis* Keifer)	Mite feeds on the upper portion of the developing nut that is covered by perianth. Feeding by mites in this zone causes physical damage to cells. The feeding sites that grow downward from the perianth appear as longitudinal patches and later develop into triangular yellow patches, turn brown, develop longitudinal fissures and finally appear as warts and develop into longitudinal splits on the surface of nuts. The liquid oozing from these patches dries and as a result dried decayed matter is noticed. The damage affects the quality of husk and dehusking becomes difficult.	Adopt phytosanitary measures in coconut plantations like crown cleaning. Collect and destroy all the fallen buttons of the affected palm. Spray neem oil-garlic-soap emulsion @ 2 per cent concentration (200 ml neem oil, 50 g soap and 200 g garlic mixed in 10 litres of water) or commercial neem formulation azadirachtin 0.004 per cent (Neemazal T/S 1 per cent @ 4 ml per litre of water) during April-May, October-November and January-February. Apply the spray solution as fine droplets on the perianth region and general surface of developing nuts of 1-6 months old bunches with hand sprayer or rocker sprayer.

Table 2.8: Diseases of Coconut and their Management

Name of the Disease	Symptoms	Management Strategies
a) Bud rot (*Phytophthora palmivora*)	☆ The first visible symptom is withering of the spindle leaf marked by pale colour. ☆ The spindle turns brown and droops down. ☆ The tender leaf base and soft tissues of the crown rot into a slimy mass of decayed material emitting a foul smell. ☆ The disease may spread to adjacent leaves, producing a dead centre with a fringe of living leaves. ☆ The disease kills the palm if not controlled at the early stages. Palms of all age are liable to be affected but normally young palms are more susceptible. ☆ The disease is more prevalent during monsoon when the temperature is low and humidity is high.	☆ Cut the palm which are in the advanced stage of disease or died due to the disease and burn the infected crown ☆ As a prophylactic measures spray 1 per cent Bordeaux mixture to all the palms in the garden in the disease endemic areas ☆ In early stages of the disease, when the spindle leaf starts withering, cut and remove all affected tissues of the crown and apply Bordeaux paste and protect it from rain by providing polythene covering till normal shoot emerges. Later remove the cover as the shoot grows ☆ Destroy infected tissues removed from the affected palm by burning. ☆ Spray 1 per cent Bordeaux mixture on spindle leaves and crown of palms around the infected area to prevent the disease spread. ☆ Provide adequate drainage in gardens and avoid over crowding.
b) Root (wilt) disease Phytoplasma. The disease is transmitted by lace bug *Stephanitis typica* and the plant hopper *Proutista moesta*.	☆ The important visual diagnostic symptoms are abnormal bending or ribbing of the leaflets (flaccidity), general yellowing and marginal necrosis of the leaflets and unopened inflorescences. ☆ The nuts are smaller and the kernel is thin. ☆ The oil content of copra is also reduced.	This disease is not lethal but only debilitating. ☆ As no curative measure is known at present, the approach will be to manage the disease in the already infected gardens. ☆ To reduce the loss due to the disease, the strategy would be to contain the disease by improving the health of affected palms and increasing the yield through proper manuring and other agronomic practices. ☆ Cut and remove all affected palms in mildly disease affected areas. ☆ In the heavily disease affected tracts, remove severely affected uneconomic adult palms (those yielding less than 10 nuts per palm per year) and all diseased palms in the pre-bearing age.

Contd...

Table 2.8–Contd...

Name of the Disease	Symptoms	Management Strategies
		☆ Adopt improved management practices in the affected gardens to enhance the yield of palms.
		☆ Organic recycling by following mixed farming system - Raising fodder crops in the interspace and maintaining milch cows and application of farm yard manure to palms.
		☆ Grow suitable inter and mixed crops.
		☆ Basin management with green manure crops.
		☆ Irrigation during summer months.
		☆ Leaf rot disease which is usually noticed in root (wilt) affected palms can be controlled by applying *Pseudomonas fluorescens* or *Bacillus subtilis* either alone or in combination @ 50 g in 500 ml water to the axil of spindle leaf
		☆ Replanting with progenies of disease free elite palms located in hot spot areas.
		☆ Follow strictly all the prescribed prophylactic measures for other pests and diseases.
c) Leaf rot (*Exserohilum rostratum* and *Colletotrichum gloesporioides*)	☆ Symptoms appear as minute water soaked angular spots on spindle leaves ☆ They enlarge, coalesce and cause spindle rot	☆ Cut and remove rotten portion of the spindle and two adjacent leaves.
		☆ Since leaf rot affected palms are prone to pest attack, filling the youngest three leaf axils with a mixture of powdered neem/marotti cake with equal quantity of sand or placing naphthalene balls (12g/palm) and covering with sand three times a year may be adopted.
		☆ Apply *Pseudomonas fluroscens* or *Bacillus subtilis* either alone or in combination as explained above
d) Stem bleeding (*Thielaviopsis paradoxa*)	☆ The disease is characterized by the exudation of dark reddish brown liquid from the longitudinal cracks in the bark, generally at the base of the trunk ☆ The bleeding patches spread throughout as the disease advances. The liquid oozing out dries up and turns black	☆ Remove water stagnation (if it is a problem) and apply 5 kg neem cake fortified with *Trichoderma* per palm along with other organics during September-October

Contd...

Table 2.8–Contd...

Name of the Disease	Symptoms	Management Strategies
	☆ The tissues below the lesions rot and turn yellow first and later black. Leaves in the outer whorl turns yellow rather prematurely, droop and dry ☆ Production of bunches is affected and nut fall also is noticed. The trunk gradually tapers at the apex and crown size becomes reduced	☆ Apply 50 kg organic manure and 5 kg neem cake fortified with *Trichoderma* per palm and provide irrigation ☆ Provide drainage channels between rows of palms ☆ Isolate the affected palm from the healthy ones by digging a trench around the affected palm ☆ Adopt phytosanitary measures (remove dead palms, bury the affected roots and bole in a pit) ☆ Intercropping of banana is desirable as the root exudates of banana are found to inhibit the growth of pathogens
e) Thanjavur wilt/ Ganoderma disease (*Ganoderma lucidum* and *Ganoderma applanatum*)	☆ Decay of root system, flaccidity of spindle leaves, browning of outer leaves, arrested fruit set and appearance of bleeding patches on the basal region on the stem are the symptoms observed. ☆ Ultimately the palm dies off. In advanced stages, brackets of fungus causing the disease are seen on stumps	
f) Leaf blight or Grey Leaf spot (*Pestalotia palmarum*)	☆ In the mature leaves of the outer whorl, yellow specks encircled by a greying band appear which later turn to greyish white ☆ The spots coalesce into irregular necrotic patches causing extensive leaf blight. When the infection is severe the leaf blade completely dries and shrivels off	☆ Cut and remove older affected leaves and spray the foliage with 1 per cent Bordeaux mixture ☆ Combined application of talc-based powder formulation of *P. fluorescens* to soil (50 g/palm/year) along with neem cake (5 kg/palm/year)
g) Mahali or fruit rot and nut fall (*Phytophthora palmivora*)	☆ Shedding of buttons and immature nuts are noticed ☆ Water soaked lesions appear on buttons near the stalk which later develop and result in the decay of the underlying tissues ☆ The disease caused by the fungus appears as whitish webby growth on the surface of the affected part	☆ Collect and burn the affected shed nuts ☆ Spray 1 per cent bordeux mixture to the bunches just before the onset of monsoon

16. Harvesting and Post-harvest Management

16.1. Harvesting

Twelve months old coconuts are to be harvested both for copra preparation as well as seed nut purpose, while for tender nut purpose; the nuts are to be harvested at 7 to 8 months stage. The maturity of nuts shall also be considered for harvest depending up on the value addition to be made using kernel, coconut water *etc*. Six to eight harvests, on an average, can be made in a year depending on the yield of palms.

16.2. Post-harvest Management

Fresh coconuts: Coconuts which are intended to be sold fresh should be harvested before they are completely ripe, as they will contain up to 95 per cent coconut water. In order to export fresh coconuts, it is recommended to remove thick fibrous husk of the coconuts immediately after they are harvested by keeping a small portion of the fibre on the upper side of the nut. Organically produced coconuts are not allowed to be treated with methyl bromide or ethylene oxide, or with ionising rays for long storage.

16.2.1. Making 'Cup Copra'

In order to make 'cup copra' the fibrous husks are removed from the freshly harvested coconut fruits. Later they are split open into two parts using a heavy knife and washed in clean, cold water to remove any foreign particles and fibres. Then they are briefly pre-dried by placing them out in the sun on racks, mats or in solar dryers. This drying facilitates easy separation of meat from the shell. The split open nuts should be kept for drying as soon as they have been opened, as any delay will result in the meat turning reddish-brown and deterioration of quality of copra. After about two days of sun-drying, the fruit meat (kernel) is usually hard enough to be removed from the hard shell. In a period of about 4-5 days, the entire drying process will be completed and copra removed from the shell. Before the copra is packed, they should be cleansed of any foreign particles (stones, sand, fibre residues *etc*.).

16.2.2. Making Ball Copra

In order to make 'ball copra', ripe coconuts are stored in the shade for about 8 to 12 months. In this way, the coconut water is gradually absorbed, and the coconut meat shrinks and dries, so that it begins to rattle around when shaken. When the meat begins to rattle, the fibres and shell are to be carefully removed. Before it is packed, the copra should be cleansed of foreign particles (stones, sand, fibre residues *etc*.). There are specific varieties suitable for making ball copra (*e.g. Kalpa Ganga*).

16.2.3. Manufacturing Dried, Grated Coconuts

In order to make dried, grated coconuts, the brown shell around the copra is removed, the meat washed with clean, cold water, then sterilised, grated, dried, and if necessary, sieved into grades. The grated, dried coconuts are to be sorted into different grades according to their grain size.

Selected References

Anonymous (2001). Organic production of coconut –guidelines Coconut Development Board. p.75.

George V. Thomas and Palaniswami, C. (eds) (2003). Recent advances in organic farming technologies in plantation crops. Central Plantation Crops Research Institute, Kasaragod p. 140.

George V. Thomas, Subramanian, P., Krishnakumar, V., Alka Gupta and Chandramohanan, R (2010). Package of practices for organic farming in coconut. Technical Bulletin No. 64. CPCRI. p.25.

George V. Thomas, Krishnakumar, V., Subramanian, P., Murali Gopal and Alka Gupta. (2012). Organic farming in coconut - feasibility, technological advancements and prospects. *Indian Coconut J.* **55** (7): 22-31.

Murali Gopal, Alka Gupta and George V. Thomas (2010).Opportunity to Sustain Coconut Ecosystem Services through Recycling of the Palm Leaf Litter as Vermicompost: Indian Scenario. *Cord* **26** (2):42-45.

Prabhu, S.R., Subramanian, P., Biddappa, C.C. and Bopaiah, B.M. (1998). Prospects of improving coconut productivity through vermiculture technologies. *Indian Coconut Journal.* **29** (4): 79-84.

Satyagopal, K., Sushil, S.N., Jeyakumar, P., Shankar, G., Sharma, O.P., Boina, D.R., Sain, S.K., Reddy, M.N., Rao, N.S., Sunanda, B.S., Ram Asre, Kapoor, K.S., Sanjay Arya, Subhash Kumar, Patni, C.S., Gangopadhyay, S., Mesta, R., Venkateshalu, Ekabote, S.D., and Rajashekarappa, K. (2014). AESA based IPM package for Coconut. p 38.

Singh, H.P. and George, V. Thomas (Eds.) (2010). *Organic Horticulture – Principles, Practices and Technologies*. Westville Publishing House, New Delhi. p.444.

Chapter 3

Organic Farming in Arecanut

☆ *Ravi Bhat, S. Sujatha*
and P. Chowdappa

1. Introduction

The arecanut palm is the source of a widely used masticatory nut, popularly known as arecanut, betel nut, or *supari*. It is mostly chewed along with the betel leaf, or in the form of value-added products like *gutka, pan masala* and scented *supari*. Arecanut is an important commercial crop of India and its industry forms the economic backbone of nearly six million people and for many of them it is the sole means of livelihood. Although the cultivation and production of arecanut is focused only in a few states of India, the commercial products are widely distributed across the country and are being consumed by all classes of people. Area expansion of arecanut in India is discouraged; however, the area increased by 70 per cent during the last two decades and the production increase was mainly due to this phenomenon.

2. Production Scenario

2.1. Global Scenario

The current production of arecanut in the world is about 1275 thousand tonnes from an area of 926 thousand hectare (Table 3.1). India ranks first in both area and production (49 per cent) of arecanut. The other major arecanut producing countries are Indonesia (16 per cent area and 15 per cent production), China (5 per cent area and 11 per cent production) and Bangladesh (20 per cent area and 8 per cent production).

Table 3.1: Country-wise Area, Production and Productivity of Arecanut
(Figures in brackets are percentage share)

Country	Area ('000 ha)	Production ('000 t)	Productivity (kg/ha)
India	453.6 (49)	632.6 (49)	1395
Indonesia	149.9(16)	187.0 (15)	1247
China	46.0 (5)	135.0 (11)	2935
Bangladesh	184.0 (20)	108.0 (8)	587
Myanmar	56.5 (6)	122.0 (9)	2159
Thailand	18.0 (2)	35.0 (3)	1944
Sri Lanka	15.9 (2)	37.7 (3)	2370
Others	2.0	17.5 (2)	—
World	925.9	1274.8	1377

Source: FAOSTAT.

2.2. Indian Scenario

In India, arecanut is cultivated in an area of about 4.54 lakh hectares with an annual production of 6.32 lakh tonnes (Table 3.2). The states of Karnataka, Kerala, Assam, West Bengal and Meghalaya are the major producers and account for more than 94 per cent of the area and production.

Table 3.2: State-wise Statistics of Arecanut in India

Country	Area ('000 ha)	Production ('000 t)	Productivity (kg/ha)
Karnataka	221.4 (49)	358.6 (57)	1620
Kerala	101.7 (22)	118.2 (19)	1162
Assam	75.1 (17)	72.6 (11)	967
West Bengal	11.4 (2)	21.2 (3)	1857
Meghalaya	16.0 (4)	23.0 (3)	1626
Others	28.0 (6)	39.0 (6)	—
India	453.6	632.6	1395

Source: NHB, 2012-13, Figures in brackets are percentage share.

3. Climatic Requirements

3.1. Geographical Position

Sub humid tropical climate suits the crop best, and therefore, arecanut thrives very well in the regions of 28° north and 28° south of the equator. Though arecanut is grown under different agro-climatic conditions, it is very sensitive to extreme climatic conditions. Although arecanut palms grow at altitude up to 1000m above MSL, at higher levels, the quality of the fruits will be adversely affected. In the high altitude areas like Wynad (Kerala) and Coorg (Karnataka), the endosperm will not develop sufficient hardness. Moreover, high altitudes will also affect the germination of seeds and quality of chali (dry kernel).

3.2. Temperature

Arecanut palms grow well in a range of temperature between 14°C and 36°C. However, the crop is being cultivated in regions with extremes of temperatures as low as 5°C (Mohitnagar, West Bengal) to high as 40°C (Vittal in Karnataka and Kannara in Kerala). Extremes of temperature and wide diurnal variations are not conducive for the healthy growth of the palms.

3.3. Rainfall

Arecanut palms require very high rainfall ranging from 300 to 450 cm per annum. However, it is grown in. areas with wide variations in rainfall such as Malnad of Karnataka where the annual rainfall may go more than 450 cm as well as in low rainfall areas like Maidan parts of Karnataka or parts of Coimbatore district in Tamil Nadu where the annual rainfall is about 75 cm. In areas of prolonged dry spell, the palms are to be irrigated.

3.4. Relative Humidity

Relative humidity directly influences the water relations of palm and indirectly affects leaf growth, photosynthesis, pollen dispersal, occurrence of diseases and finally economic yield. Therefore, very high or low relative humidity is not conducive for growth and development of arecanut. High humid conditions provide congenial conditions for the rapid spread of diseases like fruit rot, bud rot *etc*.

4. Soil Requirements

Arecanut is cultivated in a wide variety of soils; however, sandy, alluvial, brackish, or calcareous and sticky clay soils are not ideal. The largest area under the crop, however, is in regions with the gravelly laterite soils of red clay. Deep black fertile clay loam soils supports luxuriant palm growth. Under well-drained deep soil conditions, the roots of arecanut palm traverse down to about three meters and the roots confine to only about 1.40 meters under shallow soil condition and, hence the palms require deep soil, preferably not less than two meters for development of proper root system. In Malaysia and Fiji, arecanut is cultivated in the hot, moist, rich alluvial areas of the coastal belt.

5. Varieties

The areca palms could be broadly classified into tall, semi tall and dwarf types. There is wide range of variations in fruit characters, stem height, inter node length and leaf size and shape. There are also wide variations in yield, earliness in bearing, fruit number/bunch, quality, and dwarfness. Variation in plant morphology, fruit colour, shape and size could also be observed in different accessions. The varieties suited for tender nut processing may not be suitable for mature fruit drying and vice versa. Many varieties have been released by different agencies like ICAR institutes and State Agricultural Universities (Table 3.3).

Table 3.3: Characteristics and Yield Performance of Arecanut Varieties/Hybrids Released in India

Variety	Growth Habit and other Characteristics	Shape and Size of Nut	Chali Yield (kg/palm)	Areas Recommended for Cultivation	Year of Release	Agency responsible for Release
Mangala	Semi-tall, early and heavy bearer	Round, medium	3.00	Coastal Karnataka, Kerala	1972	CPCRI
Sumangala	Tall, heavy bearer	Oval to round, medium	3.28	Karnataka,Kerala	1985	CPCRI
Sreemangala	Tall, long internodes	Round, oval, bold, deep yellow	3.18	Karnataka,Kerala	1985	CPCRI
Mohitnagar	Tall, uniform nuts, sturdy stem	Oval to round, medium, deep yellow	3.67	West Bengal, Karnataka and Kerala	1991	CPCRI
South Kanara	Tall	Round, bold	2.00	Coastal Karnataka and Kerala		
CAL17(Sumrudhi)			4.37	Andaman and Nicobar Islands		
SAS-1			4.60	Valleys of Sirsi, Karnataka	1995	RARS, UAS, Dharawad
Swarnamangala	Semi tall, regular bearer	Oblong to round, bold,	3.78	Karnataka, Kerala	2006	CPCRI
Madhuramangala	Medium tall, high yielding, suitable for ripe and tender nut	Orange	3.54	Konkan region of Maharashtra, Karnataka		
Kahikuchi Tall	Tall, shorter internodes	Round, big, orange	3.70	Assam, North east hills	2010	CPCRI
Nalbari	High yielding, good processing quality for ripe nuts		4.15	Karnataka,N. Bengal and NE region		
Arecanut hybrids						
Hirehalli Dwarf x Sumangala (VTLAH 1)	Dwarf with reduced canopy, very sturdy stem, early yield stabilization	Oval, medium, yellow to orange nuts	2.55	Karnataka, Kerala	2006	CPCRI
Hirehalli Dwarf x Mohitnagar (VTLAH 2)	Dwarf with reduced canopy, very sturdy stem, early yield stabilization	Oval, medium, yellow to orange nuts	2.64	Karnataka, Kerala	2006	CPCRI

Madhuramangala

Mangala

Mohitnagar

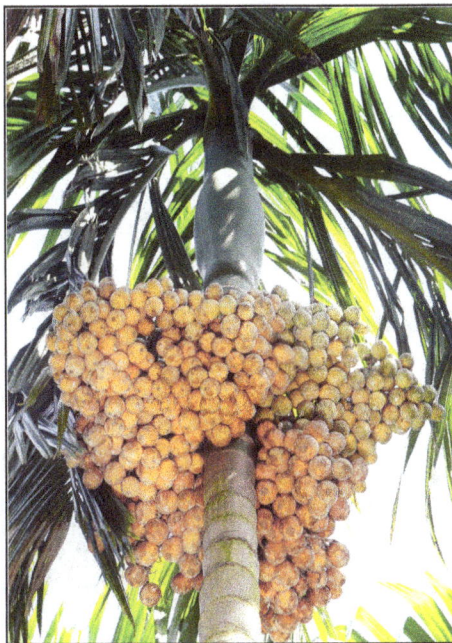

Nalbari

Figure 3.1a: Varieties of Arecanut.

Sreemangala

Sumangala

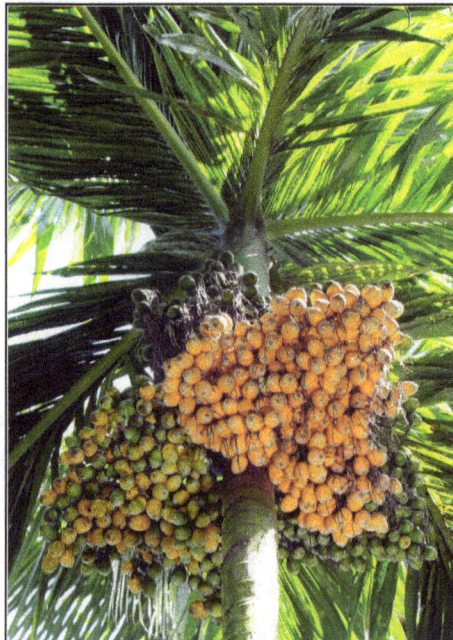

Swarnamangala

Figure 3.1b: Varieties of Arecanut.

 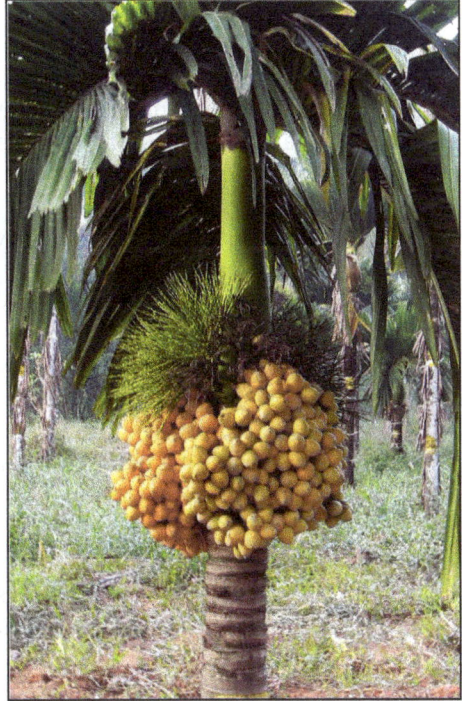

VTLAH 1 **VTLAH 2**

Figure 3.2: Hybrids of Arecanut.

Characterisitcs of Local Varieties of Arecanut

a) **South Kanara:** It is largely grown in South Kanara district of Karnataka and Kasaragod district of Kerala. It is characterised by large nuts and uniform bearing. The average yield is 1.5 chali/palm/year (7 kg ripe nuts).

b) **Thirthahali:** It is grown extensively in Maland area of Karnataka. It is preferred for tender nut processing and not as dry nut.

c) **Sree Vardhan or Rotha:** It is predominantly grow in coastal Maharashtra. The nuts are oval, in shape and the yield is 1.5 kg chali (7kg ripe nuts) per palm per year. The kernel colour when cut is marble white. Its endosperm is tastier than other varieties. It starts bearing after 6-7 years of planting.

d) **Mettupalayam:** It is grown widely in Mettupalayam area of Tamil Nadu, the nut size is very small.

6. Production of Planting Materials and Nursery Management

6.1. Selection of Mother Palms

The areca palm is a seed-propagated plant and, in general, seedlings are used as planting material. Presently, tissue culture derived plants are also available as planting material. The age at first bearing and regular bearing habit are two important characters to be considered for selection of mother palm. While selecting seeds, it must be ensured that they are obtained from trees that are already stabilized in their yielding pattern, and at least more than 10 years in maturity. More number of leaves on the crown (>10), shorter internodes, 350 to 400 fresh nuts/palm/year and high fruit set (>55 per cent) are some of the other characters to be considered while selecting mother palms. Mother palms should be selected based on the processing requirement also as all cultivars are not suitable for preparing either chali or tender processed nuts.

6.2. Selection of Seed Nut

Fully ripe nuts weighing > 35g are to be selected. Heavier nuts and those float vertically in water with calyx end pointing upwards produce more vigorous seedlings. Nuts with nine or ten months of maturity can be used for seed nut purpose. As the viability will be lost quickly, the nuts could be stored only for 3-6 days.

6.3. Raising of Seedlings

In order to produce quality seedlings, they are to be raised with adequate care. There are two steps to raise arecanut seedlings.

6.3.1. Primary Nursery

Whole fruits are to be used as seed nuts and sown immediately after harvest in soil/sand and water once in two days for early and good germination. The nuts are to be sown vertically with calyx end just covered at a distance of about 5 cm. Thick mulching using straw or arecanut leaves is to be done. The nuts commence germination by around 45-50 days and complete by about 90-95 days. The number of days required for germination increases with altitude.

6.3.2. Secondary Nursery

Seedlings are raised in secondary nursery by preparing beds of about 1.5 m width and 15 cm height. Plant three months old sprouts at 30 cm to 45 cm. Apply a basal dose of decomposed farm yard manure@ 5 t/ha in the secondary nursery. Areca sprouts and seedlings are highly susceptible to exposure to direct sun, and hence needs shading. The shade may be either of coconut or arecanut leaves spread over a pandal or by planting some fast growing green manures or banana. The commercially available shade nets can also be used for this purpose. *Sesbania aegyptica* has been found to be one of the best live shades in areca nurseries especially in Maidan parts of Karnataka, India.The concept of primary and secondary nursery can be avoided and the seed nuts can be directly sown in raised beds (15cm) of 130 cm width at a spacing of 30cm x 30cm in the sub-Himalayan region of West Bengal where the harvesting and sowing seed nuts coincide with rainy season. Seedlings can also be raised in PVC (polyvinyl chloride) bags of 25 cm x 15 cm, 150 gauge with a mixture of top soil, cattle manure, and sand used in the ratio of 7:3:2, respectively. Solarization of soil by covering with black polythene sheet and sun drying potting

Figure 3.3: Arecanut Nursery Seedlings.

mixture helps to avoid soil borne disease. Three months old sprouts can be planted in polybags. Apply water on each day during non rainy periods.

6.4. Selection of Seedlings

Arecanut, being a perennial crop committed to the land for many decades, adequate attention is to be bestowed for adoption of scientific cultivation practices right from the beginning. Disease and pest free seedlings with five or more leaves (early splitting) and minimum 90 cm height and 26 cm collar girth are to be selected from the secondary nursery for field planting. One or two years old seedlings when used for planting produce more vigorous palms with early flowering. They should have well established root system. Uproot the seedlings with a ball of earth adhering to roots for field planting. Covering the base with ball of earth with plastic sheet/ bag will help in keeping the seedlings in good condition for long distance transport.

In Assam, where soil is heavy and water stagnation could be problem, planting of 18-30 months old seedlings is ideal.

7. Planting and after Care

7.1. Site Selection and Layout

Exposure of arecanut palms to direct sunlight causes scorching effect on stems and they become weak and susceptible to wind fall, and therefore, site selected should have protection against hot sun from both southern and western side by tall growing trees. Planting methods like square, triangle and quincunx may be followed. Aligning the rows in north-south direction and planting on quincunx system with angling 35° towards west lowers the incidence of sun-scorching. The palms cannot withstand both drought and water stagnation. Thus, the site selected should have facility for irrigation during dry weather and sufficient drainage to drain away excess water during heavy rains.

7.2. Spacing

The spacing followed in different areca growing regions varies from 1.25 m x 1.25 m to 3.6 m x 3.6 m. Arecanut plants require sufficient sunlight for better growth and yield, and hence, proper spacing should be adopted. The root distribution studies in relation to individual palm yield and unit area yield indicated that 2.7 m x 2.7 m spacing is optimum for arecanut and this spacing is generally adopted by the farmers. In areas where wider spacing is adopted, the resource utilization will not be complete, and if closer spacing is followed, there will be heavy concentration of roots in the lower layers of soil resulting in reduction in yield. However, wider spacing provides ample opportunity to accommodate a number of perennial and annual crops in the interspaces.

7.3. Depth of Planting

Depth of planting is mainly decided by the soil type and the water table. In laterite soil with good drainage, the seedlings can be planted at 90 cm depth. However, in areas where natural drainage can be provided (particularly during the heavy rainfall period), deeper planting of seedlings up to 90 cm depth is preferred, as it provides firm anchorage to the roots and provides large volume of space for spread of roots. If shallow planting is practiced, the roots get exposed and the palm needs earthing up. The seedlings are to be planted with a ball of earth in the pits after filling half portion with top soil and compost mixture. The base of the seedling should be pressed properly and the pit mulched with green leaves.

7.4. Season of Planting

In areas of heavy rainfall due to south-west monsoon and in river banks, where inundation is likely, planting of arecanut seedlings can be done in September-October. In other areas planting can be done during May-June.

7.5. Drainage

Proper drainage ensures better growth and development of the plants, since arecanut cannot withstand water stagnation. The soil type will decide the number of drainage channels to be made in the plantation. In light soils, it could be less, whereas, in heavy soils, the channels should be dug in each row to drain out excess water. Prepare channels which are at least 15 - 30 cm deeper than the depth at which the seedlings are planted. The channels are to be cleaned every year for easy flow of water. The planted pits also should be provided with outlets to drain away the water.

7.6. Shading

The arecanut plants are highly susceptible for sun scorching, and hence, the seedlings should be given adequate protection against the direct exposure to sun. This can be done either by covering the plants with coconut or arecanut leaves or by planting banana in between two rows of arecanut. Such banana plants give additional income to the farmers. Sun scorching is usually noticed during October - January. The stems of young palms have to be protected during this period, as the part once lost or got damaged cannot recover. Planting quick growing shade trees on southern and western sides of the garden also helps to protect the arecanut plants from sun scorching. The palms are to be irrigated in such cases.

7.7. Irrigation Water Management

Arecanut cannot withstand drought for a long time, and being a perennial crop, once affected by water stress, it may require two-three years to regain the normal vigour and yield. Therefore, in places where such situations are likely to occur, irrigating the palms becomes essential. However, in places with high sub-soil moisture and in areas where the rainfall is well distributed, throughout the year, no irrigation is necessary. Irrigate the palms once in a week during November to February, once in 4 days during March to May. Drip irrigation @16 – 20 lit/tree/day will be economical. The water use efficiency can be enhanced through a good combination of irrigation and mulching. Mulching is very important in arecanut plantation because the highly porous soils in which the crop is planted can lead to water loss through seepage. Different types of organic mulches could be used for this purpose.

8. Arecanut Based Cropping/Farming System

Arecanut being essentially a crop of small and marginal holders, often with less than one hectare, the insufficient income will not be able to sustain dependent families. The growth habit and long pre-bearing period (5–6 years) of arecanut provide ample opportunities for increasing land and other resource use efficiency by way of adoption of multiple cropping practices. Such system ensures better utilization of basic resources, enhances income and employment opportunities. The palm with its compact crown, raised well above the ground (10–15 m), allows more sunlight to transmit to ground and maintains high humidity. Studies

Figure 3.4: Cocoa as Mixed Crop in Arecanut Field.

indicated availability of congenial microclimate and less utilization of resources for intercropping in arecanut plantations. Inter/mixed crops are to be selected based on the age of the palms, size of the crown and availability of sunlight in the garden. The initial pre-bearing phase is the ideal time to grow intercrops, especially short duration ones, whereas, in later years of growth of palms, as the canopy enlarges in height, mixed cropping with other shade tolerant perennial crops can be practiced. From the perspective of biodiversity management within the arecanut plantation, the general practice by the farmers has been that fruit crops such as papaya, citrus and banana are intercropped with arecanut without loss of soil fertility and productivity. In fact, these fruit trees provide shade and enhance the moisture retention capacity of the soil. The fruit trees also attract insects which in turn act as pollinators for the arecanut palm.

A number of annual crops, such as rice, sorghum, beans, corn, groundnut, and sweet potato, vegetables, yams, banana, and pineapple can be grown in arecanut plantation. When these crops are cultivated, all of them are also to be rasied following organic cultivation practices. Banana is the most preferred intercrop in all the arecanut gardens as it also provides good shade during early growth of arecanut palms. Black pepper can also be grown using arecanut stems as live standards. Cocoa is another important intercrop as the microclimate, especially shade, soil moisture, and temperature, in the arecanut gardens is ideal for cocoa growth. A spacing of 2.7 m x 2.7 m or a spacing of 2.7 m x 5.4 m combination could be adopted, although operational advantages are better in the latter spacing. High-density multispecies cropping in arecanut can be adopted with combination of crops such as black pepper, cocoa, coffee, or banana occupying different vertical air space levels. Population of general and function specific microorganisms will also be higher under such a system. Cropping models for heavy rainfall zone are arecanut + cocoa; arecanut-based high density multispecies cropping system with cocoa, banana and pepper; arecanut + pepper + cocoa + banana, arecanut + medicinal and aromatic plants and arecanut + vanilla.

9. Nutritional Management

Majority of the arecanut growing areas, being lateritic in nature, have low soil fertility status and exhibit wide spread micronutrient deficiency. Incidence of disorders like crown choking, crown bending, shortened internodes and oblique nodes are increasingly being noticed due to nutritional disorders. Unless these problems are taken care of, arecanut will not be a profitable crop in future. While raising nursery, apply basal dose of well decomposed FYM or vermicompost @ 2 t/acre treated with *Trichoderma* 2-3 weeks before planting seed nuts in sand bed nursery. Organic manures such as cattle manure, compost, or green leaves (*Glyricidia maculata*, which can be grown on the boundaries of arecanut gardens) are to be used to supply nutrients at the time of planting @ 12 kg/plant.

9.1. Green Manure Crops

Growing of green manure crops with the onset of monsoon will help suppress weed growth, prevent soil erosion and add large quantities of organic matter to the soil. *Pueraria javanica, Calopogonium mucunoides, Stylosanthes gracilis* and *Mimosa invisa* are good leguminous cover crops in arecanut gardens to supply green manure and nutrients. The cover crops may be sown in the months of April and May and the green matter may be cut and applied to the arecanut palms. Sesbania can be grown wherever water logging and drought are likely to occur. It can be grown in valleys of Assam, Karnataka and Kerala receiving high rainfall.

9.2. Organic Matter Recycling

Despite less canopy and leaf area, arecanut produces waste biomass comparable to other palms such as coconut (12–18 t/ha) and oil palm (14– 15 t/ha) due to a high population density per unit area. The recyclable organic biomass production from sole crop of arecanut would be around 9-10 t/ha/year. This includes arecanut leaf with leaf sheath, husk and rachis (bunch wastes) *etc*. Arecanut is intercropped with annual and biennial crops during the initial years and perennials such as cocoa, banana, clove, pepper and acid lime *etc*. in high-density multi-species cropping system (HDMSCS) during the bearing stage and therefore, the availability could be further increased to 14 t/ha by such cropping system. The husk biomass accounts for 16.5–17 per cent of the total weight of the dry nuts of arecanut, but it is largely being wasted as fuel. Taking into account the current area under arecanut cultivation, the annual recyclable biomass production could be about 9 to 11 million tonnes in the world, 50 per cent of which (4.5–5.5 million tonnes) will be contribution from India. Direct application of these wastes in the garden will result in immobilization of available nutrients due to high C:N ratio, and as it takes considerable time for decomposition, will not meet the nutrient demand of the crop immediately. The nutrient composition in these arecanut wastes is reported in the range of 0.62–1.59

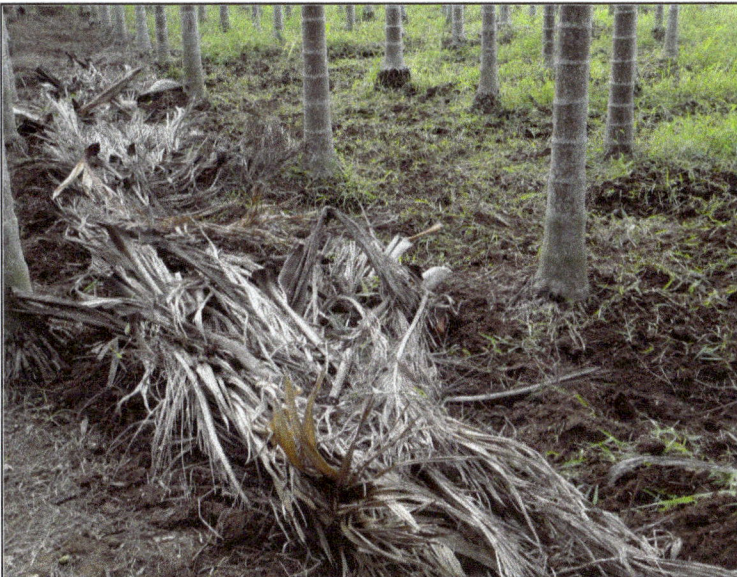

Figure 3.5: Arecanaut Leaves for Recycling.

per cent nitrogen (N), 0.07–0.16 per cent phosphorus (P) and 0.75–1.25 per cent potassium (K). Recyclable biomass in arecanut supplies approximately 95 g N, 10 g P_2O_5 and 110 g K_2O per palm per year. As husk contains higher potassium content, it can be used as a potential source of K in organic farming due to heavy K feeding nature of arecanut. Efficient recycling of these wastes supplies a substantial quantity of nutrients and meets crop nutrient demand to a great extent.

9.3. Utilization of Organic Wastes as Vermicompost

Though plantation wastes supply considerable nutrients, their direct recycling does not meet the immediate nutrient demand of the crop. Nitrogen immobilization, which occurs when plantation wastes with a high C: N ratio of 37:62, lignin and polyphenol are directly incorporated in the soil, should be avoided. The normal composting methods for management of plantation wastes is not efficient in terms of nutrient quality and time. These organic wastes could be converted into vermicompost using *Eudrilus eugeniae* earthworms, which gives a recovery of 75-88 per cent in a period of 90 days. To prepare vermicompost, areca wastes are to be chopped into small pieces of 5-10 cm and filled in tanks or pits. As the earthworms cannot eat fresh organic materials, the wastes should be in a stage of partially decomposition. Therefore the organic waste heap is to be mixed with cow dung slurry @ 10 kg/100 kg of waste and kept for two weeks with sprinkling water daily. This mixed organic waste may be watered regularly to maintain sufficient moisture (30-40 per cent) and incubated for 2-3 weeks to initiate microbial action. One or two turnings may be given to reduce the heat generated. A layer of 10-15 cm waste material is alternated with 2 cm layer of cow dung over which earthworms can be released @ 1000 numbers per square meter. The wastes are converted into fine granular, odourless vermicompost within 90 days. Application of vermicompost improves the soil organic carbon and available nutrient contents as well as helps to meet the nutrient demand to some extent.

On-farm recycling of waste biomass of arecanut-based cropping system as vermicompost produces good-quality compost and results in reduced production cost. The arecanut wastes recycled in the form of vermicompost have potential to meet about 50 per cent N, 32.5 per cent P and 26 per cent K requirement of arecanut. Under organic farming with application of vermicompost to supply 100 per cent of recommended N in arecanut the yield obtained was 1.70 kg dry nut (chali)/palm/ year which indicated that arecanut crop can be cultivated with organic inputs to realize economic yield. Leachates derived from vermicomposting are beneficial and can be used as liquid fertilizer due to high concentration of plant nutrients. Vermicompost extract obtained by mixing vermicompost and water at 1:10 ratio can be successfully used for drip fertigation.

10. Plant Protection

10.1. Pests and their Management

Important pests of arecanut and their major damage symptoms are given in Table 3.4.

Table 3.4: Pests of Arecanut and their Damage Symptoms

Name of Pest	Damage Symptom
Spindle bug (*Carvalhoia arecae* Miller and China)	Inhabit the inner most leaf axils, usually below the spindle. Both the nymphs and adults suck sap and the infested portions develop necrotic patches leading to drying. Spindle fails to unfurl and severe infestation leads to stunting of the palm.
Pentatomid bug [*Halyomorpha marmoreal* (F.)]	The later instar nymphs and adult bugs pierce the tender nuts and suck the kernel sap resulting in drying of kernels and dropping of tender nuts. There will be characteristic pinprick black marks at the feeding sites, which subsequently enter into the kernel.
White grub/root grub (*Leucopholis burmeisteri* and *L. lepidophora* Blanch)	Root/white grubs occur mostly in sandy and sandy loam soils and are voracious feeders of arecanut roots. The early instar grubs feed on the roots of grasses and other humus, while the second and third instar grubs feed on tender and mature roots of the palm. In severe cases, the bole of the palm is also eaten up. Feeding on arecanut seedling roots results in dropping and drying of leaves. The affected seedlings come off easily since the entire root system is usually eaten up. The palms with continuous infestation show a sickly appearance, with yellowing of leaves, tapering of stem, and reduction in yield. In case of severe attack and loss of roots, the palms may topple down. Root grubs also feed on roots of intercrops such as banana, cocoa, tapioca, yams *etc*.
Inflorescence caterpillar (*Tirathaba mundella* Walker)	The caterpillars feed on the inflorescences especially the tender female flowers and rachillae, web them into a wet mass with silken threads and take shelter in it. Burrowing and feeding activities produce visible damage symptoms in the form of frass production and a sticky, gummy exudate. Mature caterpillars can damage newly opened inflorescences also. In severe cases, they bore into the tender buttons and tender nuts as well. Delayed spathe opening, yellowing of spadices, presence of small holes with frass and drying patches on the spathe are the external symptoms of attack.
Red mite (*Oligonychus indicus* Hirst. and *Raoiella indica* Hirst.)	The reddish mites are easily seen against green leaves. Heavy infestations of the mites are typically on the lower surface of the leaves, and yellow speckles and blotches on the leaves are seen from the feeding damage. Yellowing of the leaves may often be severe. In severe infestations yellowing of leaves is quite prominent.

The farmers can adopt an integrated approach for managing pests. Use as many different control measures as possible *viz.*, cultural, mechanical, physical, biological *etc*. (Table 3.5).

10.2. Disease and their Management

The important diseases of arecanut, their symptoms and management under organic production are given in Table 3.6.

11. Harvesting and Post-harvest Management

11.1. Harvest

Harvesting of nuts at correct stage is very important for obtaining quality product. The stage of harvesting depends on the type of produce to be prepared for the markets. Two types of final product are seen in arecanut. Ripe nuts are

Table 3.5: Cultural/Biological Control Measures for Pests of Arecanut

Name of Pest	Control Measure
Phytophagous mite	Cultural control: Collect and destroy the heavily infested and drying leaves of young palm in the initial foci of colonization
Scales	Biological control: Release *Chilocorus nigritus* periodically @ 4-5 beetles/palm
	Conserve predators such as coccinellid beetles (*C. nigritus* and *C. circumdatus*)
Spindle bugs	Cultural control: Digging and forking of the soil before and after the monsoon will help in eliminating the various developmental stages of the beetle
Root grub	Cultural control: Deep summer ploughing to expose the immature stages for avian predation.
	Mechanical control: Collection and destruction of beetles emerging from the soil during pre-monsoon showers in the evening hours. Install light traps @ 1 trap/acre and operate between 6 pm and 10 pm
	Biological control: Conserve and augment entomopathogenic nematodes such as *Heterorhabditis* spp. and *Steinernema* spp.
	Application of neem cake @ 2 kg/palm/year at the base of the plant during June-July
Inflorescence caterpillar	Mechanical control: Affected spadices may be opened and if all the female flowers have been damaged the inflorescence should be removed and burnt

Figure 3.6: Harvested Mature Nuts.

harvested for production of dry nuts called 'chali', while green nuts at 6-7 months age are harvested for tender nut processing.

Table 3.6: Diseases of Arecanut and their Management

Disease	Symptoms	Management
Koleroga/mahali/fruit rot (*Phytophthora palmivora* and *P. meadii*)	Initial symptoms appear as dark green/yellowish water-soaked lesions on the nut surface near the perianth (calyx). The infected nuts lose their natural green lustre, quality and, hence, have a low market value. The lesions on the fruits gradually spread covering the whole surface before or after shedding which consequently rot. As the disease advances the fruit stalks and the axis of the inflorescence rot and dry, sometimes covered with white mycelial mats. Infected nuts are lighter in weight and possess large vacuoles. When infection occurs later in the season, it leads to rotting and drying up of nuts without shedding (known as 'Dry Mahali'). The fruit bunches infected towards the end of rainy season may remain mummified on the palm and such nuts provide inoculum for bud rot or crown rot or the recurrence of fruit rot in the next season. The disease spreads through heavy winds and rain splashes. The severity, persistence and spread of fruit rot are related to the pattern of rain. The disease appears usually 15 to 20 days after the onset of regular monsoon rains and may continue up to the end of the rainy season. Continuous heavy rainfall coupled with low temperature (20 to 23 °C), high relative humidity (>90 per cent) and intermittent rain and sunshine hours favour the outbreak of fruit rot.	Collect all the infected nuts and other plant parts and destroy. Spray 1 Bordeaux mixture on fruit bunches. Also cover the bunches with poly bags
Inflorescence die back (*Colletotrichum gloeosporioides*)	Disease appears on rachillae of the male flowers and then in the main rachis as brownish patches which soon spreads from tip downwards covering the entire rachis causing wilting. The female flowers of the infected rachis shed and the whole inflorescence shows 'die back' symptom. The fruiting bodies of the fungus (conidia) appear as concentric rings in the discolored areas. The disease is severe mostly during dry condition (February-March). Button shedding followed by die-back of inflorescence is a severe problem in arecanut plantations during monsoon periods. The spread is through air borne conidia.	Remove the fully affected inflorescence and destroy them to prevent spread.
Yellow leaf disease	Yellowing of tips of leaflets in 2 or 3 leaves of outermost whorl. Brown necrotic streaks run parallel to veins in unfolded leaves. The yellowing extends to the middle of the lamina. Tips of the chlorotic leaves dry up. In advanced stage all the leaves become yellow. Yellowing of leaves is conspicuous during October to December. Finally the crown leaves fall off leaving a bare trunk. Root tips turn black and gradually rot. The disease is caused by Phytoplasma and transmitted by plant hopper (*Proutista moesta*).	Remove and destroy the diseased palms in the mildly affected areas to prevent the spread. Adopt biomass recycling and excess application of phosphorus 100g/palm in the form of rock phosphate.

Contd...

Table 3.6–Contd...

Disease	Symptoms	Management
Basal stem rot/foot rot/ anaberoga/Ganoderma wilt (*Ganoderma lucidum*)	The leaflets in outer whorls become yellow, which spreads to the whole leaf and the leaves droop down covering the stem. Later, the inner whorl leaves also become yellow. Subsequently all the leaves droop, dry up and fall off, leaving the stem alone. Then the stem becomes brittle and easily broken by heavy wind. The base of the stem shows brown discoloration and oozing of dark fluid. Bracket shaped fructifications of the fungus called 'anabe' appears at the base of the trunk. Roots become discoloured, brittle and dried. When infected trunk is cut open brown discoloration can be seen up to one metre from ground level. The disease is severe in neglected, ill-drained and over-crowded gardens especially with hard, black loamy acid soils of higher iron and calcium contents. The disease is soil borne, but secondary spread is through air-borne spores.	Improve drainage. Avoid dense planting. Avoid flood irrigation and water flowing from infected palms to healthy palms. Avoid repeated ploughing and digging in the diseased gardens. Balanced manuring. Cutting and burning of dead palms along with the bole and roots should be followed strictly.
Bud rot or crown rot (*Phytophthora meadii*)	Initial symptom is the characteristic change of spindle leaf colour from green to yellow and then brownish. The leaves rot and the growing bud rots causing death of the palm. The affected young leaf whorl can be easily pulled off. The outer leaves also become yellow and droop down one by one leaving a bare stem.	Remove and destroy the diseased palms in the mildly affected areas to prevent the spread.

11.2. Post-harvest Management

Arecanut is processed by two methods in different states: 'chali' fully ripe dehusked graded nuts accounting for about 80 per cent of production, and saraku, semi-ripened, dehusked, boiled, coloured and dried nuts accounting for about 15 per cent. The most popular type of arecanut is dried whole nuts.

Figure 3.7: Spreading Harvested Nuts for Drying.

11.2.1. Processing of Ripe Nut for Making Chali (Dry kernel)

After harvesting, fully ripened fresh nuts are sun dried for 40-45 days. Spread the nuts in single layer and turn them once in a week for uniform drying and better quality of produce. Ensure proper drying to avoid fungal infections. The dry nuts are dehusked manually or mechanically, graded based on the size and quality, and marketed. Good quality chali should be free from immature nuts, surface cracking, husk sticking, fungal and insect infection.

11.2.2. Processing of Tender Nut (Immature nut)

To prepare tender processed nuts, the nuts to be are harvested at 6- 7 months maturity when they are green and soft. The processing consists of dehusking, cutting nut into halves and boiling with water dilute extract from previous boiling. After boiling, arecanut pieces are coated with 'kali', which is a concentrated extract after boiling 3-4 batches of arecanut, to get good quality processed nuts. These nuts are generally sun dried, though occasionally oven drying is followed. The well-dried product should be crisp, dark brown in colour, glossy in appearance and well-toned astringency.

Selected References

Ananda, K.S., Ravi Bhat, Chandramohanan, R.and Sujatha,S.(2011). Calendar for arecanut. Technical Bulletin No.67. CPCRI, Directorate of Arecanut and Spices Development, Kozhikode. p.43.

Ravi Bhat and Sujatha, S (2004). Crop management, In: *Arecanut* Balasimha, D and Rajagopal, V.(Eds.) CPCRI, Kasaragod, pp: 76-102.

Satyagopal,K., Sushil, S. N., Jeyakumar,P., Shankar, G., Sharma, O. P., Boina,D.R., Sain,S.K., Srinivasa Rao,N., Sunanda, B.S., Ram Asre, Kapoor, K. S., Murali, R., Sanjay Arya,. Subhash Kumar, Patni, C. S., Joseph Raj Kumar, A., Patnaik, H. P., Sahu, K.C., Mohapatra, S. N., Surajit Khalko, Nripendra Laskar,and Ayon Roy T.K.H.(2014). AESA based IPM Package for Arecanut. 51p.

Sujatha, S. Bhat Ravi and Chowdappa, P. (2015). Recycling potential of organic wastes of arecanut and cocoa in India: A short review. *Environmental Technology Reviews.* **4**(1):91-102.

Sujatha, S. and Bhat Ravi. (2015). Resource use and benefits of mixed farming approach in arecanut ecosystem in India. *Agricultural Systems.* **141**: 126-137.

Sujatha, S. Bhat Ravi and Chowdappa, P. (2016). Cropping systems approach for improving resource use in arecanut (*Areca catechu*) plantation. *Indian Journal of Agricultural Sciences.* **86**(9):1113–20.

Chapter 4

Organic Farming in Cocoa

☆ *Ravi Bhat, S. Sujatha, V. Krishnakumar*
and P. Chowdappa

1. Introduction

The cocoa tree (*Theobroma cacao* L.), belonging to the genus *Theobroma*, meaning "food of the god's in Greek, is believed to be originated from several localities in the area between the foot of the Andes and the upper reaches of the Amazon forests (South America) (the subspecies Forastero Amazónico) as well as the rainforests of central America (the white seed subspecies Criollo). Cocoa arrived in Europe through the former colonial powers Portugal and Spain, and was later to be found in Africa. Today, cocoa is cultivated in all of the humid, tropical countries. Though there are several species in this genus, but only one, *Theobroma cacao*, is grown commercially. World over, the vast majority of cocoa (70 per cent to 90 per cent) is grown by smallholder farmers. In cocoa producing countries, it is grown mainly for its beans, processed into cocoa powder, cake and cocoa butter. These products are largely used in the manufacture of chocolates, soaps, cosmetics, shampoo and other pharmaceutical products. The cocoa seeds are highly rich in fat content, and therefore, provide an energy-rich and delicious foodstuff. The natural habitat of cocoa tree is in the lower storey of the evergreen rainforest. Climatic and site requirements place cocoa in the tropical regions of the equator. The climatic factors, including the ambient weather conditions have a direct effect on the morphology, growth, fruiting and general health of cocoa plants.

Over the past few decades, cocoa has increasingly gained considerable attention in the international market as it continues to become one of the most lucrative and heavily traded food commodities in the world. This has paved the way for taking efforts to continuously increase cocoa production across the world, most especially by the four main growing countries in West Africa—Côte d'Ivoire, Ghana, Nigeria and Cameroon—now together providing about 75 per cent of the global cocoa market. The cocoa market is one of the largest food commodities exported from the developing countries to the rest of the world. The consumers across the globe are aware of the environment an ecological issues and prefer organic products even in chocolates. The certified organic cocoa market represents a very small share of the total cocoa market, estimated to around 0.5 per cent of total production. However, the demand for organic cocoa products is growing at a very strong pace, as consumers are increasingly concerned about the safety of their food supply along with other environmental issues.

The International Cocoa Organization (lCCO) estimates production of certified organic cocoa at 15,500 tonnes. Seventy per cent of the organic cocoa is produced in South America with Dominican Republic as the leading producer. The other countries are: Madagascar, Tanzania, Uganda, Belize, Bolivia, Brazil, Costa Rica, El Salvador, Mexico, Nicaragua, Panama, Peru, Venezuela, Fiji, India, Sri Lanka and Vanuatu. Indian organic cocoa production is largely confined to Kerala and it is done mostly in Idukki and Kottayam districts. It is estimated that about 10-20 per cent of the total cocoa production of over 12,000 tonne in the country is organic. The organic cocoa market is expected to be largely driven by the health consciousness among consumers. However, lack of proper supply of organic cocoa restrains the global organic cocoa market which also leads to increase in price of organic cocoa. There is a huge opportunity in the North America and Western Europe and Japan. Asia Pacific excluding Japan is an untapped market which is also a potential market for organic cocoa. This is attributed to increasing inclination of consumers towards organic products, rapid urbanization, strengthening supply chain for organic cocoa and rising health consciousness among consumers.

Currently, Dominican Republic is dominating the organic cocoa market in terms of production that holds around 70 per cent of the total market share; Peru, Ecuador and Mexico together hold around 20 per cent of the market share in terms of production; rest around 10 per cent is held by Bolivia, Ghana, Brazil and others. Majority of organic cocoa is exported to Western Europe followed by North America. U.K and U.S are the largest manufacturing countries of organic cocoa products in the world, since organic chocolate is more popular in U.K, U.S and Germany and consumers from those countries owe a significant inclination towards organic chocolates irrespective of high pricing of organic products. Organic cocoa has many nutritional benefits such as more fiber, iron, magnesium, copper, manganese and many other minerals. Moreover, it also contains antioxidants, which helps to protect skin. Organic cocoa also helps to improve blood flow and lower blood pressure. Certified organic cocoa producers must comply with all requirements associated with the legislation of importing countries on production of organic products. The benefit for cocoa farmers is that organic cocoa commands a premium price than conventional cocoa, usually ranging from US$ 100 to US$ 300 per tonne.

2. Production Scenario

2.1. Global Scenario

Cocoa is grown in 58 countries in around 10 million hectares with an estimated production of 4.36 million tonnes during 2013-14 (Table 4.1). Among the major countries, Côte d'Ivoire has the highest productivity of 660 kg/ha, while the world productivity is 504 kg/ha. The four West African countries *viz.*, Côte d'Ivoire, Ghana, Cameroon and Nigeria contributed for 73.3 per cent of worldwide cocoa production, whereas, Côte d'Ivoire alone accounted 40 per cent. If the production of Indonesia is added to the output, the five countries reach a market share of 80 per cent. Latin America, where the cocoa plant originated, presently accounts for only 16 per cent of worldwide cocoa production. With the exception of Brazil, cocoa production is mainly concentrated in small-scale farms. Cocoa production is therefore highly important for many households, as it is a key source of income and livelihood.

Table 4.1: Production of Cocoa Beans ('000 tonnes)

Region	2012-13	Estimates (2013-14)	Forecasts (2014-15)
Africa	**2836**	**3197 (73.3 per cent)**	2984
Cameroon	225	211	220
Côte d'Ivoire	1449	1746	1740
Ghana	835	897	696
Nigeria	238	248	235
Others	89	95	93
America	**622**	**708 (16.2 per cent)**	729
Brazil	185	228	215
Ecuador	192	220	250
Other	246	260	264
Asia and Oceania	**487**	**454 (10.5 per cent)**	455
Indonesia	410	375	370
Papua New Guinea	41	40	42
Others	36	38	43
World total	3945	4359	4168

Source: ICCO Quarterly Bulletin of Cocoa Statistics, Vol. XLI, No. 2, Cocoa year 2014/15.

Published: 29-05-2015.

Note: Totals may differ from sum of constituents due to rounding.

2.2. Organic Cocoa

According to SSI Report (2014), organic cocoa production accounted for about 2.5 per cent of the world's cocoa production in 2011, with 103,554 metric tonnes certified from an area of 176,880 hectares. Production is highly concentrated in the Dominican Republic (69.5 per cent), Peru (9.2 per cent), Ecuador (7.5 per cent), and

Mexico (3.1 per cent). About 75 per cent of all production of cocoa (77,539 metric tonnes) was sold as certified during the same year, which is well above the average of 33 per cent across all standards involved in the cocoa sector. The volume of organic cocoa production have almost doubled since 2008 (from 53,730 metric tonnes to 103,554 during 2011), with 60 per cent of the growth coming from the Dominican Republic (Table 4.2.). Demand for Organic cocoa continues to grow, particularly within markets for fine aroma cocoa.

Table 4.2: Area, Production and Sale of Organic Cocoa in Leading Countries of the World (2011)

Name of the Country	Area (ha)	Production (mt)	Sales (mt)
Dominican Republic	94,000	72,000	50,000
Peru	12,000	9,500	9,500
Ecuador	20,000	7,800	5,400
Mexico	13,000	3,200	2,200
Brazil	7,100	1,800	1,500
Ghana	6,700	2,000	1,400
Sao Tome and Principe	3,800	300	200
World Total	176,880	103,554	77,539

Source: SSI Report (2014) page 143.

2.3. Indian Scenario

Cocoa, a beverage crop having high commercial potential, is mostly grown in India as a mixed crop in arecanut and coconut gardens. In the global production scenario, India is a very small player with the production share of a meagre 0.3 per cent. The cocoa industry in the country has expanded to a considerable extent in recent years, with a production of 15,133 tonnes of cocoa from an area of 71,245 hectares and contributes about Rs.2,000 million annually to the GDP of the nation. Although the per capita cocoa consumption in India is very less (0.04 kg/head) in comparison with major cocoa consumers, the consumption is continuously increasing over the last one decade, indicating a bright prospect for the cocoa sector. Taking into consideration the present day consumption patterns and growth of confectionery industry in India at around 15 per cent, the demand for cocoa is likely to increase in the coming years. In India, cocoa is cultivated mainly in the states of Tamil Nadu, Andhra Pradesh, Kerala, and Karnataka (Table 4.3). At present, demand for cocoa beans is higher than the domestic production, necessitating large scale imports to meet the national requirements. India produced 15,133 tonnes of cocoa from an area of 71,245 hectares. Tamil Nadu has the highest area under cocoa (33.6 per cent), followed by Andhra Pradesh (31.1 per cent), and while in the case of cocoa production, Kerala has the major share (41.8 per cent) followed by Andhra Pradesh (37.0 per cent). Indian productivity is 236 kg/ha whereas, the world productivity is

504 kg/ha. The projected demand of cocoa by 2050 is 212 thousand tonnes against the estimated supply of 121 thousand tonnes. With the projected supply, there would be a demand supply gap of 101 thousand tonnes of cocoa beans in 2050. To achieve this target, the cocoa production in the country should increase at an annual growth rate of 7.68 per cent considering the market growth at 20 per cent and the cocoa sector has a great potential to develop in future years.

Table 4.3: Cocoa Area, Production India 2013-14

	Area (ha)	Production (mt)	Productivity (kg/ha)
Kerala	13,483 (18.9)	6,320 (41.8)	750
Tamil Nadu	23,959 (33.6)	1,071 (7.0)	500
Karnataka	11,683 (16.4)	2,142 (14.2)	250
Andhra Pradesh	22,120 (31.1)	5,600 (37.0)	450
Total	71,245	15,133	

Figures in the bracket are the percentage share.

ICAR-CPCRI is recognized as the national active germplasm site. The institutes in the traditional zones through three to four decades of research have developed high yielding elite clones and hybrids, with yield range of 1 to 2 kg dry bean/tree and with varying processing qualities. These varieties can contribute effectively as a mother source for the improvement of cocoa productivity in the country. Cocoa clones, specifically suitable to grow under arecanut and coconut canopies and suitable for both the shades have been identified. Of late, the area expansion has occurred in non-traditional zones of Tamil Nadu and Andhra Pradesh, therefore, the challenge is to identify high yielding clones suitable for such locations and also for different cropping systems.

3. Environmental Benefits

A multi-layered forest system continues to be the optimum environment for organic cultivation and, cocoa grown in this type of system, holds enormous potential for environmental and cultural conservation. The environmental advantages would include soil conservation, increased diversity of plants and animals, utilization of local and renewable resources, reduced soil and groundwater pollution and can contribute towards specific habitat conservation.The cocoa tree naturally prefers the shade, especially when young, and can grow in harmony with and in support of the local ecosystem when thoughtfully cultivated by small-scale family farmers using organic production methods. Planting cocoa among other trees, such as fruit trees, provides many benefits such as shade for the cocoa tree, increased biodiversity on the farm, inhibited growth of weeds (reducing the need for chemical herbicides), and additional food and income for the farmer's family. Organic farming also offers opportunities for the diversification of farms and has the potentials to contribute towards rural development.

4. Botany of Cocoa

Theobroma cacao belongs to the family of Sterculiaceae. Cocoa trees attain a height of 8m to10 m. The blossoms appear in the many year old wood of the leaf axil, on the trunk and on the branches (Kauliflorie). The flowering in cocoa is throughout the year, provided no extreme drought periods or seasonal temperature fluctuations occur. The berries develop within 5 to 6 months from flowering and are pollinated by insects, mainly of the species Forcipomyia and Lasioshelea. The cocoa fruit has a cucumber-like shape, and is about 25 cm long, 8-10 cm thick and weighs 300-400 g. The shell, which can be up to 20 mm thick, surrounds the sugar-rich, bitter-sweet, acrid pulp. The fruit contains 25-50 almond-shaped, bitter tasting beans or seeds, in 5-8 long rows.

5. Varieties of Cocoa

Three large groups of cocoa *viz*. Criollo, Forastero and Trinitario can be distinguished, each with several varieties and strains: Criollo and Forastero are the types naturally evolved at the centre of origin of this crop, whereas, Trinitarios are considered to be the derivative of a natural cross between Criollo and Forastero. It is hardier and more productive than Criollo. It has a share of roughly 10 to 15 per cent of the total world production and can fertilize self-incompatible species of other groups.

5.1. Criollo

With its strong and fine flavor, the Criollo group produces the highest cocoa quality. This group has characteristic white cotyledons, and originated mostly in Mexico and Venezuela. Unfortunately, the yields are low and, therefore, it is rarely cultivated. The unripe pods of criollo are purplish red, red, turn to yellow or red when ripe. It has pointed fruits, with large seeds, plump and almost round, white or pale violet cotyledons, which are less astringent. Criollo is superior in fruit possessing generally elongated and with distinctly ridged pods. The beans ferment quickly and yields 'fine' cocoa. Additionally, the white seeded Criollo cocoa (in some Latin American countries the "forastero amazónico" is also called "criollo") is much more demanding in terms of its habitat requirements, and improper production practices thus render it much more susceptible to pests and diseases. It can be sub divided into Central American *criollo* and Venezuelan *criollo*.

5.2. Forastero

Due to its high yield, the Forastero group, which is native to the Amazon region, is by far the most widely grown with around 80 per cent of total area under cocoa. It constitutes bulk of world's cocoa and almost all of the production currently coming from Brazil, West Africa and South East Asia is of Forastero. Though it gives high yields, its taste, however, is relatively weak. Forastero cocoa plants are stronger, vigorous and more productive. Unripe pods of Forastero are green, and turn to yellow on ripening. They are inconspicuously ridged and furrowed, with smooth surface, and ends rounded or bluntly pointed. The pod wall is thick. The seeds are flattened, fresh cotyledons deeply pigmented and dark violet giving an astringent product. The trees are hardy and give high yields. Quality is not comparable with Criollo. The beans take 5-6 days for fermentation.

5.3. Trinitario

The Trinitario is a hybrid of the Forastero and Criollo types originated in Trinidad. These are very heterogenous and exhibit a wide range of morphological and physiological characters. Individual clones show a wide range of characters, from Criollo type to Forastero type. It is difficult to specify the characters of Trinitarios as they have pod and bean characters ranging from those of typical Criollos to those of Forasteros. Usually hardier and more productive than Criollo, the flavour of the best reaches that of Criollo. Of total world production, Trinitario has a share of roughly 10 to 15 per cent. This variety has the capacity to fertilize the species of the other groups which generally face the constraint of being self-incompatible. In addition to these three types, several other subgroups fall under *forasteros*. The best known *forasteros* are the *"Amelonados"*, the most cultivated cocoa type in West African countries since the 19th century. The *Amelonados* are self-compatible and the pods have a melon shape with nearly smooth pod surface. Each country has its own unique cultivar. The cultivars vary with the country from where they are introduced and the extent of hybridization. The details of cocoa varieties released from ICAR-CPCRI and Kerala Agricultural University are presented in Tables 4.4 and 4.5. There are thousands of clones in the field gene banks of different cocoa research institutions in the world. Selection and hybridization work in the cocoa growing countries of the world have yielded a large number of clones under each category with distinct and unique features.

The clones are named after the place or the river in the region in which they have been traditionally harvested or the research station where it was evolved:

Some of the important clones coming under each type are:

1. Criollo

   a) RIM 189(MEX)- Rosario Izapa Mexico
   b) ICS39, 40, 45, 47,60,84,89 and 91- Imperial College Selections

2. Forastero

   a) AMAZ 2/1,3/2 (Amelonado) -Amazonas
   b) UF 4,29,36,700,701,703, 705, 710 (United Fruit Company)
   c) SCA 6,8,9, 12 (Scavina)

3. Trinitario

   a) BL Z 9/R (MEX), BL Z 23/R (MEX), BL Z 34/R (MEX), BL Z 52/R (MEX), BL Z 56/R (MEX)- BeLize
   b) UF1,10,168,221,296,613,650,654,666,667,668,676,677,688,707,708,709,711, 712 (United Fruit Company)
   c) ICS 6, 8,16,43,61,95 (Imperial College Selections)

Table 4.4: Details of Varieties Released by ICAR-Central Plantation Crops Research Institute

Name	Tree Characteristics	Pod Characteristics and Yield	Bean Characteristics
VTLCH-1 (Vittal Cocoa Hybrid-1)	Vigorous, early, heavy bearer	50 yellow pods/tree/year Yield: 1014 kg/ha	No. of beans/pod- 42 Single dry bean weight- 1.00g Dry bean yield/tree/year- 1.48 kg Shelling- 13 per cent, Fat content- 53.6 per cent
VTLCH-2 (Vittal Cocoa Hybrid-2)	Early, heavy bearer, BPD tolerant	70 yellow pods/tree/year Yield: 800 kg/ha	No. of beans/pod- 40 Single dry bean weight- 1.15 g Dry bean yield/tree/year- 1.15 kg Shelling- 11 per cent, Fat content- 54 per cent
VTLCH-3 (Vittal Cocoa Hybrid-3)	Early, heavy bearer, suitable for water limited conditions	45 yellow pods/tree/year Yield: 993 kg/ha	No. of beans/pod- 45 Single dry bean weight- 1.07 g Dry bean yield/tree/year- 1.45 kg Shelling- 13 per cent
VTLCH-4 (Vittal Cocoa Hybrid-4)	Early, heavy bearer, suitable for water limited conditions	40 red pods/tree/year Yield: 856 kg/ha	No. of beans/pod- 43 Single dry bean weight- 1.01 g Dry bean yield/tree/year- 1.25 kg Shelling- 12 per cent
VTLCC-1 (Vittal Cocoa Clone-1)	Early, heavy bearer, both self and cross compatible	75 yellow pods/tree/year Yield: 911 kg/ha	No. of beans/pod- 37 Single dry bean weight- 1.05 g Dry bean yield/tree/year- 1.33 kg Shelling- 12 per cent, Fat content- 52.5 per cent
Cocoa selection 1 (VTLCS 1)	Stable high yielder both under arecanut and coconut, withstands biotic and abiotic stress	55 red attractive pods/tree/year Yield: 1700 kg/ha	No. of beans/pod- 42 Single dry bean weight- 1.13 g Dry bean yield/tree/year- 2.52 kg Shelling- 11 per cent, Fat content- 52.1 per cent
Cocoa selection 2 (VTLCS 2)	Stable high yielder both under arecanut and coconut, less incidence of pests and diseases	55 yellow pods/tree/year Yield: 1850 kg/ha	No. of beans/pod- 42 Single dry bean weight- 1.21 g Dry bean yield/tree/year- 2.7 kg Shelling- 15 per cent, Fat content- 53 per cent

Table 4.5: Details of Varieties Released by the Kerala Agricultural University

Name of Variety	Pod Characteristics	Pod/Bean Weight	Yield Potential
CCRP I	Medium, green, turn to yellow on ripening. Constricted at the base, blunt beak at the tip with moderately deep ridges and furrows	Mature pods weigh about 385 g with 46 beans and oven dry bean weight is 0.8 g.	The average yield is about 56 pods per tree per year going up to 72 pods per year on reaching steady bearing
CCRP 2	Medium, smooth, green, turn to yellow on ripening. Spherical pods without any basal constriction and the apex is almost obtuse. Ridges and furrows are almost absent and the rind is very smooth with a thin pericarp	Mature pods weigh 311 g with 45 beans and 0.9 g oven dry weight. Average bean weight per pod is 96.5 g.	The average yield of pods per tree is 53.9, going up to 90 pods per year under favourable conditions
CCRP 3	Medium, smooth, green, turn to yellow on ripening. Pods are elliptic with moderate ridges and furrows. A slight constriction is present at the base. The apex is slightly acute	Mature pods weigh 240 g with 42 beans and 0.8 g oven dry weight. Average bean weight per pod is 94.8 g.	The average yield per tree is 68.5, going up to 80 pods per year under favourable conditions
CCRP 4	Large, purple tinged, turn to deep yellow on ripening. Pods are beaked with a prominent acute tip. Basal constriction is shallow or absent and peri carp is deeply rugose having deep ridges and furrows	Mature pods weigh about 402 g with 45 beans and oven dry bean weight is 1.1 g.	The average yield per tree is 66, going up to 93 pods per year under favourable conditions
CCRP 5	Large, elliptical, green, turn to yellow on ripening. Pods have moderately deep ridges and furrows and apex is almost acute	Mature pods weigh about 425 g with 42 beans and oven dry bean weight is 0.8 g.	The average yield per tree is 38, going up to 55 pods per year on reaching steady bearing
CCRP 6	Very big, green, turn to yellow on ripening. Thick rind, elliptical without basal constriction, with shallow ridges and furrows	Mature pods weigh about 895 g with 48 beans and oven dry weight is 1.9 g.	The average yield per tree is about 50 pods per year, going up to 180 pods per year on reaching steady bearing.
CCRP7	Large, elongated, green, turn to yellow on ripening. Beaked with slightly acute apex and a slight basal constriction, pod surface is rugose with moderately deep ridges and furrows	Mature pods weigh about 526 g with 47 beans and oven dry bean weight is 0.9 g.	The average yield per tree is about 78 pods, going up to 95 pods per year on reaching steady bearing.

Contd...

Table 4.5—Contd...

Name of Variety	Pod Characteristics	Pod/Bean Weight	Yield Potential
CCRP 8 (Hybrid between CCRP 1 x CCRP 7)	Medium, green, turn to yellow on ripening. Almost spherical without any basal constriction and the apex is attenuate	Mature pods weigh 389 g, with 48 beans and oven dry weight is 0.9 g. Average bean weight per pod is 126 g.	The average yield per tree is 90 pods, going up to 131 pods per year under favourable conditions
CCRP 9 (Hybrid between CCRP 1 x CCRP 4)	Medium, smooth, green, turn to yellow on ripening. Almost spherical without any basal constriction and the apex is attenuate	Mature pods weigh 370 g, with 36 beans and oven dry weight is 0.8 g. Average bean weight per pod is 84.9 g.	The average yield per tree is 105, going up to 358 pods per year under favourable conditions
CCRP 10 (Hybrid between CCRP 3 x GVI 68)	Medium, smooth, green, turn to yellow on ripening. Almost spherical without any basal constriction and the apex is attenuate	Mature pods weigh 332 g. with 41 beans and oven dry weight is 1.1 g. Average bean weight per pod is 102 g	The average yield per tree is 79, going up to 154 pods per year under favourable conditions

VTLCC 1

VTLCH 1

VTLCH 2

VTLCH 3

Figure 4.1a: Cocoa Varieties.

VTLCH 4

VTLCS 2

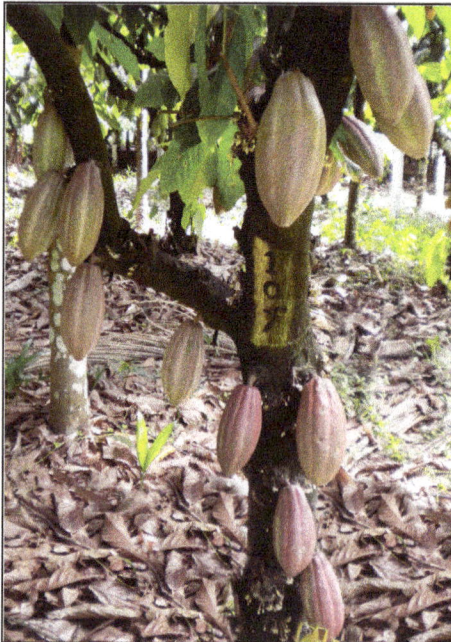

VTLCS 1

Figure 4.1b: Cocoa Varieties.

6. Climatic Requirements

6.1. Geographical Position

Though cocoa grows between 20°N and 20°S latitude, over 75 per cent of the area is distributed within 8°N and S of the equator. Cocoa is grown from sea level up to an elevation of about of 1100-1200 m, wherein most of the cocoa growing areas lie below 300 m MSL.

6.2. Temperature

Cocoa is a crop adapted to the moist, lowland tropics and thrives in a climate where the average temperature of the coldest month is not less than 16 °C and is not colder than 10 °C. If the absolute minimum temperature falls below 10°C for several consecutive days, the yield is likely to be reduced. Defoliation and dieback occur between 4-8°C. Low temperatures below 10 °C damage the sprouting seedlings, while long periods of high temperatures above 30 °C affect the physiology of the cocoa trees, bean characteristics and yield. When the temperature goes beyond 38°C, the pod size gets reduced with smaller beans and lesser butter content. Bean abnormalities like caking, occurrence of high vivipary and low pulp content are common in pods developing at high temperatures especially in summer season. It grows within a temperature range of 15-39°C and optimum temperature range is from 25 to 28 °C. The temperature in most of the cocoa growing areas lies between a maximum of 30- 32°C and a minimum of 18-21°C.

6.3. Rainfall

Rainfall and its distribution is one of the most important climatic factors deciding the cocoa cultivated area. The yield of cocoa trees from year to year is found to be influenced more by rainfall than by any other climatic factor. To ensure good growth, rainfall distribution is more important than the total amount received. In other South American, African, and Southeast Asian cocoa producing countries, rainfall distribution is more or less even. It is so well distributed that about 10 cm of rainfall is received every month, in particular, in Brazil, Ghana, and Malaysia. Ideally, cocoa requires a minimum of 90-100 mm rainfall per month with an annual precipitation of 1500-2000 mm. In most of the South-East Asian cocoa-producing countries, rainfall distribution is more or less even with minor peaks and around 100 mm of rain is received in almost every month. If rainfall is less than 1,250 mm, the crop needs irrigation during the rainless period. Cocoa plants can tolerate three to four months of low rainfall site conditions. In such cases, the cocoa plants display a more distinct rhythm of flowering and fruiting. Shortage of water leads to leaf fall and dieback. Annual rainfall in excess of 2,500 mm may lead to problems such as black pod disease due to more humid conditions. The incidence of vascular streak dieback will be more when rainfall exceeds 2,500 mm. Cocoa is susceptible to longer periods of water logging and poor aeration of soils. Heavy rainfall is the prime cause for the severity of VSD in Papua New Guinea and its rapid spread in India. In India, concentration of the rainfall to 2–3 months aggravates the spread of the disease. High rainfall during pod ripening tends to reduce rate of recovery of cured beans to as low as 23 per cent. In Kerala, the distribution of rainfall in Idukki and Kottayam is so even and, hence, the pod and bean qualities are very good.

6.4. Relative Humidity

Cocoa needs high humidity throughout the year, and hence, a hot and humid atmosphere provides suitable condition for the optimum development of cocoa trees. Such conditions are prevalent in the tropical lowland between 15° north and south of the equator. The plants grown at lower relative humidity produce small leaves and tend to be curled and withered at the tip. Though the growth of cocoa is favoured by high relative humidity, storage of cocoa beans is affected. When the relative humidity exceeds 80 per cent, the beans absorb moisture and turn mouldy. At higher altitudes (Uganda 1,400 m elevation), cocoa can be grown successfully only within close range to the equator. Strong and steady winds can damage cocoa severely. Areas exposed to such winds are to be avoided.

7. Soil Requirements

The plant can grow in a wide range of soils, and it is predominantly grown on clay loam and sandy loam soils. Very coarse sandy soils are not suitable for cultivation of this crop as its water holding capacity is very poor. To develop a good root system, cocoa requires a deep, well-drained soil, having sufficient water-retaining capacity as well as sufficient organic matter content. Soil containing coarse particles and sufficient quantity of nutrients to a depth of 1.0 m to 1.5 m will be ideal. Avoid heavy soils in high rainfall areas, whereas, light soils in drier areas. Cocoa can grow in soils with pH ranging from 4.5-8.0 with optimum being 6.5-7.0. Excessive acidity (pH 4.0 and below) or alkalinity (pH 8.0 and above) must be avoided. Virgin, freshly cleared forest soils are used for the cultivation of this crop in major producing countries. Hence, soils of most of the cocoa growing regions of the world are rich in organic matter and nitrogen, well-drained. Exchangeable bases in the soil should amount to at least 35 per cent of the total cation exchange capacity. Cocoa is susceptible to longer periods of water logging and poor aeration of soils. One of the most important measures for the improvement and maintenance of soil fertility is the continuous addition of woody (ligneous) organic material, of which large amounts become available every year as a result of pruning measures.

8. Conversion of Old Cocoa Farms into Organic

The existing cocoa planted under the canopy of trees in either a primary or secondary forest, could be converted to an organic farm. This conversion should take place within a period of three years during which organic farming practices are carried out and an organic management plan agreed upon between the farmers or their representatives and the certification body, and is strictly adhered to.

9. Establishment of New Cocoa Plantation

When choosing the site for a new plantation, the natural site requirements of cocoa should be followed. Ideal sites are those with alluvial soils, which are not susceptible to water-logging. When creating a new plantation, care should be taken to reproduce as closely as possible the natural structure of forests. This means that all of the varieties that are to be cultivated along with cocoa in the agro eco-system should be planted either simultaneously or even beforehand. The best method is to leave an area free for natural growth, and to plant tall-growing trees which will rapidly provide cover, such as bananas and to plant the cocoa in-between them at a later date. In this way, the biological activity of the soil is maintained, and the mycorrhiza of the cocoa can begin to develop immediately.

9.1. Time of Planting

Time of planting depends on the climatic condition of the place where cocoa is intended to be cultivated. The young seedlings need moisture for better establishment. The seedlings cannot withstand excess water also. Cocoa can be planted with the onset of Southwest monsoon (May-June) in areas where rainfall intensity and amount is less or at the end of the monsoon (September) in high rainfall areas. If the moisture can be maintained by irrigation, planting can be done at any time of the year.

9.2. Land Preparation

Cocoa can be planted in forest lands by thinning and regulating the shade suitably. It can also be grown in arecanut and coconut gardens as a mixed crop. Since cocoa is a shade-loving crop, all other crops to be mixed with cocoa should either be planted beforehand or at the same time as that of cocoa. If the area had any natural growth, then some trees should be left standing during land preparation. On the other hand, fast-growing plants, which will rapidly provide cover, such as bananas should be planted before cocoa is planted. Different land preparation practices are to be adopted depending on the slope, the preceding crop or previous use of the site, existing vegetation and other factors. Burning the vegetation for field preparation is not recommended. Instead the site should be cleared by slashing and chopping or shredding the hard plant materials and by distributing them homogeneously on the soil surface.

9.3. Preparation of Good Quality Planting Materials

Good quality planting material is an important input to ensure high yield and quality of cocoa. The selected varieties should be good yielding under local climatic conditions, with limited susceptibility to common pests and diseases, and produce the required quality according to the market demand.

Adopt the following for production of good quality cocoa seedlings

1. Identify cocoa trees that give consistent yields. The trees should preferably be selected in the same region in which the plantations are going to be established.

2. Harvest healthy and mature pods only. Good quality hybrid seeds obtained from local research stations can also be used for seedling production.

3. For raising cocoa nursery, enough shade, ample water and protection from wind are essential.

4. Fresh beans should be used for sowing as cocoa seeds lose their viability faster.

Figure 4.2: Cocoa Nursery Seedlings.

5. Rub the seeds with dry sand or wood ash to remove the mucilage and plant with their pointed ends upwards.

6. Sow the seeds in plastic bags (25 cm x 15 cm size with 150 gauge thickness).

7. Prepare soil mixture for the nursery consisting of 40 per cent top soil, 30 per cent compost and 30 per cent sand. A fertile, loam topsoil is ideal for filling the bags.

8. Relatively dense initial shade is recommended (more than 50 per cent). But shade must be decreased, as the seedlings grow.

9. Apart from watering, the plants do not need much attention in the nursery. However, too much watering may promote fungus attack.

10. Seedlings can be kept in the nursery for up to 6 months.

The following points are to be considered for selection of seedlings

1. Only vigorous seedlings based on height and stem girth are to be used.

2. When seedlings are grown under heavy shade, hardening for 10 days by exposing to higher sun light becomes necessary before transplanting.

3. Watering of the nursery beds should be done before lifting seedlings for transplanting to avoid breakage of roots and should be taken along with little earth around the roots.

4. If seedlings are raised in polythene bags, the cover should be removed before planting.

5. The seedling/graft/budded plant should be planted with ball of earth into the centre of the pit, not too deep.

6. While planting grafts, polythene strip tied over graft joint should be removed and the joint should be above the soil.

7. The planting material may be of 4-6 month old seedling or grafted or budded plant.

8. Avoid planting seedlings with twisted or damaged tap root or pot bound plants.

9. For long distance transportation, seedlings can be packed with moisture retaining materials like coir compost.

Softwood grafting is found successful and suitable in cocoa to breed true to type planting materials. These grafts can also be planted in the pits as in the case of seedlings after three months of hardening.

9.4. Spacing

The optimum spacing between cocoa trees is the distance which will give the optimal economic return of cocoa per unit area, always considering the stability of the organic production system. Factors such as labour requirements, establishment costs for the plantation and cost of inputs, possible losses due to pests and diseases *etc.* will also play significant role in finalization of spacing in cocoa plantation. In addition, the spacing is determined by factors such as the vigour of the trees, the soil and climate or the selected planting system.Providing optimum spacing is one of the important factors, which has a direct bearing on the yield of the crop. The optimum spacing for cocoa is one, which will give maximum economic returns. The spacing adopted for cocoa in the major cocoa-producing countries is highly variable and each country has adopted certain spacings which have become traditional. The spacing followed varies from as low as 1.0 m x 1.0 m in Trinidad to as high as

Figure 4.3: Cocoa as Mixed Crop in Coconut Garden.

5.0 m x 5.0 m in Sri Lanka. Although closer spacing usually produces higher yields in the first years after planting, once the canopy forms, the plantation becomes dense.

For the African situation where the less vigorous Amelonado type, which has a long pre-bearing period, is predominantly cultivated and where practically no costly input is used, a closer planting with 1.7 m x 1.7 m or 2.7 m x 2.7 m is beneficial. This will also mean a better crop in the early bearing period. The closer spacing is advantageous in the early years, especially for the unshaded cocoa. For the Amazonian types, a wider spacing in the range from 2.7 m x 2.7 m to 3.3 m x 3.3 m is recommended in Ghana. When cocoa is cultivated as mixed crop, the spacing of cocoa largely depends on companion crops. Among the other cocoa producing countries like Philippines, Papua New Guinea and Malaysia, where cocoa is cultivated along with coconut, the spacing followed in Malaysia is rather a close one, with two rows of cocoa at a plant-to-plant distance of around 2 m in between the 2 rows of coconut.

When coconut is spaced at 8 to 10m, the cocoa population in a hectare would be about 1000. For the space- planted coconut plantation, a row of cocoa at a plant to plant distance of 3 m is recommended for Indian cocoa. When cocoa is to be raised as a mixed crop with coconut, single hedge or double hedge system of planting can be adopted. In single hedge system, cocoa can be planted 2.7 m apart in a single row between two rows of coconut, while in double hedge system it can be planted 2.5m apart in paired rows between two rows of coconut palms. Cocoa is also planted in the interspaces of arecanut. Arecanut is usually planted at spacing for 2.7 m x 2.7 m and the usually adopted spacing of cocoa also is the same. Cocoa can be planted at the centre of four areca palms. The general experience is that such spacing results in crowding of cocoa canopy. Hence, a row of cocoa may be planted in the interspace of alternate rows of arecanut, and then the spacing for cocoa will be 5.4m x 2.7m. While establishing new garden a spacing of 3.3m x 3.3m can be followed for both areca and cocoa. The ultimate spacing and population level will depend upon the extent of canopy development, the variety used and the type of management.

9.5. Planting

If soils are naturally deep enough and are fertile, no significant advantage is seen for making planting pits. However, if soils are gravelly or if hard pan exists within the depth of penetration of roots, the practice of taking pits is advantageous. Cocoa seeds can be sown directly or seedlings planted at any time of the year if soil moisture conditions are suitable. Under Indian conditions, the most ideal time for planting seedlings in the main field is with the onset of pre-monsoon showers in May-June. If the soils are of low fertility, incorporation of manures during filling of pits may be advantageous. In areas where soils are of low fertility and gravelly laterite zones occur at varying depths, it is better to dig pits of 50 cm^3 to 60 cm^3 and fill with compost. A point to be noted is that cocoa seedlings are to be planted on the soil surface as the feeding roots of cocoa get concentrated in the top 10-15 cm layer of soil, irrespective of the zone at which seedlings are initially planted. Except for India, Malaysia, and the Philippines, cocoa is planted as a mono crop under natural or planted shade trees.

9.6. Shade Requirement

Cocoa is a plant that has originated under shade and has been traditionally cultivated under shade. The shade levels at which this crop was cultivated had been, however, highly variable. The shade requirement of this crop varies widely depending on stage of growth. Cocoa requires as much as 75 per cent shade in the early stages, which is to be gradually brought down to about 25 per cent when it comes to production. The role of the shade is not just to reduce light intensity but also to buffer the micro-environment so that excessive moisture stress in the young plants is avoided. The shade also helps to protect the young plants from the wind. It is possible to cultivate cocoa without shade under Kerala conditions and the productivity could also be higher under shade - free situations. Though the yield of cocoa under open condition, without any shade, may be good in the early years, the yield declines drastically in the long run. The shade in young plants has a major role to play as it decides the early growth of the crop. The shade crop should provide good shade throughout the dry season and not compete with the cocoa roots for moisture and soil nutrients and should be easy to remove later on without damaging the cocoa canopy. Such a plant should not be an alternative host to any insect pests of cocoa and if possible it should be of commercial value.

Young cocoa plants require shading to ensure the right form of growth, and 50 per cent shade will be sufficient. Heavy shade is needed initially, but this must be adjustable in the first few years leaving in the end a small number of trees as shade for the mature cocoa. Low light intensities with heavy shade leads to long internodes and few side branches and the high light intensity make the plant bushy. So appropriate shade is essential for the young cocoa plants for better canopy formation. In traditional cocoa growing countries two types of shade trees are available *viz.* temporary shade trees and, permanent shade trees. Temporary shading is essential during the initial years of planting cocoa. They should be relatively dense allowing not more than 50 per cent of the total light at least for two years after planting. It should be progressively reduced to 25 per cent as the cocoa tree develops, but never before the jorquettes have been properly developed. Temporary shade trees are planted at the same spacing as cocoa, alternating with it. The common temporary shade plants in African countries are banana, cassava or cera rubber (*Manihot glazeovii*) and cocoyams (*Colocasia esculenta*). In pure crop situations, permanent shade trees are planted or left without removal at a wider spacing of about 13 to 15 m. The permanent shade tree commonly used for planting in Ghana is *Terminalia ivorensis*. Other trees grown are *Leucaena leucocephala* in Papua New Guinea, Indonesia, *Glyricidia sepium* in Central America, West Indies, Malaysia and Indonesia.

With the combination of permanent and temporary shade plants, the shade level will be high resulting in best vegetative growth of young cocoa plants. Once the cocoa canopy develops and comes to bearing stage, the temporary shade is removed and the increased light stimulates the production. Apart from these, multipurpose trees are also being used as shade trees especially in Asian countries. Coconut, arecanut and rubber are used as shade trees in Malaysia, India and Brazil.

Because of the high proportion of light, which penetrates through the canopy of palms, they are considered most suitable as shade tree for cocoa.

9.7. Weed Management

Weeds compete with the crop for water, nutrients and light. Therefore, the cocoa planted field should be kept weed free. Weeding also increases air circulation and reduces relative humidity and thereby reduces the incidence of black pod disease. The weed control measure commonly adopted in major cocoa producing countries, where cocoa is grown as a sole crop, is manual slashing twice a year at a height of 5 to 15 cm. It was found that near-complete removal of weeds has no additional advantage over slashing. Slashing will have the advantage of less labour intensive compared to clean cultivation. Once the canopy closes and plant start shedding their leaves which normally occurs in 3-4 years, weed growth will be restricted. Under intercropping situations weeding is done two to three times a year.

9.8. Irrigation

As a general management practice, irrigation is not done in any of the major cocoa producing countries because of the well-distributed rainfall in these regions. As cocoa is sensitive to water stress, irrigation is essential for performance of the crop during post monsoon season. In countries where the rain-free period extends from four to six months, irrigation in the summer months will be beneficial. Cocoa is mainly grown as an intercrop in the irrigated coconut and arecanut gardens in South India, where the dry period extends from 4-6 months a year. In such cases, the crop is to be irrigated once in 6 days during January-March, once in 4-5 days during April-May, and once in a week during November- December. Any of the methods of irrigation *viz.* basin, sprinkler, drip may be adopted depending upon the availability of water. Among these, daily one hour drip irrigation using emitters supplying 20- 24 litres of water is the most efficient method in terms of water use efficiency, supplying moisture to the root zone with maximum rhizosphere.

10. Cocoa as an Intercrop

Cocoa is generally grown as an intercrop under other plantation crops like coconut, arecanut and oil palm in Asian countries. Inter planting of coconuts with cocoa is popular in Papua New Guinea, Malaysia and India. In Papua New Guinea, where the coconuts are usually spaced at 9.0 m, cocoa is planted at 4.5 m intervals between and within the rows of palms, which gives 360 trees per ha. In Malaysia, coconut palms are planted at 8.0 m or 9.0 m and two rows of cocoa seedlings are planted between the rows of palms. The cocoa spacing is maintained to have population of 1040 trees per ha. In India the general recommendation is one row of cocoa in between two rows of coconut planted at 7.5 m x 7.5 m. If the spacing of coconut is wider (9.0 m x 9.0 m), two rows of cocoa can be planted in double hedge with a spacing of 3.65 m.When oil palm is planted at a spacing of 9.9 or 10.5 m in triangular method, the cocoa planted with a spacing of 2.4 m in triangular method produces better growth and yield. In India cocoa is extensively grown under arecanut also. The microclimate under arecanut is congenial for cocoa cultivation.

11. Nutritional Management

In general, most of the major cocoa-producing countries do not use manures or chemical fertilizers on a regular basis, as cocoa is cultivated in cleared forest soils rich in humus. The presence of shade trees, the dense canopy development of cocoa, and the large recycling of cocoa litter prevent any substantial soil erosion or loss of nutrients. The demand of shaded cocoa for fertilizer applications is considerably lower than that of unshaded cocoa. Hence, in organic agriculture which does not permit the application of synthetic fertilizers, the cultivation of cocoa under shade is essential. The creation of organic material through mulching and pruning activities is sufficient for an economically viable production. Nutrient requirements for cocoa plants can depend on the stage of growth of plants such as young or mature plants, which are yielding or non-yielding. With respect to nutrient requirements, organic farming practices should be such that the use of synthetic fertilisers under any circumstances should be avoided.

One of the most important measures for the improvement and maintenance of soil fertility is the continuous addition of organic material, both woody and fresh plant materials such as mulches. Part of this material can come from pruning of trees and from harvest residues, for example when composted cocoa pods are returned to the plantation and distributed evenly over the soil surface. By ensuring consistent use of prunings and pods within the plantation, soil fertility can generally be maintained for successful organic cocoa production. Thus regular pruning of trees and maintenance of a multitiered, diverse and densely populated agro-ecosystem is generally sufficient for profitable cocoa production.Organic manures or mulches can be used but their application in large quantities will not be very economical. However, this can be accomplished when leguminous plant species are grown in the border areas of plantation itself. Application of animal manure and compost is very beneficial in cocoa plantations, as they provide nutrients and improve the soil structure and its capacity to hold water and nutrients.Through mycorrhiza-symbiotic association, many palm varieties are in a position to actively break down phosphorus and other nutrients. In addition, they are capable of binding heavy metals in soil, so that their uptake through cocoa is reduced. This is important, since in many cases the heavy metal content of cocoa beans reaches critical levels. Therefore, suitable palms are to be integrated into the cocoa cultivation system.

11.1. Nutrient Recycling in Cocoa Garden

Cocoa tree, which produces leaves throughout the year, is an exhaustive crop that uses large quantities of nutrients, especially K and the nutrient mining is 43.8 kg N, 8 kg P and 64.3 kg K/ha. The biomass production varies with age and growing conditions. The pattern of biomass accumulation of cocoa is different in arecanut and coconut due to differences in population density and growth habits. In India, cocoa intercropped in coconut plantation adds organic material through litter fall to the extent of about one to two tonnes/ha/year. Pruning of cocoa plants could contribute about 8–10 kg/tree of organic biomass and litter fall will be about two to three kg/tree/year. The available recyclable biomass from cocoa (which includes pruned biomass, leaf litter and pod husk) could be around 0.7 to 0.8 million tonnes. In Brazil, where cocoa plantations are under different agro-ecosystems, the recyclable biomass production is about 6 to 8 tonnes/ha under shading and 8.5 tonnes/ha under open conditions. By recycling organic wastes available from the cocoa gardens, it could supply around 540 t, 72 t and 244 t of N, P and K every year. The cocoa leaves from the garden can be converted into vermicompost using earthworms. The composted leaves were found rich in nutrients, including micronutrients, which was more than the normal compost. A recovery of 74 per cent could be obtained in case of vermicomposting. Though cocoa waste biomass such as pruned and fallen leaves, as well as pod husk contain relatively higher K concentration of 1.2 per cent to 3.0 per cent, the vermicompost made out of them has lower K content. In addition to this and as mentioned above, it is essential to return the (composted) cocoa pods to the plantation after removing the beans.

12. Maintenance of Mature Cocoa Farms

For any organic cocoa farm to be economical, there should be suitable conditions for growth and yield, based on the organic farming practices. Proper maintenance of cocoa trees during the early stages of growth improves later yields of the plantation. For good development, young cocoa plants need favourable growing conditions with the soil protected from the sun, little competition from weeds, proper pruning, adequate shade and improved soil fertility. Maintenance of existing cocoa farms will be economical only if optimal conditions for growth and yield are provided. A farm with complete cocoa canopy rarely needs more than occasional attention to clear the vegetation at the field's edges and in pockets within the farm. This would also require that weeds, pests and diseases are effectively controlled, cocoa trees regularly pruned accordingly, shade correctly adjusted and, appropriate manures are applied. A practice such as the regular pruning of cocoa trees and the correct adjustment of shade and adequate organic nutrients would lead to optimal yields in an organic cocoa farm and should be encouraged.

12.1. Pruning

Cocoa trees produce more branches and leaves than they need in order to be strong enough to compete with other trees. The more branches a tree grows, the more energy and "food" it must provide to these branches which reduces the size and number of pods that reach maturity.The best cocoa tree has one stem only and two or three main branches, with enough side branches and leaves to capture most of the sunlight. Removing unnecessary branches by pruning is, therefore, important for increasing production and reducing pests and diseases. The basic aim of pruning cocoa trees is to encourage a tree structure that allows sunlight to filter through to the main branches and trunk (what is known as a jorquette) to stimulate flowering, facilitate harvesting, more productive and efficient. Pruning is an important operation in cocoa especially when it is grown as an intercrop. It is a regular practice in all the cocoa growing countries except in West Africa. When cultivated as mixed crop under palms, a maximum of two-storey canopy architecture may be maintained. Cocoa trees should be pruned regularly to maintain a good canopy shape. It is optimum to have a canopy area of 15-20 m². Cocoa plants are grown under the shade of arecanut and coconut plantations. It is, therefore,

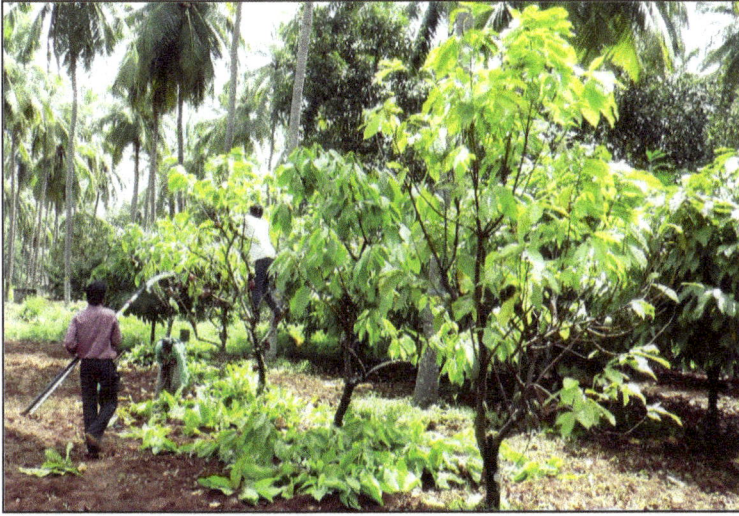

Figure 4.4: Pruning of Cocoa Plants.

necessary to regulate the canopy size and shape of plants so that the main crop is not affected. Proper and systematic pruning is essential in cocoa cultivation. The proper pruning of cocoa ensures adequate ventilation in garden; maintain tree height, makes spraying and harvesting operations easier. Pruning is usually done annually in July-August. Pruning reduces the number of unnecessary branches, and allows more light and wind to pass through the branches which reduces pest and disease levels and also amenable for easy harvest of pods.

Formation pruning and maintenance pruning are the two types of prunings generally practiced in cocoa.

12.1.1. Formation Pruning

Formation pruning is done in young plants, mainly to adjust the height of first jorquette. It is mainly to shape the canopy for desired shape, which should be umbrella-shaped. Normally the height at which the jorquette is formed depends upon the shade condition in the garden. Low shade intensity leads to jorquette formation at lower height. When the jorquette is formed at lower height, it has to be removed at an early stage to facilitate upward growth. This pruning will decide the number of jorquettes per tree, fan branches per jorquette and height of first jorquette. For pruning the seedling material, first adjust the height of first jorquette for easy cultural operations. A low jorquette will make it difficult to carry out cultural operations. Hence, a jorquette at 1.5 m to 2.0 m height is preferable. However, the jorquette-height varies significantly from tree to tree. The decision to control vertical growth depends upon the cropping system and the convenience of the farmer. Generally the vertical height is restricted to first jorquette. It has been found that increasing light intensity decreases the jorquette-height.

Presently, emphasis is being given to planting of graft materials obtained from soft wood grafting method using high yielding cocoa clones. For pruning the graft material after first year of planting, primary pruning should be done to obtain a supporting framework of one or more upward growing main stems. Then drooping or inward growing branches are to be removed. Secondary pruning is suggested to develop well shaped canopy and desired canopy should be maintained in umbrella shaped form with about 3.8m to 4.2m spread and 2.7m height depending upon the space and main crop in which cocoa is under planted/grown. All the chupons arising from below the jorquette have to be cut regularly to maintain the height. If there is any damage to the jorquette, then only the chupons is left for the development of next jorquette. It is important to note that the maximum leaf area should be maintained with pruning practices to avoid self-shading. The strongest of the re-growing chupon can be selected and all others removed. In due course, this chupon will produce a jorquette at a higher level.

Vegetatively propagated plants generally form a jorquette at ground level. Fan branches should be limited to 3 to 4 to allow more light to enter and decrease the humidity within the canopy. Basal chupons should be removed at regular intervals and all lower branches that form or bend below the jorquette should be trimmed off. Furthermore all branches within 60 cm of the jorquette, all old and diseased branches and branches growing into the centre of the tree canopy should be removed. This should be done at regular intervals through maintenance pruning. All prunings should be left in the field to rot down, except the diseased ones.

12.1.2. Maintenance Pruning

Removing new shoots and new branches that are not needed for the health and strength of the tree throughout the year is called maintenance pruning. Maintenance pruning can be done at any time of the year. This pruning is done on mature trees to maintain the health and vigour of the tree by cutting all the diseased and unproductive branches, which is called sanitary pruning and to maintain the structure of the tree, which is called structural pruning. Sanitary pruning also includes removal of all unnecessary chupons, dead branches, epiphytes, climbing plants, ant nests, diseased and rodent damaged pods, and over ripe pods. Chupons must be removed very often. It is also necessary to prune infected branches with diseases like witches broom or vascular streak dieback. Structural pruning is done to shape the canopy to desired size and architecture. In either case care must be taken, when removing large branches to ensure that exposed wood surface is not damaged, to prevent the entry of fungi or insects. Apply recommended organic fungicides immediately after the pruning.

13. Plant Protection

Cocoa can be affected by many pests and diseases, which thrive well in the warm and humid climates where cocoa is commonly grown. However, with proper understanding and implementation of a natural agro-ecosystem, pests and diseases can be effectively managed. The loss due to pest and disease in cocoa could be as high as 30 per cent to 40 per cent of global cocoa production. Diseases are usually caused by different kinds of fungi or may be viral, while pests affecting cocoa are predominantly insect pests but can also include birds and mammals. Combating of cocoa diseases and pests will be the major challenges for the success of organic cocoa farming in many countries.In an organic production system, the application of synthetic pesticides is not permitted, and hence, the conditions in the cocoa plantation have to be developed in a manner wherein the infestations of pests and diseases can be largely prevented. Organic practices would require that diseases and pests are to be managed using methods that would avoid the use of synthetic chemicals. However, in exceptional circumstances where all non synthetic methods have been used and are unable to remedy the situation, chemicals within a list agreed upon between the certifiers and the farmers could be used to solve the situation. The manner in which such chemicals are used and the reason for them being used should be well documented through a method agreed upon between the farmers or their representatives and the certifying bodies.

Accurate pest management, dependent on a strong scouting, often referred to as monitoring or surveying of pests, is essential in any pest management programme. The cocoa farmers should know the conditions of their crop and the pests in the field so that they can determine the best actions to adopt for their management. Most pest and disease infestations have been found to occur under the following conditions:

a. Cultivation of cocoa as mono crop with a small number of shade trees and species (in conventional cocoa production only 25 to 40 trees per hectare of mostly the same species are recommended).

b. High density of vegetation due to very close planting between the different varieties in a system and failure to thin and prune the trees.

c. Unsuitable site (water-logging, too dry, no possibility for deep root development of plants)

d. Poor and degraded soils, which lack organic material

e. Unsuitable shade management.

f. Deficiencies in plantation hygiene: Non-removal of diseased pods, branches and leaves.

g. Unsuitable harvest practices: Irregular harvesting.

Lack of air, excess moisture as well as physical disorders of the cocoa plant (inadequate nutrition) often cause fungal diseases. In many cases, effective and sustainable control can only be achieved through improvement of the entire plantation system, especially shade management.

Generally proper management of pests and diseases can be achieved by:

a. **Using disease resistant and pest tolerant varieties** - Cocoa varieties with tolerance to black pod disease and swollen shoot virus disease are available.

b. **Ensuring field hygiene** - This is probably the single most important method for managing key cocoa diseases. All diseased or infected plants, pods and other plant parts should be removed from the plantation and destroyed. Regular removal of diseased pods can suppress the black pod disease. To ensure healthy planting material, shoots should be taken from non-infested trees and plantations only.

c. **Regulating cocoa tree height, pruning and shade management** - Removal of some branches of cocoa and shade trees by pruning and proper maintenance of the height of the cocoa trees will allow light to penetrate to the centre of the tree, and will increase air circulation. Such cultural operation makes the conditions unfavorable for the black pod disease. Removing shade trees with a shorter life cycle than cocoa at the end of their life cycle is an important measure to be undertaken in this aspect.

d. **Maintaining soil fertility** - To ensure general health of trees, efforts are to be taken for improving soil fertility, particularly where cocoa is grown on poor soils with low nutrient levels.

e. **Proper weeding** - Weeding increases air circulation and reduces the humidity in the plantation and thereby reduces the incidence of diseases, particularly the black pod disease.

Table 4.6: Pests of Cocoa and their Management Practices

Name of Pests	Symptoms of Pest Infestation	Control Measures
Mirids or capsids (*Sahlbergella singularis Distantiella theobroma Helopeltis* spp. *Monalonium* spp.)	Most important insects in many of the cocoa cultivating areas of the world Most significant effect on cocoa farming in the West African Region Sucking insects and damage young shoots and pods thereby reduce yield Brown or black sap lesions that are later infested by disease Young cocoa trees are very vulnerable to pest attack when they are grown without shade	Keep the canopy of cocoa complete and protect nests of predatory ants
Mealy bugs (*Planococcoides njalensis P. citri* Kokoo sumor)	Adult females and young ones feed on the tender shoots, flower cushions, flowers, cherelles and pods Colonization of seedlings and young plants cause retarded growth and excessive branching at undesired height Causes cushion abortion and wilting of cherelles Act as vectors of cocoa swollen shoot virus (CSSV)	The population build up of the bugs is more during the summer months. Avoid unscrupulous cutting of trees
Tea mosquitoes (*Helopeltis antonii*)	Damage the pods and the infested pods develop circular water soaked spots around the feeding punctures They later on turn pitch black in colour Deformation of pods also occurs because of multiple feeding injuries	Mechanical control
Thrips (*Heliothrips selenothrips*)	Brown spots on dry or silvered leaves	Avoid nutritional imbalance, poor soil conditions and sudden change of shade level

13.1. Pests and their Management

Though many insects and non-insects are known to feed on cocoa, only a few are of economic importance and many of them have geographical isolation and, thus, the number of major pest species in any one area tends to be low. Cocoa is affected by more pests in West Africa than any other country. The most common pests of cocoa are mirids or capsids and mealy bugs. The details of pests of cocoa and their management are given in Table 4.6.

Some of the other pests of cocoa are:

The red stem borer of coffee is found to bore into the branches and trunks of cocoa trees. The portion of branch above the point of entry of the pest dries up. Control of the pest is best achieved by pruning off and destroying the attacked branches.

Aphids colonise on the underside of tender leaves, succulent stem, flower buds and small cherelles. Heavy infestation brings about premature shedding of flowers and curling of leaves.

The larvae of Stem girdler beetle tunnel the bark first and penetrate deeper making galleries. On younger trees, the pest attack occurs at the jorquette, which normally results in the drying or breaking of the portion above. Mechanical extraction of the larvae is suggested as control measures under organic cocoa cultivation.

Leaf eating caterpillars and leaf eating beetles are some of the other pests of cocoa.

13.2. Diseases and their Management

Details of important diseases of cocoa and their management practices are given in Table 4.7.

14. Harvesting

The most essential quality characteristics of cocoa will be decided based on correct processing, which begins with the harvesting process and ends with the storing of the processed product. A wrong step in any of the post-harvest operation leads to poor quality beans. The harvest can begin when the fruits are completely ripe and turn yellowish or reddish orange in color. In many Trinitario types, with their red and dark violet fruits, this can be recognized by an orange discolouring of the pod. Since the amount of flowering and pods set are higher at periods of high temperature, the main harvest will take place several months after such a period. Depending on the temperature, pod ripening can take between 4.5 to 7 months from flowering. Depending on the region and weather conditions, there are usually one or two harvesting phases, which are spread out over several months. In order to achieve a uniform ripeness of the fruits harvested, to avoid overripe pods in the trees, harvest all the ripened fruits every 2-3 weeks. The best way to avoid harming the bark is to cut off the fruits at the base of the blossom with a sharp knife or other suitable instrument.

Table 4.7: Diseases of Cocoa, Symptoms and their Management

Name of Disease	Disease Symptoms	Control Measures
Swollen-shoot virus	Virus is a major problem in Ghana and Nigeria Swellings on roots, chupon and jorquette shoots Red vein-banding of young flesh leaves, leaves develop interveinal chlorosis; trees look generally yellow Pods become mottled, smoother and rounded in shape with fewer beans Virus is transmitted by mealybugs	Inoculating trees with a mild virus strain Use resistant varieties Control of mealybugs Removal of infected plants and of adjacent trees
Black pod disease (*Phytophthora palmivora*, *P. megakarya, P. capsici*)	Occurs during rainy season when humidity is high and the temperature is constantly optimum. Fungus infects seedlings, flower cushions, pods, shoots, leaves and roots Initially small, chocolate brown circular lesion(s) anywhere on the pod surface Spots develop into brown patches which spread over the whole pod surface and turn black, and hence the name White or yellow sporulation over infected areas Fishy smell On some varieties cankers are formed; pink-red discoloration below diseased bark Root infection is important part of annual cycle	Reduction of overhead shade to reduce relative humidity and to improve aeration Regular weeding Regular harvesting of mature pods Removal of infested plant parts, particularly fruits or pods In emergency cases: spraying copper,sulphur or bentonite compounds before disease builds up Periodical spraying of 2 per cent epiphytic bacterium (*Pseudomonas fluorescens*) Cut off infested bark
Moniliasis/ Moniliophthora pod rot (*Moniliophtora roreri*)	Fungus mainly appears in South America Infections on young pods Dark brown spots appear one month after infection and gradually cover whole pod White sporulating mycelium	Reduction of shade Frequent removal and destroying infected pods Use of more resistant varieties Application of lime to stem application of copper fungicides Applying epiphytic bacterium (*Pseudomonas aureoginosa*)
Witches'-broom disease (*Marasmius perniciosus, Crinipellis perniciosa*)	Fungus mainly appears in South America and some West Indian islands Major symptom being the brooms; thicker branches with short lateral shoots Abnormally thick stalks of flowers Distorted young pods; black speckles on old pods Small pink mushrooms on dead brooms	Regular removal and disposal of diseased material Identification and removal of susceptible trees Using resistant trees

Contd...

Table 4.7–Contd...

Name of Disease	Disease Symptoms	Control Measures
Vascular streak dieback (*Oncobasidium theobromae*)	Fungus mainly appears in South East Asia. First yellowing of one or two leaves on the second or third flush behind the growing tip. Falling of diseased leaves. Short shoots grow from leaf axils after fall of leaves. Bundles of black vascular streaks inside diseased stems. The disease is spread by spores produced on diseased branches, which are released only at night under certain specific climatic conditions and are dispersed by wind	Use of Amazon type of cocoa. Removal of unwanted branches. Prepare nurseries away from diseased cocoa field. Prune diseased branches around 30 cm below diseased xylem and remove. Regular pruning of chupons on the trunk
Black root disease (*Rosellinia pepo*)	Fungus appears mainly in the West Indies. Infected roots are covered with grey mycelium which later turns purplish black. Trees wilt and leaves die. Root diseases usually arise from residues of felled trees	Removal of the infected tree with all the roots. Removal of adjacent trees
Charcoal pod rot (*Botryodiplodia theobromae*)	Pods of all ages are susceptible. The infection appears as dark brown to black coloured spot and the affected spots turn black and remain on the tree as mummified fruit. The internal tissues become rotten and the affected beans turn black. Spores appear in masses forming a soot. Infection takes place through wounds	Spraying with one per cent Bordeaux mixture is recommended to control this disease.
Pink disease (*Pellicularia salmonicolor*)	The presence of a pinkish powdery coating on the stem. It causes wilting of shoots, shedding of shoots, shedding of leaves and finally drying up of the branch. The disease persists from season to season through dormant mycelium inside the bark and in the cankerous tissues.	It could be checked by pruning the affected branches and swabbing the cut ends with Bordeaux paste. Regular spraying of 1 per cent Bordeaux mixture is also recommended for management of the disease.
Canker (*P. palmivora*)	The cankers appear either on the main trunk, jorquettes or fan branches. The initial symptom is the appearance of a greyish brown water soaked lesion on the outer bark. A reddish brown liquid oozes out from these lesions, which later dries up to form rusty deposits. The tissues beneath the outer lesion show reddish brown discoloration due to rotting. When these cankers girdle the main stem or branches, dieback symptoms appear and ultimately the tree dies. The infection may also spread from the infected pod to the peduncle and then to the cushion and bark.	The disease can be managed in the early stages by removing the infected tissues and applying Bordeaux paste. All infected pods should also be removed and destroyed. The drainage system is to be improved in the garden, if the disease appears.

Contd...

Table 4.7–Contd...

Name of Disease	Disease Symptoms	Control Measures
Cherelle rot (*Colletotrichum gloeo-sporioides/other factors*)	Large number of two to three months old and developing fruits (pods) known as 'cherelle' dry up and remain on the tree as mummified fruit and this type of drying of pod is commonly referred as cherelle wilt. The shrivelling and mummifying of some young fruits are a familiar sight in all cocoa gardens. In the early stages the fruits lose their lusture and in four to seven days the fruits shrivel. The fruits may wilt but do not abscise. Many factors are associated with this malady. The most important factors are: insects, diseases, competition for nutrients, over production etc.	The management practices depend upon the nature of the causative factors involved.

Figure 4.5: Harvested Mature Pods.

In case the fruits are harvested when overripe, germination can already have begun in the pod. A large number of already germinated cocoa seeds will not pass a quality control, and for this reason, overripe and diseased fruits should not be mixed together with healthy cocoa fruits, but are to be processed separately. Ripe pods should be harvested as soon as possible in order to minimise attack by fungal diseases or animal pests such as squirrel and rodents. Harvesting too late leads to overripe pods with the pulp drying up, and in extreme cases, the beans may start to germinate. Lack of pulp will not give a good fermentation. Also, germinated beans will not ferment well, and the hole caused by the emerging shoot will allow mould inside the bean. Pods that are still green or partly green have more solid pulp (with less sugar content) and the beans may be hard to break up. Unripe pulp gives rise to clumps of beans and leads to poor fermentation. Therefore, harvesting unripe pods produces beans of low quality. After harvesting, unopened pods can be kept for about 5 - 7 days before opening, as storage allows the pulp to increase in sugar content, which causes faster fermentation. Storing pods for longer than 7 days may allow mould to damage the beans and/or encourage the beans to germinate. Pods are to be opened for the removal of the beans. To reduce the risk of damaging the beans, the pods are to be cracked on a hard surface (stone or wood) or by hitting them with a piece of wood. It is important to separate the beans from the placenta. Either the pods are opened in the field and only the beans moved for fermentation or the pods are transported and opened near the fermenting kegs.

15. Post-harvest Management

The following are some of the very important aspects to be considered in the post-harvest handling of cocoa beans:

1. Avoid injuring the beans during pod-splitting (pods cut open to extract the beans);

2. Wet cocoa beans should undergo fermentation for 7 days in order to kill the seeds and enhance the chocolate flavor;

3. Dry under the sun or by a mechanical dryer;

4. Dried beans are kept in gunny sacks and stack on raised platforms. Avoid damp conditions to control fungus attack which lower the bean quality.

15.1. Fermentation

The beans must be fermented as soon as they are removed from the pod. Fermentation has the following objectives:

1. Remove the mucilage (pulp) attached to the beans.

2. Kill the embryo so that the beans cannot germinate and give the beans a good taste when roasted.

3. Encourage chemical changes within the bean, which produce the substances responsible for developing the cocoa/chocolate flavour needed by chocolate manufacturers.

4. Reduce the moisture content of the beans.

Fresh cocoa seeds are enveloped in a white, fruity-sweet pulp that composes up to 15-20 per cent of the fruit's weight. They consist to 80 per cent of water, 10-15 per cent glucose and fructose, as well as 0.5 per cent non-volatile acids (mainly citric acid) and pectin, and have a pH-value of 3.5. The seeds have a strong, bitter taste, which is caused by the dark violet-coloured anthocyan constituent in the seeds. It is only with the beginning of the fermentation that these bitter parts are chemically transformed, when seeds begin to take on a chocolate brown colouring, and the first signs of the typical aroma begin to develop. The fruit pulp of the cocoa beans offers excellent living conditions for microorganisms that play an important role during the entire fermentation process. As the process commences, the fermentation is dominated by yeast fungi which produce alcohol. The fruit pulp begins to disintegrate, and flows away. The fermentation tanks, therefore, are to be constructed and set up so that the fruit pulp juices can drain away.

Large harvests also produce large quantities of juice, which is not allowed then to flow directly into the environment. It must either be processed, or disposed of in a soakage or sewage pit. This is very important in an organically cultivated cocoa plantation.Fermentation can be carried out in two ways.

1. Traditionally the beans are heaped or wrapped in banana leaves. Every second day, the banana leaf packages are turned over to ensure even fermentation. The size of heaps is determined by the need for a sufficiently high temperature of 40 °C to 50 °C to permit liquid to drain out and air to circulate freely around the beans. Flavour development begins when the temperature of the beans is raised to a high enough level during fermentation.

2. A second way of fermenting the cocoa beans is by placing them in wooden tray stacked on top of each other and covered. This ensures better fermentation. Waste water from the fermentation process should

Figure 4.6: Fermentation of Beans.

be properly disposed off to avoid environmental pollution of the organic farm.

The temperature does not get high enough to start flavor development in a fermentation heap if small quantities of less than about 70 kg are used, whereas, while in heaps of more than 150 kg, aeration becomes restricted. The first part of the fermentation is done by bacteria that do not need air. Therefore, the first two days the heap is not mixed. The second part of the fermentation process is done by bacteria that require air. Therefore, after the first two days the heap must be turned/mixed thoroughly every day to allow air to reach all the beans. The end of the fermentation process is reached when most of the beans are brown. When 75 per cent of the beans have pale cotyledons in the centre with a brown ring, the fermentation process should be stopped. Fermentation usually will be completed within 3 to 5 days for Criollo cocoa, whereas, it takes 6 to 8 days in the case of Forastero cocoa. The relationship between colour, degree of fermentation and flavor is given in Table 4.8.

Table 4.8: Relationship between Colour, Degree of Fermentation and Flavor in Cocoa

Colour of Beans	Degree of Fermentation	Flavour of Roasting
Brown	Fully fermented	Strong cocoa flavour, balance of acidity, astringency and bitterness
Brown/purple(mauve)	Partly fermented	Good cocoa flavour, higher acidity, astringency and bitterness
Purple	Low fermentation	Low cocoa flavour, strong acidity, astringency and bitterness
Greyish or black	Unfermented	Absence of cocoa flavor, predominantly acid, astringent and bitter. Overall sour flavor

15.2. Drying

After the fermentation process is finished, any unripe or damaged beans are to be sorted out. The remaining fermented beans must be dried to prevent deterioration, reduce the growth of mould and to improve the flavour. Dried cocoa beans are easier to be stored and transported. The initial moisture content of around 55 per cent must be reduced to around 6-7 per cent before the beans are stored. Enzymatic transformation processes continue to take place during the drying process, whereby the contents are oxidised, and the cotyledon inside the seeds' shells turns a brown colour, and the typical aroma of chocolate develops. In addition, the surplus acetic acid evaporates. It is important to dry the cocoa beans carefully, in order to maintain a certain stability and storability. Drying is done by spreading them out in a thin layer on the mat in the sun on concrete floors or raised mats. Any foreign matter found should be removed from the beans while they are spread out. Sunlight will increase the browning process and also the development of aroma. Slow, careful drying in the sun can take up to 7 days.

Figure 4.7: Drying of Beans.

It is important to turn the cocoa beans by often raking to ensure that the beans are dried uniformly and carefully. Drying apparatus that utilise warm air are recommended for use in those regions where it is often cloudy during the harvest season. It is important to note that the cocoa should not come into contact with the fumes from the fuel – as this would adversely affect their taste and smell, and therefore, their quality. Well-dried beans should have a moisture content of about 6 to 7 per cent. Beans with a moisture content of more than 8 per cent become mouldy, while beans with moisture below 5 per cent become brittle. A properly fermented and dried bean should be brown in colour when the bean is cut into half. All flat, broken and poorly fermented beans should be removed before packing dried beans into air permeable bags.

15.3. Bagging of Dried Beans

The beans are usually bagged in bags of 60 to 70 kg.

For exporting, the bags should provide the following information:

☆ Name and address of producer/packer, country of origin

☆ Designation of product, quality class

☆ Date of harvesting

☆ Weight

☆ Lot number

☆ Destination with address of trader/importer

☆ Clear information on organic certification (standards applied, certifying agency, year of conversion or full organic status).

The bags are to be kept in dark, dry and well ventilated stores at low temperatures.

Short term: ca. 16 °C and 55 per cent humidity;

Long term: ca. 11 °C and 55 per cent humidity.

If the organic cocoa is being stored with conventional cocoa (mixed store), confusion has to be avoided by carrying out suitable measures such as:

1. Training of and instructions to storekeepers
2. Clear labelling in the store (*e.g.* green colour for organic)
3. Keeping of store book in which arrivals and departures of goods can be clearly distinguished.

15.4. Storage

Dried cocoa beans can be stored for years in temperate climates; on the other hand, due to the high temperature and humidity in the tropics, stored cocoa rapidly gets attacked by storage pests and moulds, because dried cocoa easily absorbs moisture. In regions with 80 to 90 per cent humidity, the moisture content of cocoa beans often increases to more than 10 per cent and as a result, cocoa loses its storage capacity. Therefore, the storage area should always be well-ventilated and the inside temperature should remain below the outside temperature. The cocoa should be stored in air-permeable sacks on the production site for only a short time, whereby the sacks should be stacked on wooden planks or boards. The use of sacks made of organic material (jute) should be avoided, if these have been treated with pesticides. Organic production does neither allow the treatment with methyl bromide nor the application of any synthetic storage insecticides.

Selected References

Ananda, K.S (2006). *Cocoa cultivation practices*. Technical bulletin. CPCRI, Kasaragod. p14.

Bhat, R and Sujatha, S. (2002). Climate, soil and agronomy.In: *Cocoa*.D.Balasimha (Ed), CPCRI, Kasaragod. pp:48-69.

Elain Apshara,S. (2010). Cocoa planting material production. Technical Bulletin No. 66. CPCRI and Directorate of Cashew nut and Cocoa Development, Kochi. p.24.

Elain Apshara,S.,Sujatha, S and Hubballi, VN. (2012)(Eds.).*Cocoa: towards sustainability.* p.102

Jason Potts., Matthew Lynch., Ann Wilkings., Gabriel Huppé., Maxine Cunningham., and Vivek Voora (2014). The State of Sustainability Initiatives Review 2014, Standards and the Green Economy. 7.*Cocoa Market* pp:131-154.

Prasannakumari, S *et al.* (2012). *Cocoa in India*. Kerala Agricultural University p.72.

Chapter 5

Organic Farming in Cashew

1. Introduction

The cashew tree (*Anacardium occidentale*) originated from South and Central America (from Brazil to Mexico) was introduced in other parts of the world during the early 16th century mainly for the purpose of afforestation and soil conservation. However, from its initial beginning as a crop intended to check soil erosion, cashew has come out now as a major foreign exchange earner in most of the countries where it is cultivated. Though cashew had its origin from Brazil, it gained greater popularity and attention in India, Vietnam and other African countries than Brazil itself. The production of raw cashew nut (RCN) has grown from 0.29 million tonnes in 1961 to 2.60 million tonnes in 2013, registering an 804 per cent increase with a growth rate of 4.13 per cent (CAGR). Cashew tree is short, stocky, low spreading evergreen tropical tree which flowers once a year between November to January. The fruits ripen fully within two months of flowering. There are two main parts for the cashew fruits *viz.* cashew apple and cashew nut. The raw cashew nut is the main commercial product of the cashew tree. These cashew nuts while processing releases the by-product cashew nut shell liquid (CNSL) which has industrial and medicinal applications. Cashew apple is generally processed and consumed locally by making juice from it which is very high in Vitamin C. It can also be fermented to give a high proof spirit.

2. Production Scenario

2.1. Global Scenario

The significant cultivators and producers of raw cashew nut are India, followed by Cote'd'Ivoire, Brazil, Vietnam, Mozambique, and Tanzania. Other countries in the Asian and African continents also grow the crop, on a smaller scale. The global cashew cultivated area during 2013 was 5.02 million hectares with the production of about 2.60 million tonnes. The country-wise raw nut production data is given in Table 5.1. India stands first in area and production of raw cashew nut in the world. India's share of the world raw nut area under cultivation and production accounts to about 20 per cent and 29 per cent, respectively. The highest productivity is registered in Tanzania with 1209 kg/ha, followed by Vietnam (804 kg/ha). Raw nut production in South East Asian Countries has registered approximately 10 fold increase since 1980. Latin American countries have registered approximately three fold increase during the same period. In recent times, India is facing stiff competition from Vietnam and Brazil in international cashew trade.

Table 5.1: Major Country-wise Area, Production and Productivity of Raw Cashew Nut during 2013

Name of Country	Area ('000ha)	Production ('000 tonnes)	Productivity (kg/ha)
India	1009 (20.10)	757 (29.08)	751
Cote'd'Ivoire	882 (17.58)	480 (18.44)	544
Brazil	726 (14.48)	138 (5.30)	777
Vietnam	324 (6.45)	252 (9.68)	804
Mozambique	77 (1.53)	50 (1.92)	653
Tanzania	80 (1.59)	120 (4.61)	1209
Others	1920 (38.26)	806 (30.97)	–
Total	5018	2603	

Source: FAO STAT; DCCD, Cochin; MARD, Vietnam; BGE Brazil; MoA, Indonesia.

Figures in the brackets are percentages.

2.2. Indian Scenario

Cashew has assumed an important place in the Indian economy and its cultivation is confined mainly to the peninsular India. The major cashew producing states are Karnataka, Kerala, Maharashtra along the West Coast and Odisha, Andhra Pradesh and Tamil Nadu along the East Coast. It is also grown to a limited extent in Goa, West Bengal, Andaman and Nicobar Islands, Madhya Pradesh, Manipur, Meghalaya, Assam and Tripura. The area, production and productivity of cashew are showing an increasing trend due to making available superior clones, standardization of vegetative propagation techniques and near self sufficiency in quality planting material production and supply to the growers. The area under raw cashew nut cultivation during 2013 was 1.01 million hectares with the production being around 0.8 million tonnes (Table 5.2). Andhra Pradesh has largest area under cultivation, however, Maharashtra ranks first in production of cashew in India. India's raw nut production has increased from 0.079 million tonnes in 1955 to 1.01 million tonnes by 2013 in around six decade's time, which is quite remarkable. The productivity is the highest in Maharashtra (1317 kg/ha) followed by West Bengal (1168 kg/ha), though their total area under cultivation is comparatively low than other states.

Table 5.2: Area, Production and Productivity of Raw Cashew Nut in different States of India during 2013

Name of State	Area ('000ha)	Production ('000 tonnes)	Productivity (kg/ha)
Andhra Pradesh	185 (18.33)	100 (13.21)	543
Maharashtra	184 (18.24)	243 (32.10)	1317
Odisha	167 (16.55)	86 (11.36)	514
Tamil Nadu	139 (13.78)	67 (8.85)	483
Karnataka	124 (12.29)	81 (10.70)	650
Kerala	85 (8.42)	83 (10.96)	979
Goa	58 (5.75)	32 (4.23)	558
West Bengal	11 (1.09)	13 (1.72)	1168
Others	56 (5.55)	52 (6.87)	=
Total	1009	757	751

Source: DCCD, Cochin.

Figures in the brackets are percentages.

3. Botany of Cashew

The cashew plant can grow as high as 14 m, but the dwarf types that grow up to 6 m has proved more profitable with earlier maturity and higher yields. The true fruit of the cashew tree is a drupe (kidney or pear-shaped), (sometimes called as hypocarpum or pseudocarp or false fruit) which develops from the pedicel and the receptacle of the cashew flower (called the cashew apple). The drupe develops first on the tree, and then the pedicel expands to become the cashew apple. During

Figure 5.1: Cashew Nut in different Maturity Stages.

maturity, the cashew apple turns into a yellow and/or red structure about 5–11 cm long. It is edible, and has a strong sweet smell and a sweet taste. The pulp of the cashew apple is very juicy, but the skin is fragile, making it unsuitable for transport. Within the true fruit is a single seed, which is the cashew nut.

4. Varieties of Cashew

Many varieties of cashew have been released by different research organizations in India and the details of some of them are given in Table 5.3.

Commercial Varieties of Cashew

High yielding varieties recommended for the region is to be selected for new planting and replanting. By principle, preference is to be given for pest and disease resistant/tolerant varieties. Unfortunately, none of the improved varieties are found to be resistant to major pests of cashew. However, Damodar is apparently tolerant to Tea Mosquito Bug. Likewise, Goa-11-6 (Bhaskara), released by National Research Centre for Cashew, Puttur is reported to perform well even under unsprayed situations. Due to increased temperature during flowering and fruiting in the mid and late varities, the tea mosquito population comes down, thereby, the crop damage due to TMB is minimum or nil. Some of the most promising mid season flowering varieties are Bhaskara, Dhana (H1608), Amrutha (H1597), Priyanka (H1591), BPP-8, Vengurla-4, Vengurla-7, and late flowering varieties are Ullal-1, Chintamani-1 and Madakkathara-2.

5. Climatic Requirements

5.1. Geographical Position

Cashew is essentially a tropical crop, grows best in the warm, moist and typically tropical climate. The distribution of cashew is generally occurs in altitudes below 700 m where the temperature does not fall below 20°C for prolonged periods, although it may be found growing at elevation up to 1200 m MSL. The best production is noticed up to the altitude of 400 m with at least 9 hr sunlight/day from December-May. It is well suited to the coastal regions. The flowering time of cashew depends on the latitude. In Brazil and Tanzania, peak flowering is between August and September. The highest flowering occurs in October in Mozambique, while in Philippines, it is March. In the west coast of India, flowering is from October to March, whereas, in the east coast of India, flowering is delayed by about 2-3 months. However, the crop is ready for harvest in summer, both in the north and south of the equator.

5.2. Rainfall

Cashew is a hardy, drought resistant crop and comes up well in areas receiving annual precipitation of 1000 - 2000 mm. It requires well distributed rainfall during growing and pre-flowering phase (from September to November) for higher productivity. However, any unusual heavy rains during November and December inordinately delay the reproductive phase of late-season varieties. Cashew needs a clearly defined dry season of at least 4 to 5 months. A dry spell from January to

Table 5.3: Cashew Varieties Released from different Cashew Research Institutes of India

Name of Variety	Salient Features	Recommended for
Madakkathara1 (BLA-39-4)	It is a selection of Bapatla having compact canopy. It comes to flowering during November and fruiting in January-March. It has yellow apple with 72 per cent juice. Average nut yield 14 kg/tree having nut weight of 6.2 g, kernel weight of 1.64 g with 26.8 per cent shelling rate and its export grade is W280.	Kerala
Madakkathara-2 (NDR-2-1)	It is a selection of Neduvellur having open canopy. It comes to flowering during January-March and fruiting in February-May. It has red apple with 68 per cent juice. Average nut yield 17 kg/tree having nut weight of 7.25 g, kernel weight of 2.87 g with 26 per cent shelling rate and its export grade is W210.	Kerala
Kanaka(H-1598)	It is a hybrid (BLA-139-1×H 3-13) having open canopy.It comes to flowering during November-December and fruiting in December-March. It has yellow apple with 70 per cent juice. Average nut yield 13 kg/tree having nut weight of 6.8 g, kernel weight of 2.08 g with 30.58 per cent shelling rate and its export grade is W280.	Kerala
Dhana(H-1608)	It is a hybrid (ALGD-1×K-30-1) having compact canopy. It comes to flowering during December-January and fruiting in January-March. It has yellow apple with 72 per cent juice. Average nut yield 11kg/tree having nut weight of 8.2 g, kernel weight of 2.44 g with 29.8 per cent shelling rate and its export grade is W210.	Kerala, Odisha, Karnataka and Assam
Amrutha(H-1597)	It is a hybrid (BLA-139-1 x H-3-13) having spreading canopy. It comes to flowering during December- January and fruiting in January-March. It has yellow apple with 72 per cent juice. Average nut yield 18kg/tree having nut weight of 7.18 g, kernel weight of 2.24 g with 31.58 per cent shelling rate and its export grade is W210.	Kerala
Priyanka(H-1591)	It is a hybrid (BLA-139-1 x H-30-1) having open canopy. It comes to flowering during December-January and fruiting in February-May. It has yellow red apple with 67 per cent juice. Average nut yield 17kg/tree having nut weight of 7.40 g kernel weight of 2.87 g with 26.57 per cent shelling rate and its export grade is W180.	Kerala
K-22-1	It is a selection of Kottarakkara 22 having compact canopy. It comes to flowering during December-February and fruiting in February-March. It has red apple with 67.5 per cent juice. Average nut yield 13kg/tree having nut weight of 6.2 g kernel weight of 1.6 g with 26.5 per cent shelling rate and its export grade is W280.	Kerala

Contd...

Table 5.3–Contd...

Name of Variety	Salient Features	Recommended for
Vengurla-4	It is a hybrid of Midnapur red x Vetore-56 having open canopy. It comes to flowering during November-December and fruiting in February-May. It has red apple with 76 per cent juice. Average nut yield 17kg/tree having nut weight of 7.7 g kernel weight of 1.91 g with 31 per cent shelling rate and its export grade is W210.	Maharashtra, Andhra Pradesh, Karnataka, Goa, Odisha, Gujarat, Jharkhand, Assam, Chhattisgarh, Meghalaya, Tripura, Andaman and Nicobar Islands, Puducherry and West Bengal
Vengurla-6	It is a hybrid of Vetore-56 x Ansur-1 having compact canopy. It comes to flowering during November-December and fruiting in February-May. It has yellow apple with 85 per cent juice. Average nut yield 14 kg/tree having nut weight of 8 g kernel weight of 1.91 g with 28 per cent shelling rate and its export grade is W210.	Maharashtra and Goa
Vengurla-7	It is a hybrid of Vengurla-3 x M-10/4(Vri-1) having compact canopy. It comes to flowering during November-December and fruiting in March-May. It has yellow apple with 86 per cent juice. Average nut yield 18 kg/tree having nut weight of 10 g, kernel weight of 2.9 g with 30.5 per cent shelling rate and its export grade is W180.	Maharashtra, Karnataka, Goa, Assam, Meghalaya, Tripura, Gujarat, Andaman and Nicobar Islands and Puducherry
BPP-4	It is a selection of Epurupalem having open canopy. It comes to flowering during February-April and fruiting in April-May. It has yellow apple with 64 per cent juice. Average nut yield 10kg/tree having nut weight of 6 g, kernel weight of 1.15 g with 32.3 per cent shelling rate and its export grade is W400.	Andhra Pradesh
BPP-6	Trees having open/spreading canopy. It comes to flowering during February-May and fruiting in April-June. It has yellow apple with 74 per cent juice. Average nut yield 10 kg/tree having nut weight of 5.2 g, kernel weight of 1.44 g with 24 per cent shelling rate and its export grade is W400.	Andhra Pradesh
BPP-8 (H2/16)	It is a hybrid (T1 x T3) having compact canopy. It comes to flowering during February-April and fruiting in April-May. It has yellow apple with 64 per cent juice. Average nut yield 14kg/tree having nut weight of 8.2 g, kernel weight of 1.89 g with 29 per cent shelling rate and its export grade is W210.	Andhra Pradesh, West Bengal, Odisha, Jharkhand and Chhattisgarh

Contd...

Table 5.3–*Contd...*

Name of Variety	Salient Features	Recommended for
Vridhachalam-3 (M 26/2)	It is a selection of Edayanchavadi having compact canopy. It comes to flowering during January-February and fruiting in February-May. It has red apple with 72.8 per cent juice. Average nut yield 12kg/tree having nut weight of 7.18 g. Kernel weight of 2.16 g with 29.1 per cent shelling rate and its export grade is W210.	Tamil Nadu, Kerala, Goa, Andaman and Nicobar Islands and Puducherry
Ullal-1	It is a selection of Thaliparamba having medium spreading canopy. It comes to flowering during November-April and fruiting in February-May. It has yellow apple with 64.2 per cent juice. Average nut yield 16kg/tree having nut weight of 6.7 g, kernel weight of 2.05 g with 30.7 per cent shelling rate and its export grade is W210.	Karnataka
Ullal-2	It is a selection of 3/67 Guntur having medium spreading canopy. It comes to flowering during December-March and fruiting in February-April. It has red apple with 64.1 per cent juice. Average nut yield 9kg/tree having nut weight of 6 g, kernel weight of 1.83 g with 30.5 per cent shelling rate and its export grade is W320.	Karnataka and Assam
Ullal-3	It is a selection of 5/37 Manchery having open canopy. It comes to flowering during November-January and fruiting in January-March. It has dark red apple with 66.1 per cent juice. Average nut yield 15kg/tree having nut weight of 7 g, kernel weight of 2.1 g with 30 per cent shelling rate and its export grade is W210.	Karnataka, Assam, Meghalaya and Tripura
Ullal-4	It is a selection of 2/77 Tuni having open canopy. It comes to flowering during November-January and fruiting in January-March. It has yellow apple with 65.4 per cent juice. Average nut yield 9kg/tree having nut weight of 7.2 g, kernel weight of 2.15 g with 31 per cent shelling rate and its export grade is W210.	Karnataka, Assam, Meghalaya and Tripura
Chintamani-1	It is a selection of 8/46 Thaliparamba having open canopy. It comes to flowering during January-April and fruiting in February. It has yellowish red apple with 65.4 per cent juice. Average nut yield 7 kg/tree having nut weight of 6.9 g, kernel weight of 2.1 g with 31 per cent shelling rate and its export grade is W210.	Karnataka
Chintamani-2	Selection from ME 4/4 Average nut yield 12.4 kg/tree with nut weight 7.9 g. Shelling percentage 30, kernel weight 2.35 g, its export grade is W210.	Karnataka
UN-50	It is a selection of 2/27 Nileswar having medium canopy. It comes to flowering during November-January and fruiting in February -May. It has yellow apple with 65.2 per cent juice. Average nut yield 10kg/tree having nut weight of 9 g, kernel weight of 2.24 g with 32.8 per cent shelling rate and its export grade is W180.	Karnataka

Contd...

Table 5.3–Contd...

Name of Variety	Salient Features	Recommended for
NRCC-2	It is a selection of 2/9 Dicherla having medium compact canopy. It comes to flowering during November-January and fruiting in February-March. It has pink apple. Average nut yield 9 kg/tree having nut weight of 9.2 g, kernel weight of 2.15 g with 28.6 per cent shelling rate and its export grade is W210.	Karnataka
Jhargram-1	It is a selection of T.No 16 of Bapatla having medium compact canopy. It comes to flowering during February-April and fruiting in April-May. It has yellow apple with 63.5 per cent juice. Average nut yield 8kg/tree having nut weight of 5 g, kernel weight of 1.5 g with 30 per cent shelling rate and its export grade is W320.	West Bengal and Odisha
Bhubaneswar-1	It is a selection of WBDC-5 (V36/3) having medium compact canopy. It comes to flowering during January-March and fruiting in March-May. Cluster bearing, has reddish yellow apple. Average nut yield 10kg/tree having nut weight of 4.6 g, kernel weight of 1.47 g with 32 per cent shelling rate and its export grade is W320.	Odisha
Goa-1	It is a selection of Balli-2 having semi spreading compact canopy. It comes to flowering during December-February and fruiting in March-May. It has yellow apple with 68 per cent juice. Average nut yield 7kg/tree having nut weight of 7.6 g, kernel weight of 2.2 g with 30 per cent shelling rate and its export grade is W210.	Goa
Bhaskara(Goa 11/6)	Tree of seedling origin spotted in cashew plantation of Gaodengrem Forest Department, Canacona Taluk, South Goa. Average nut yield 10.7 kg/tree, with nut weight of 7.4 g. Shelling percentage 30.6. Its export grade is W 240.	Coastal Karnataka

Amrutha

Dhana

K22-1

Kanaka

Figure 5.2: Varieties of Cashew.

May with occasional light summer rains ensures better cashew production. If there is heavy rainfall during these periods, it will encourage high incidence of pest like tea mosquito bug (TMB) and reduce yield and quality. Coincidence of excessive rainfall and high relative humidity with flowering may result in flower/fruit drop and heavy incidence of fungal diseases. Prolonged dry spells, frost, foggy weather and heavy rains during flowering and initial fruit setting also affect production. Cashew is highly sensitive to high relative humidity (> 80 per cent) during the harvesting season. The humidity of suitable region should range from 60-80 per cent. High relative humidity will adversely affect the nut quality.

5.3. Temperature

Cashew is a tropical plant and, though it can tolerate wide range of temperature, the optimum monthly temperature is between 24°C and 28°C. Cashew grows in the semi-arid regions like northern Mozambique where a daily maximum temperature exceeds 40°C and in Assam, cashew survives up to 7°C. It has been reported that cashew cultivation is not economical in regions where annual temperature falls below 20°C for prolonged periods. Cashew grows at reasonably high temperatures and does not tolerate prolonged period of cold and frost especially during the juvenile period. However, temperature above 36°C between the flowering and fruiting period could adversely affect the production. Year to year variation in time of flowering of a variety is common even under uniform cultural and management practices. It signifies the influence of weather factors on flowering behavior of cashew. The summary of climatic factors that influence the growth and production of cashew are the following:

☆ Bright sunshine (greater than 9 h/day) with moderate dry weather is good for flowering.

☆ Dry spell during flowering and fruit setting ensures better harvest.

☆ Cloudy weather during flowering enhances scorching of flowers due to tea mosquito infestation.

☆ Heavy rains during flowering and fruit set adversely affects production.

☆ High temperature (39-42°C) during fruit set development stage causes fruit drop.

6. Soil Requirement

It can be grown in almost all types of soils from sandy to laterite as well as in poor soils, however, its performance would be much better on good soils such as well-drained laterite soil with high organic matter content. The most ideal soils for cashew are deep and well-drained sandy or sandy loams without a hard pan. Cashew can also thrive well in hard degraded laterites, red sandy loam, and coastal sands. Water stagnation and flooding are not congenial for cashew. Cashew prefers slightly acid soil of pH 4.5 to 6.5, with low Ca content. Soil pH of >8.0 is unsuitable for cashew cultivation. Excessive alkaline and saline soils also do not support its growth.

7. Production of Planting Materials

Choosing the right type of planting material is very important in cashew cultivation, as cashew is a cross pollinated crop and exhibits wide variation in respect of nut, apple and yield of seedling progenies. Therefore, vegetative propagation is recommended to obtain true to type progeny. Farmers can identify high yielding mother trees in their own gardens and use them for production of planting materials. Further multiplication should be done by vegetative methods *viz.*, grafting or air layering.

The following criteria are to be adopted while selecting a high yielder:

☆ The plant should be fast growing and have a spreading nature.

☆ It should give 10-30 flower panicles per square meter canopy area. Higher the number of flower panicles, more is the yield potentiality.

☆ The male flowers should be 200-300 per bunch and should have 40-80 bisexual flowers in it. Higher the number of bisexual flowers, more will be the ultimate yield.

Figure 5.3: Soft Wood Grafts.

☆ The mean weight of the nut should be 7-9 g.

☆ The tree should start flowering in the mid season (January-February) or late in the season (February-March) to escape the attack of tea mosquito bug. Further, the flowering season should be as short as possible (1-3 months).

Figure 5.4: Cashew Nursery Plants Ready for Planting.

☆ The tree should be resistant to pest and disease attack.

☆ The kernel yield should be more than 28 per cent of the total nut weight.

☆ A well matured tree (15 years and above) should give 15-25 kg nut yield.

☆ When early selection method is adopted, tree of about 8 years should yield more than 10 kg.

Cashew can be propagated by seedlings, air layers and softwood grafts. Air layering has been quite successful but survival percentage is low and it has been reported that the plantations raised from air layers are more susceptible to drought and the life of such plantation is shorter as compared to that of grafted or seedling ones. The anchorage has also been observed to be poor, especially in cyclone prone areas. Epicotyl grafting and softwood grafting are found to be successful because it is easy to produce large number of grafts in a short time. However, epicotyl grafting also has limitations due to high mortality at transplanting and incidence of collar-rot at the nursery stage. Soft wood grafting is the only commercially viable and feasible method of propagation and gives a success rate of about 70 per cent. The potting mixture used for raising grafted plants should contain organic manure (cow dung), rock phosphate and biofertilizers. The potting mixture should have $1/3^{rd}$ top fertile soil, $1/3^{rd}$ cow dung and the remaining $1/3^{rd}$ sand and 5g each of *Azospirillum, Pseudomonas* and *Aspergillus* biofertilizers and 5g of rock phosphate.

8. Preparation of Land and Planting

8.1. Land Preparation

Organic cashew orchards, whether planted new or converted from existing orchards, should be isolated from the conventional orchards by a minimum distance of 500 m. A minimum period of three years is required for converting an existing cashew plantation into organic. In the case of new plantations, as the newly planted trees take 2-3 years for yielding, the nuts collected from the first harvest itself can be considered as organic.

For developing a new plantation, wild growth particularly forest tree growth should be cleared. The area identified for cashew planting should be ploughed thoroughly and leveled. In case of forestlands, the jungle should be cleared well in advance and the debris burnt. The roots of the weeds and bushes are to be completely uprooted around 2 m radius of the planting pit, which ensures competition-free environment for the newly planted cashew grafts. In the absence of inter crops or cover crops, the space in between the plants is also to be cleared in phased manner in the subsequent years. In sloppy areas, after clearing the growth of other plants, the land is to be terraced or bunds constructed. In order to ensure better moisture conservation, trenches are to be made across the contours. Complete the land preparation work before the onset of monsoon season *i.e.* during May-June.

8.2. Season for Planting

Since cashew is a rainfed crop, the ideal time for planting is usually during monsoon season. Grafts are best planted in July–August. However, in heavy rainfall areas, the planting is to be taken up only once the heavy rains are over. If irrigation facilities are available, planting can be done throughout the year except winter months.

8.3. Pit Size

Cashew grafts are to be planted in pits of 60 cm x 60 cm size in light to medium soils. If a hard substrate like laterite is present in the subsoil, pits may be 1 m x 1 m to compensate for the lesser depth of soil. Take the pits at least 15–20 days prior to planting to expose planting holes to direct sunlight which can help remove termites and other harmful insects that can damage young plants, if present. Keep the top soil and sub-soil separately and allow withering under sun. Later on, fill the pits with a mixture of top soil, 5 kg of compost or farmyard manure or 2 kg of poultry manure, 20 g biofertilizers (nitrogen fixers and phosphate solubilizers) and 200 g rock phosphate. Contour planting is to be followed in sloping areas. Standard conservation measures need to be followed on steep lands when cashew plantations are established.

8.4. Spacing

For planting cashew, spacing of 7.5 m x 7.5 m or 8 m x 8 m is recommended in square system of planting, which gives a tree density of 175 and 156 trees per hectare, respectively. In level sites, however, it would be advantageous to plant cashew at a spacing of 10 m x 5 m to have a tree density of 200 trees per hectare, which provides sufficient space to plant intercrops during the initial years of establishment. In case of sloppy lands, the triangular system of planting is recommended to accommodate 15 per cent more plants without affecting the growth and development of the trees. In undulating areas, the planting should preferably be done along the contours, with cradle pits or trenches provided at requisite spacing in a staggered manner to arrest soil erosion and help moisture conservation.

8.5. High-Density Planting

High density planting is a recent technique recommended for enhancing the productivity of cashew plantations. This technique involves planting more number of grafts per unit area and thinning at later stages. High density planting at 4 m x 4 m giving a tree density of 625 trees per hectare in the initial years and subsequently thinning in stages to reach a final spacing of 8 m x 8 m by the 10th year after planting can also adopted. This helps to realize higher returns during the initial years and as the canopies grow in volume, alternate trees are to be removed to achieve the desired final spacing. The benefits of high-density planting are effective soil conservation and checking weed growth, especially in forest lands.

9. Management of Cashew Plantations

9.1. Training and Pruning

As an orchard management technique to improve the sanitation, removal of water shoots, lower branches, criss-cross branches and dry branches are necessary to enhance flowering and the yield.Training and pruning of cashew plants during the first three to four years is essential to provide better frame work and proper

Figure 5.5: Cashew Plantation.

shape to the trees. During first year of planting, the sprouts coming from the rootstock portion of the graft that is from the portion below the graft joint should be removed periodically to ensure better health of the plant. The trees are to be shaped by removing lower branches and water shoots coming from the base during subsequent growth period of three to four years. The tree should be allowed to grow by maintaining a single stem up to 0.75-1.0 m from the ground level and stems are allowed to grow thereafter. Since cashew trees have a tendency to spread their canopies and lodge easily, proper staking is also essential. New and longer stakes are to be provided after removing old and weaker ones during the second and third year after planting. Proper training of trees and maintenance of single stem at base enables easy undertaking of cultural operations such as terracing, weeding, manure application, nut collection and stem/root borer infestation control *etc*. When large sized shoots are pruned, care should be taken to see that the cut surface is as smooth as possible and hence sharp blades should be used. The cut portion should be swabbed with Bordeaux paste. The training and pruning of cashew plants is done during August – September.

The flowers appearing during first and second year of planting should be removed (de-blossoming) and plants should be allowed to bear fruits only after third year. After three to five years, the main branch which is growing vertically should be beheaded at a height of 2.5 m to 3.5 m, which helps to reduce over shading effect of higher branches on the lower branches. This encourages better spread of the canopy as well as uniform distribution of light on all the branches. The ideal period for pruning would be after the nut harvesting and before the onset of new shoots. Thereafter, regular removal of dried/dead wood, criss-cross branches and water shoots once in 2-3 year is required to keep the plant healthy.

9.2. Cover Cropping

Raising cover crops in cashew plantation prevents soil erosion and help in moisture conservation. When leguminous plants are used, they have the additional benefit of enriching soil fertility by adding fixed nitrogen, in addition to building up the carbon base through biomass incorporation. The popular cover crops for cashew plantations are *Pueraria javanica, Calopogonium muconoides* and *Centrosema pubescens.* Cover crop seeds are generally sown in the inter space of cashew plants with the beginning of the monsoons at a seed rate of about seven kg/ha. On degraded steep lands, cover crops are usually established on seed beds between tree rows.The seeds of these cover crops may be sown in the beginning of rainy season. The seed beds of 30cm x 30cm size are to be prepared in the inter space in slopes by loosening soil and mixing a little quantity of compost. The seeds of these crops are sown in the beds and covered with a thin layer of soil. Presoaking of the seeds in water for six hours ensures better germination. Before harvest of the nuts, the cashew basins must be fully cleared of the cover crops to ensure easy harvest to gather all the fallen fruits with nuts intact.

In China, natural grass and leguminous crops are usually maintained at the time of land clearance to conserve soil. During initial years after planting, green manure crops are also grown. Creeping cover crops, such as *Pueraria phaseoloides* and *Centrosema pubescens,* and bush cover crops such as *Glyricidia maculata* and *Leucaena leucocephala,* and nitrogen fixing trees such as *Acacia mangium* are the principal cover crops grown in cashew plantations in Sri Lanka.

9.3. Mulching

As cashew is generally planted on the wastelands and in very dry areas where other crops are seldom grown, availability of soil moisture is always low, and hence, mulching is essential to conserve available soil moisture. In low rainfall areas, mulching around the base of trees with organic matter or residues helps in the control of weeds, retention of moisture and modulation of soil temperature, especially in the hot summer months and prevents soil erosion. Locally available materials like green or dry grass or weeds are ideal for mulching the basins. Mulching with black polythene is also beneficial to increase the growth and yield of cashew. Small pebbles or stones can also be used for mulching of the basin. The plastic or stone mulch does not improve soil health but ensures better moisture retention in the soil and also prevents attack of soil borne insects and pests.

9.4. Weed Management

Use of any chemical herbicides is prohibited in organic farming and hence, only manual weeding is to be done. Until the tree canopies grow enough to shade out the weeds, weeding is essential around the tree trunks up to a radius of about 2 m. In general, two weedings are to be made in cashew plantation; the first weeding in August during manure application and the second weeding just before the start of flushing and flowering during October or November. Weeding with a light digging should preferably be done before the end of rainy reason. Hoeing, cutting the weeds off underground is more effective than slashing. The weed biomass can be effectively recycled as mulch cum green manure by applying around the plant basin. Mature trees on attaining full canopy can smother weed growth to an appreciable extent.

9.5. Irrigation and Drainage

Cashew cultivation is generally carried out under rainfed conditions. However, it is preferable to give supplementary irrigation during summer months, especially during January-March at fortnightly intervals @ 200 litres per plant which has proved to double the yield. Cashew responds well to irrigating @ 60-80 l water per tree once in four days through drip after initiation of flowering till fruit set and development (December to March). In China, supplementary irrigation is only provided during the early stages of the orchard establishment. Other Asian countries rarely practice supplementary irrigation. The mono crop orchards or the adult orchards rarely receive supplementary irrigation, whereas, interior mixed crops are raised, in order to achieve their best yields, irrigation will be needed. Cashew cannot withstand water stagnation. In high rainfall areas, plantations located in low lying areas should have trenches to drain out excess water.

9.6. Soil and Water Conservation Methods

Since the cashew plant is normally grown as rain fed crop and along the slopes, arresting soil erosion and runoff water during monsoon for proper soil and moisture conservation can be achieved by planting on terraces along the contour and opening pits to catch running water at the lower end. In sloppy area, terracing should be taken up around each plant within second year. Initially terrace with inward slope may be made and a catch pit of 2 m x 0.3 m x 0.45 m at a distance of 1.8 m -2.0 m away from the base of the plant on the upper side of the slope prepared. A small channel connecting catch pit sideways/water ways is to be made to drain out excess water during rainy season. The terrace and trench could also be made in semicircular pattern. On a level land, square, circular or staggered trench of 0.3 m depth should be dug and the soil is spread around plant basin.

Individual tree trenching with crescent bund is the best soil and water conservation measure in sloppy lands. Terraces are to be prepared by removing soil from the elevated portion of slope and spread on the lower side to form a flat basin of 1.5 m to 2 m radius depending upon the age of the plant. Terraces may be crescent with inwardly sloping, so that the top soil which is washed off from the upper side due to rain water is deposited in the basin of the plant. Modified crescent bunds made at 2 m radius having a crescent shaped bund on the upstream of the plant or staggered trenches with coconut husk burial are superior. Coconut husks buried in trenches of 1 m width, 0.5 m depth and 3.5 m length per plant opened across the slope between two rows of cashew plants helps in better soil and water conservation. Generally three to four layers of coconut husks can be buried one above the other and about 100 husks are needed to bury in a trench of 3.5 m length. The first layer is laid with the convex surface of the husk touching the ground. After spreading a layer of soil on the husks, the second layer of husks is laid in the same position and the last layer is to be covered with soil up to 10 cm thickness.

10. Inter/Mixed Cropping

The wider spacing adopted for establishing cashew plantation offers scope for inter cropping and thereby utilizing the available space to the maximum extent possible. Different intercrops are to be established early in the plantation as delay would lead to a smothering effect due to the spreading cashew canopy. Various leguminous crops like horse gram, cowpea; spice crops like turmeric and ginger as well as other crops such as ground nut, tapioca, vegetable and fodder crops, elephant-foot yam can be grown as intercrops in cashew plantations. Besides the annual crops, arid zone fruit crops having less canopy especially annona, phalsa,

Figure 5.6: Turmeric as Intercrop in Cashew Plantation.

Figure 5.7: Pineapple as Intercrop in Cashew Plantation.

sapota *etc*, can also be grown depending on the suitability. All these crops should also be grown organically.

Pineapple is highly economic intercrop that can be grown in the inter space between two rows of cashew during the first seven years. Open three trenches (1 m x 0.5 m (width x depth and of any convenient length) in between the rows across the slope and plant two rows of pineapple suckers in each trench at 60 cm between rows and 40 cm between two suckers within the row. Add half basket of compost for one meter length and mixed with soil before planting suckers. In this way, one hectare of cashew plantation can accommodate 15,000 suckers. Growing pineapple in trenches across hilly slopes will also help check soil and water erosion. Pineapple starts yielding from second year and after fourth year, it should be replanted in a new trench dug out by the side of existing trench or the same could be retained till seventh year of cashew plantation. After seven years, because of heavy shade of cashew tree over the pineapple and due to difficulty in picking raw cashew nuts fallen over pineapple plants, it may not be feasible to grow this intercrop economically. Casuarina and acacia are unsuitable as intercrops due to adverse effects on soil structure and soil moisture, as the former requires high quantity of soil moisture and its root system spread almost throughout the plantation.

Glyricidia maculata can be grown in the inter space between two rows of cashew or all along the border. If it is grown in the inter space, it may be spaced at one meter distance. Three rows of Glyricidia can be grown in the inter space of two rows of cashew. Glyricidia may be grown by sowing seeds or planting stem cuttings of one meter length during rainy season. Nearly 60 kg leaf and tender branches can be collected and applied to each cashew plant as green leaf manure. Glyricidia grown as an intercrop during initial years is found to contribute around 5.75 t/ha dry matter equivalent to 186, 40.8, 67.8 kg N, P_2O_5 and K_2O. Growing glyricidia and sesbania as green manure crops will be helpful in increasing yield of cashew plants. Medicinal and aromatic plants can also be planted in the inter space. In Indonesia, sweet potato and peanut are popular intercrops. Recently, watermelon and sweet melon and chilli or hot pepper has also been tried as intercrops. Vegetables can only be grown as intercrops when facilities for supplemental irrigation are available. When melons are cultivated, a lot of biomass after the harvest of the fruits is available for incorporation into the soil, which will help build up the organic carbon content. In Myanmar, several intercrops, predominantly annuals, such as sweet potato, sesame, peanut, maize, cassava, pigeon pea *etc.* are grown. In Sri Lanka, banana is a popular intercrop. Pineapple, papaya, pomegranate, and coconut are also grown as intercrops. In Sri Lanka, the common annuals grown in cashew plantations are legumes (cowpea, black and green gram), oil crops (sesame, ground nut), and condiments such as hot pepper and onion.

Mixed cropping should be adopted in organic cultivation of cashew to achieve species diversity and sustainability of the production system. Therefore, irrespective of the size of the holding, growing a variety of crops within the available land contributes to the ecological balance. Farmers are advised to analyze various aspects of cultivating mixed crops in their locations such as suitability and compatibility with the main crop, water management options, manuring requirement, marketing avenues *etc.* before venturing into mixed cropping in their plantations. Mango, sapota, kokum, amla, jack fruit *etc.* could be grown as mixed crops in cashew plantations. In countries like Malaysia, Indonesia and Cambodia, rambutan,longon, dwarf bamboos are grown with cashew by providing proper spacing for each crop according to canopy coverage.

11. Top Working

Top working is a technique evolved to rejuvenate unproductive and senile cashew trees. Top working can successfully rejuvenate poor yielders in the age group of 5-20 years. The unproductive trees are to be beheaded at a height of 0.75 to 1.0 m from ground level. The stem should be cut with a sharp saw to avoid stump splitting. The best season for beheading trees is May-September. Soon after beheading, the stumps and cut portions should be applied with Bordeaux mixture and 5 per cent neem oil to prevent fungal attack.Sprouts emerge 30-45 days after beheading. Sprouting will be profuse in young trees. New, 20-25 days old shoots should be grafted with scions of high yielding varieties.

Figure 5.8: Rejuvenation after Top Working.

12. Nutritional Management

In view of increasing demand for organically grown cashew, strategies are to be developed for organic farming for different agro-ecological zones because organic manure availability may vary according to different locations. In organic farming system, the nutrients should be given in an organic way. Cashew is generally grown in soils with low fertility status and water holding capacity. To ensure supply of sufficient nutrients leading to optimum growth and yield in organic cashew, an integrated approach consisting of growing leguminous green manure/cover crops, recycling of crop residues, application of organic manures and bio-fertilizers is to be followed. Growing of leguminous cover/green manure crops is highly beneficial particularly in young plantations where intercrops are not raised. When organic manures are used, around 25 kg poultry manure, 60 kg FYM or 30 kg vermicompost may be used per adult tree. Addition of bio-fertilizers such as N-fixers and P-solubilisers @200 g/plant also gives benefits. Apply 1/5th dose of the organic manure during the first year, 2/5th dose during second year and progressively reaching full dose from fifth year onwards. Application of organic manure should be done in the beginning of monsoon (June) in low rainfall areas and during mid monsoon (August) in high rainfall areas. After the application of organic manure around the trees, it should be covered by a thin layer of soil and properly mulched. If all the organic materials available in the orchard are fully utilized, it can meet a major portion of the nitrogen and a part of other macro and micro nutrient needs.

Cashew plantations have vast potential of organic biomass available for recycling. Around 1.4 to 5.2 tonnes/ha of leaf litter is available from cashew plantations of 10 to 40 years. Converting available cashew biomass waste (about 5.5 tonnes per hectare) produces around 3.5 t of compost or vermicompost, and its application can meet 50 per cent nutrient requirement of cashew. The amounts of nutrient elements recycled from canopy fallout may partially meet the nutrient requirements of cashew. From a six years old cashew trees in Australia, about 15 to 38 per cent of tree total requirements of macronutrients could be met through recycling from canopy biomass fallout of leaves, cashew apples and flowers. When there are plants of more than 20 years of age in a plantation, pits of 1.0 m length, 1.0 m breadth and 0.5 m depth should be dug at the centre of the four trees and all the cashew biomass with fruits are to be incorporated into the pits along with organic manure and bio-fertilizers. Periodic spraying cow's urine (1: 1 dilution) or compost tea (1: 40 dilution) facilitates better growth. They also prevent pest and disease attack.

13. Plant Protection

13.1. Pests and their Management

It is observed that there are about 30 species of insects infesting cashew. Out of these, tea mosquito, flower thrips, stem and root borer and fruit and nut borer are the major pests, which are reported to cause around 30 per cent to 40 per cent loss in yield. The details of major pests of cashew, their damage symptoms and control measures are given in Table 5.4.

Of the several species of defoliating hairy caterpillars recorded as cashew pests, *M.hyrtaca* assumes serious proportions occasionally. Other species include: *Circula trifenestrata* Helfer; *Bombotelia jacosatrix* Guen, *Lymantria* sp., *Thalassodes quadraris* Guen, *etc.* which cause damage by defoliation on isolated trees in certain localities. The caterpillars are nocturnal feeding. They are gregarious and congregate on the lower surface of leaves. Initially, they feed on fresh leaves starting from petiole and continue half way on a leaf, leaving the midrib intact. In later stages, the caterpillars feed voraciously on mature leaves and tender twigs leading to complete defoliation leaving only the bare branches. After feeding, they hide at the base of the tree trunk, shaded branches and on the ground under dry leaves during day time. Dark green,

Table 5.4: Major Pests of Cashew, their Damage Symptoms and Control

Name of the Pest	Damage Symptoms	Control Measures
Tea mosquito bug (*Helopeltis antonii* Signoret)	This pest is considered to be the most serious pest of cashew in India, and causes more economic loss to the crop than any other pest. It is estimated that this pest alone is responsible for damage of nearly 25 per cent of shoots, 30 per cent of inflorescence and 15 per cent of tender nuts. It causes more than 30 per cent economic loss by inflorescence blight and immature nut fall.	Cultural control: Remove alternate hosts such as neem, guava, cocoa, mahogany, cinchona, cotton, apples, grapes, drumstick, black pepper, jamun etc.
	It attacks the tree in all the seasons during flushing, flowering and fruit setting period but the peak period of infestation is from October to March.	Remove the volunteer (self-sown) neem plants in and around cashew plantations.
	Nymphs and adults of this mirid bug suck sap from the leaves, young shoots, inflorescence, developing young nuts and apples.	At the out-break situation, the management programme against this pest should be launched on large scale community basis as the efforts made by an individual farmer may not be of much use.
	The injury made by the suctorial mouth parts of the insect results in exudation of a resinous gummy substance from the feeding punctures.	Monitor crop regularly for signs of damage.
	The tissues around the point of entry of stylets become necrotised and black scab formed, due to the action of the phytotoxin present in the saliva of the bug, infesting the tender shoots/inflorescences at the time of feeding.	Avoid interplanting cashew with other crops which are hosts for *Helopeltis* bugs such as and cotton.
	These lesions turn pinkish brown in 24 hours and become black in 2-3 days. Feeding on tender leaves causes crinkling.	Spray either neem oil (0.5-1 per cent) or Pongamia oil (2 per cent) during flushing, flowering and fruiting phases. Add teepol/soap. Repeated sprayings at fortnightly intervals may be required in specific situations such as heavy infestations or young plantations
	Affected shoots show long black lesions and may cause die-back in severe cases.	
	Infested inflorescence usually turns black and die, immature nuts may drop off Heavily infested trees show scorched appearance, leading to the death of shoots and growing tips	Trees which harbour large populations of predator ants and spiders being natural enemies of TMB and other pests can provide protection. So promote predator ant and spider colonies.

Contd...

Table 5.4—Contd...

Name of the Pest	Damage Symptoms	Control Measures
		The plants can escape pest attack if the new flush is delayed. Planting mid season or late season flowering varieties would be the right strategy to escape TMB.
		Even early flowering varieties also flower 10-15 days late if the plants are grown organically compared to chemical fertilizer applied ones.
		The delay in flowering naturally minimizes the incidence by escaping multiplication of the pest population.
		The pest could be repelled by smoking the garden by burning organic residues three times during flushing, flowering and fruiting. Care must be taken to see that small heaps of organic wastes in several places on the ground below the canopy of the tree is burnt slowly. This can be achieved by putting a thin layer of soil on the heaps and setting fire. At any chance the burning should not produce too much of heat lest flowers and shoots gets affected.
		Natural enemies of mosquito bug: Parasitoids: *Trichogramma* spp., *Telenomus* spp., *Chaetostricha* sp, *Erythmelus helopeltidis*
		Predators: Red ant, dragon fly, ladybird beetle, spider, praying mantis, black ant, anthocorid bug

Contd...

Table 5.4—Contd...

Name of the Pest	Damage Symptoms	Control Measures
Stem and root borer (*Plocaederus ferrugenius* L.)	This is the most serious pest of cashew as its damage results in death of trees. It is an internal tissue borer and hidden dreaded enemy of cashew tree as it is capable of killing the tree outright. The infestation by the pest is more severe in neglected plantations.	Cultural control: Phytosanitary measures such as removal of dead and dried branches of trees, dead trees and trees at advanced stages of infestation at least once in 6 months
	Stem borer infestation could be identified by the presence of small holes at the collar region, gummosis, extrusion of frass through the holes at the collar region, yellowing and shedding of leaves, drying up of twigs and gradual death of the tree	Roots should not be left exposed in the field
		Avoid injuring the plants by sickle and other garden tools, which otherwise will attract the adult for egg laying
	Adult beetles lay eggs in the crevices of bark on the trunk	
	The grubs that hatch out, bore into the bark and feed on the sub-epidermal and vascular tissues and the tissues are tunneled in irregular fashion	The affected bark should be removed along with the grubs
	As a result of the injury to the bark tissues, gum oozes out and gets hardened subsequently resulting in gummosis	Smear lime on the bark crevices
	When the vascular tissues are damaged the ascent of plant sap is arrested and the leaves become yellow and start shedding	Apply wood ash (15-20 kg/tree) and common salt at the base reduces the pest infestation
	In the advanced stages of infestation twigs dry up and the tree dies	Mechanical control: Mechanical removal of the immature stages (grubs) of the pest during initial stages of infestation
	Cashew trees more than two years of age are prone to be attacked by this pest	Identify the borer hole (alive) and extract mechanically by chiseling out the damaged area of the tree and swab mud slurry or coal tar and kerosene (1:2) for adult trees or neem oil 5 per cent (50 ml neem oil in 1 litre of water + 0.5 ml teepol or 5 g of bar soap) on the tree trunk up to 1.0 m height, thrice in a year
	However, the infestation is severe in older and neglected plantations. Even though the infestation is noticed throughout the year, the peak period of infestation will be during summer months	

Contd...

Table 5.4–Contd...

Name of the Pest	Damage Symptoms	Control Measures
Leaf miner (*Acrocercops syngramma* Meyrick)	This is a major pest causing serious damage to the tender foliage of post-monsoon flushes. Young caterpillars soon after hatching, start mining the epidermal layer on the upper surface of the tender cashew leaves, leaving tortuous markings Later on, the thin epidermal mined areas swell up. As a result, the affected areas form blistered patches of greyish white colour When the infested tender leaves mature, big holes are manifested in the damaged areas. The result of injury is the permanent damage to the young leaves which are shrivelled, dried and shed prematurely Nursery seedlings and young plantations are more prone to the infestation of this pest than the older ones Normally 3 to 8 blisters and as many as eight caterpillars are observed on a single leaf	Natural enemies of leaf miner: Parasitoids: *Chelonus* spp., *Sympiesis* sp. Predators: Lacewings, robber fly, Coccinellids, spiders, red ants, dragon fly, praying mantis etc.
Leaf and blossom webber (*Lamida moncusalis* Walker)	Leaf and blossom webber is reported to be a major pest in East-Coast, particularly in Tamil Nadu, Andhra Pradesh and Odisha of India. Another species, *Orthaga exvinacea* Hamps (Noctuidae) has also been recorded as pest on cashew, but it is of minor importance Leaf and blossom webbers attack new flushes and inflorescences. The caterpillars of this pest web the shoots and inflorescences together, remain inside and feed on them Subsequently the webbed portion of the shoots and blossom dry up. Hence, it is called shoot and blossom webbing caterpillar	Natural enemies of leaf and blossom webber: Parasitoids: *Tetrastichus* sp, *Trichogramma* sp, *Bracon* spp., *Goniozus* sp, *Trichospilus pupivorus* Predators: King crow, common mynah, wasp, reduviid bug, big eyed bugs, pentatomid bug, earwigs, red ants, ground beetle, rove beetle, lacewing, ladybird beetle, spiders, robber fly, praying mantis, dragon fly, etc.

Contd...

Table 5.4–Contd...

Name of the Pest	Damage Symptoms	Control Measures
	The galleries of silken webs reinforced with castings and scraps of plant parts are indicative of the presence of caterpillars inside the webbed portion	Follow the common cultural, mechanical and biological practices
	The incidence is found severe mostly on young trees	
Flower thrips (*Rhynchothrips raoensis* Ramakrishna Ayyar)	Thrips attack cashew inflorescence.	Cultural control:
	The rasping and feeding injury made by these thrips results in scab on floral branches, apples and nuts, forms corky layers on the affected parts and subsequent shedding of flowers, improper filling of kernel, malformation of nuts	Inter crop with *Sesbania grandiflora*, to provide barrier which regulate the thrips population.
	Adults and nymphs are seen in colonies on the lower surface of leaves and suck the sap from leaves, inflorescence and apples and nuts	Do not follow chilli or onion as intercrop crop—both the crops are attacked by thrips
	As a result of their rasping and sucking activity the leaves become pale brown, scab on floral branches, apples and nuts, forms corky layers on the affected parts	Sprinkle water over the seedlings to check the multiplication of thrips
	In severe cases there will be shedding of leaves and stunting of growth of trees	
Foliage thrips (*Selenothrips rubrocinctus* Giard, *Rhipiphorothrips cruentatus* Hood, and *Retithrips syriacus* Mayet)	Three species of thrips have been reported as foliage pests of cashew of which *S. rubrocinctus* is of economic importance. This thrip is widespread throughout the tropics, however, it is not considered to be a major pest because its infestation is restricted to a small area	Natural enemies of flower thrips: Predators: Predatory mite (*Amblyseius swirskii*), predatory thrips (*Aeolothrips* spp.), insidiosus flower bugs (*Orius insidiosus*), ant lion, lygaeids, ladybird beetle, anthocorids etc.
	Adult and young thrips are seen in colonies on the lower (abaxial) surface of leaves and suck the sap	Natural enemies of foliage thrips:
	As a result of rasping and sucking activities, the leaves of infested young trees become pale-brown and crinkled with roughening of upper surface	Predators: Predatory mite, predatory thrips, insidious flower bugs, ant lion, lygaeids, ladybird beetle, anthocorid bugs etc.
	In severe cases, there will be shedding of leaves and even stunting of growth	

Contd...

Table 5.4—Contd...

Name of the Pest	Damage Symptoms	Control Measures
	The population of thrips varies from tree to tree in the same area and the insects do not feed indiscriminately, some trees even remain uninfested while others are heavily infested. The trees which produce leaves after the onset of monsoon, are found free from attack because leaves become old during summer months and hence are not normally infested	
	The leaves of post-monsoon flushes about full size but not mature, are mostly invaded by thrips, but as the leaves become old, a number of newly hatched nymphs decline to accept the old leaf tissue and finally the remaining population disperse	
	In most of the young plants, thrips are found to appear on the foliage and cause heavy damage, particularly in summer months. The population increases during the dry season from December to January to a peak in April-May and then rapidly declines during wet season	
Apple Fruit and nut borer (*Thylocoptila panrosema*)	The apple and nut borer causes 10 per cent yield loss during years of severe infestation in certain tracts	Natural enemies of fruit and nut borer:
	The young larva of move to the joints of nut and apple, scrape the epidermis and then bore into them	Parasitoids: *Trichogramma* spp., *Bracon* spp.
	In later stages, they bore into tender apples and nuts and feed on them	Predators: *Chrysoperla carnea*, ladybird beetle, King crow, common mynah, wasp, dragonfly, spiders, robber fly, reduviid bug, praying mantis, red ants, big eyed bugs (*Geocoris* sp), pentatomid bug (*Eocanthecona furcellata*), earwigs, ground beetles, rove beetles, shield bug, anthocorids etc.
	The borer affected nuts do not develop, become shrivelled and dried up resulting in pre mature fall of nuts and apples	
	Usually, the borers tunnel near the junction of apples and nuts, and the entry holes are plugged with excreta	

Contd...

Table 5.4—Contd...

Name of the Pest	Damage Symptoms	Control Measures
Mealy bug	The mealy bug is a serious pest of cashew in all cashew growing areas. Two species of mealybugs infesting cashew includes *Planococcus lilacinus* and *Planococcus citri*. They are called *chappathi poochi, maavupoochi, kallipoochi* in different localities in Tamil Nadu	Follow the common cultural, mechanical and biological practices
	The nymphs and adults colonise on the lower surfaces of tender leaves, twigs, inflorescence panicles and fruit peduncles and suck the sap. The insects being sessile remain stationary while feeding on plant parts. The terminal leaves turn pale yellow and curl downwards due to continuous drainage of sap by large mealy bug populations	Cultural control: Regular monitoring and early detection of infestation are essential to combat this menace
	Besides causing direct damage, the bugs excrete a sweet and sticky substance	The plantation and neighbouring areas should be free from weeds and alternate hosts
	It falls on the upper surface of lower leaves, twigs and fruiting parts	The infested portion of the plant with mealy bug colonies may be pruned and destroyed
	This clogs stomata and makes the plant surface shiny on which black sooty mould fungus develops	Fallen leaves under the tree canopy should be collected and burnt to avoid further spread of the pest
	This black coating, covering the upper surface of leaves and twigs impedes the photosynthetic activities	Natural enemies of mealybug: Parasitoids: *Aenasius advena, Blepyrus insularis, Anagyrus* spp.
	Nut qualities also suffer	Predators: Mirid bug, dragonfly, spiders, robber fly, praying mantis, red ants, lacewings, big-eyed bugs (*Geocoris* sp.)., Coccinellids such as *Cryptolaemus montrouzieri, Cheilomenes sexmaculata, Rodolia fumida, Scymnus coccivora, Nephus regularis*
	Ants are attracted to the sweet excreta and aid in spreading the bugs	
	Nut yield is lost heavily in affected trees	

pea size faecal pellets littered in mass, on the soil under tree canopy reveals the presence of the pest in the orchard. *M.hyrtaca* is a polyphagous pest known to infest drumstick, jamun, sapota, cocoa and garden plants. The pest appears during early monsoon and continues up to December. The natural enemies of hairy caterpillar are predators: Coccinellid, spiders, praying mantis, dragonfly *etc*. Nearly six species of leaf folders attack cashew. Of these, the pyralid leaf folder *Sylepta aurantiacalis* Fisch is a major pest. Its activity is seen from September to January synchronising with flushing and flowering seasons of cashew. The caterpillars, after emerging from eggs, roll up the tender leaves from tip downwards or longitudinally towards dorsal side throughout the breadth or length to form a tubular niche. The leaves are tightly fastened with silken thread of salivary secretion of the caterpillars. Normally one caterpillar is present in each roll. The caterpillar remains inside the leaf roll and feeds on the green tissues by scraping. The fed green matter of the leaf is visible through the transparent integument and the caterpillar appears green in colour. As a result of the damage, the terminal and tender leaves dry up. This affects the terminal growth in root stock and young grafts. On regular flushes, the panicle emergence is arrested. Natural enemies of leaf folder are parasitoid: *Cotesia flavipes* and predators: *Chrysoperla carnea,* Coccinellids, King crow, common mynah, wasp, dragonfly, spiders, robber fly, reduviid bug, praying mantis, red ants, big eyed bugs (*Geocoris* sp), pentatomid bug (*Eocanthecona furcellata*), earwigs, ground beetles, rove beetles *etc*.

13.2. Disease and their Management

The important diseases of cashew and their management are detailed in Table 5.5.

14. Harvesting

The flower panicles emerging from the graft during 1st and 2nd year should be removed in order to allow the plant to put good vegetative growth and better framework. Economic bearing in cashew commences after 3rd year of planting. Normally, about 92 per cent of the trees yield by the third year from planting. It reaches full bearing during tenth year and continues to give remunerative yields for another 20 years. The ripened fruit will fall down and nuts from fallen fruits have to be collected. The land should be clean during the harvesting season to facilitate collection of cashew. The main harvesting season is from February to May. Normally, harvesting consists of picking of nuts that have dropped to the ground after maturing. Cashew nuts when fully mature look greyish brown. A simple test of maturity is to float nuts in water when mature nuts will sink while the immature and unfilled nuts will float. Nuts are to be gathered every week during the harvest season. Sometimes farmers harvest their crop before they drop to prevent pilferage. This very often results in poor quality of the kernels. The average yield per tree increases from about 2 kg at 3-5 years to 4 kg at 6-10 years as the canopy develops and 5-10 kg when trees are 11-15 years of age. Thereafter, trees yield in excess of 10 kg as the trees get older and depending on management of the plantation. However, in plantations of unknown origin or seedling progenies with conventional methods of cultivation, the yield could be less than one kg of raw nuts per tree.

Table 5.5: The Important Diseases of Cashew and their Management

Name of the Disease	Damage Symptoms	Management
Die–back or pink disease (*Corticium salmonicolor* Berk. and Broome)	It is a very common disease of cashew, often assuming great importance during the south-west monsoon period	Chisel out the affected parts and apply Bordeaux paste
	Whitish or pinkish growth of the fungus can be seen on the affected branches	
	The fungus penetrates into the deeper tissues and causes the death of the shoots from the tip downwards and hence the name dieback	
	After heavy rains a film of silky thread of the fungus is seen on the branches	
	In advanced stages, the bark splits and peels off. Sometimes only one branch is affected, but often many branches turn yellow and shed giving a barren appearance to a portion of the tree	
Damping off, seed rot, seedling blight and root rot (*Phytophthora palmivora* (Butler)	The disease occurs in nurseries where drainage conditions are poor	These diseases can be effectively managed with the following integrated control measures:
	The organisms attack the roots or collar region of seedlings or both the regions and cause their death	Provide proper drainage facilities in the nursery
	When seedlings are infected by *Phytophthora palmivora*, they become pale	Provide enough drainage holes on the bags used for raising seedlings
	Water-soaked lesions can be observed at the collar region which turns dark and girdles the stem	Raise seedlings in solarised potting mixture. Potting mixture has to be solarised for one month using 150 gauge transparent polythene sheets
	The seedlings droop and ultimately the plants die	After filling the potting mixture in the polythene bag, use Trichoderma enriched manure for potting mixture
	On leaves, water-soaked lesions can be observed in severe cases	Incorporate Mycorrhiza @ 10g/kg and PGPR mix I.5g/kg potting mixture before sowing the seeds
	These lesions enlarge and coalesce, often covering the entire leaf lamina	
	All the organisms in combination or alone may cause the disease	

Contd...

Table 5.5–*Contd...*

Name of the Disease	Damage Symptoms	Management
		Remove and destroy the disease affected seedlings
		Never re-use contaminated potting mixture
		Provide sufficient spacing in the nursery to ward off excess humidity
		Never raise cashew nurseries in heavily shaded areas
		Drench the nursery bag with 1 per cent Bordeaux mixture. While drenching, care should be taken to drench sufficient quantity of fungicide to soak the entire potting mixture in the polythene bag
		Spray the seedlings with 1 per cent Bordeaux mixture as a prophylactic measure to prevent aerial infection
Anthracnose (*Colletotrichum gloeosporioides* (Penz.) Penz. and Sacc.)	It is known to cause severe loss in Brazil.	Remove and destroy the affected plant parts
	The disease affects all young tissues, *viz.,* tender leaves, twigs, inflorescence, shoots, nuts and apples	Give a prophylactic spray of 1 per cent Bordeaux mixture
	Appearance of reddish brown, shiny water soaked lesions followed by resin exudation on the affected parts is the initial symptom	Plant wind breaks like Casuarina or Eucalyptus to check the spread of disease from one plantation to another
	In severe cases the lesions coalesce resulting in defoliation	
	The tender leaves are crinkled and fruits shriveled	
	The infected inflorescences turn black	

Contd...

Table 5.5–Contd...

Name of the Disease	Damage Symptoms	Management
	Repeated Infection of the terminal shoots leads to the death of the tree in course of time	Remove and destroy the affected plant parts
Inflorescence blight (*Colletotrichum mangiferae* Kelker and *Phomopsis anacardii* Early and Punith)	This disease is occurring during the monsoon period	
	The characteristic symptom is the drying of floral branches	
	The symptoms appear as minute water soaked lesions on the main rachis and secondary rachis	
	The lesions are pinkish brown, enlarge and soon turn scabby. Gummy exudates can be seen at the affected regions	
	The lesions develop into bigger patches and result in drying up of the inflorescences	
	The incidence is very severe when cloudy weather prevails	
Shoot rot and leaf fall (*Phytophthora nicotianae* var. *nicotianae* Breda de Haan)	During the south west monsoon months of June - August extensive leaf fall and shoot rot symptoms are observed	Spray Bordeaux mixture @ 1 per cent before the onset of monsoon and again during the break in the monsoon to check the spread of the disease
	Black elongate lesions are first developed on the stem with exudation of gum	
	Later, infection spreads up and down, causing the tender stem to collapse and tender leaves to shrivel up	
	The lower mature leaves are also infected with black elongated lesions on mid rib, which later spread to the main lateral veins and the leaf blade	
	The infected leaves are soon shed	

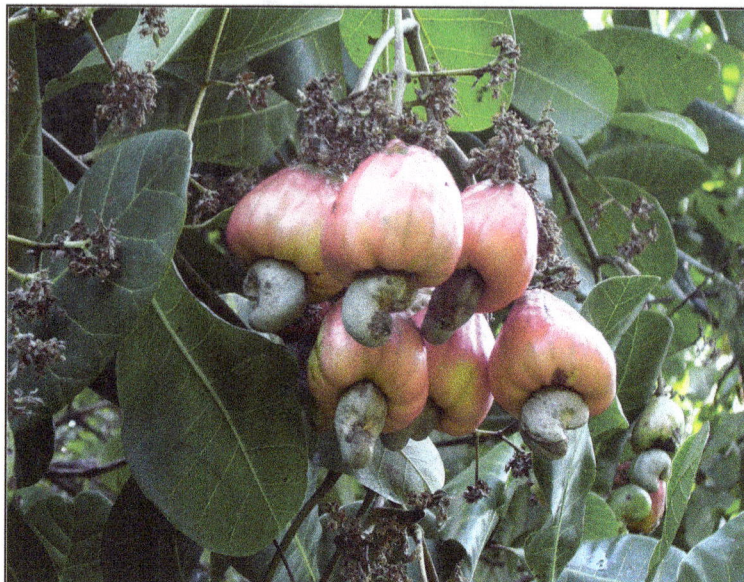

Figure 5.9: Mature Nuts Ready for Harvest.

15. Processing of Cashew

The cashew fruits produce cashew apples and cashew nut which can be used as foodstuffs. In addition, a by-product, Cashew Nut Shell Oil, is used in the paint and brake-liner industries. The actual fruit, the cashew nut, requires a special kind of procedure in order to break open the hard shell and extract the nut. Cashew nuts are traded in a variety of different quality grades, and are used in the snack, confectionery, chocolate and baking industries. Raw nuts, immediately after harvest, are to be separated from cashew apple. If the raw nuts are not dried properly, during subsequent storage, they get spoiled due to microbial infestation. The drying process also helps to retain flavor and quality of the kernels. High quality nuts are obtained when freshly fallen nuts are separated from the cashew apples and dried in sun on cement floor for 2-3 days to bring down the moisture from about 25 per cent to below 9 per cent. Nuts after drying can be stored in gunny bags well protected from rodents and stacked on a platform above the ground level leaving space on all the sides of the room. In this form, cashew nuts can be stored for up to 2 years, in ideal storage conditions (dry, dark, cool, well-ventilated). However, they are usually processed within the same year of harvesting.The processing of cashew nuts involves preliminary cleaning, roasting and shelling and separation.

Manual picking of large objects and by sieving is done during preliminary cleaning of cashew nuts. Later, the cleaned cashew nuts are roasted in open pan or earthen ware or rotary cylinder or hot oil bath. The first two methods are simple and cheap, but they are time consuming and lead to poor recovery of CNSL (Cashew Nut Shell Liquid). The rotary cylinder method is more hygienic and efficient, but results in loss of major portion of the CNSL. The hot oil bath process helps in good roasting and high recovery of shell liquid. The most common method adopted is roasting by rotary cylinder method. After roasting, the shells are removed and the nuts extracted manually. In manual shelling, recovery of whole kernels is more compared to the mechanical shelling. The kernels are dried in hot air chambers which facilitates peeling of the outer coating or testa. To prevent breakage, the kernels are to be handled very carefully, as they are brittle at this stage. The shelling percentage of cashew varies between 20-25.

16. Grading and Packing

Grading is done for export purposes based on "counts" or number of kernels per pound. Sound kernels are named as "wholes" and broken ones as "splits". The wholes are again classified as whole white kernels, whole scorched kernels, whole dessert kernels (a) and whole dessert kernels (b). The splits are also further graded into white pieces, scorched pieces, dessert pieces (a) and dessert pieces (b) based on certain physical characters. The wholes are packed in several grades *viz.*, 210, 240, 280, 320, 400, 459 and 500; the popular grade is 320. The specifications for graded kernels are that they should be fully developed, ivory white in colour and should be free from insect damage and black and brown spots. Packing is done in time by Vita pack method (exhausting the air inside the packing tin, pumping in carbon dioxide and sealing). The cashew nuts are not allowed to be treated with methyl bromide, ethylene oxide or ionising rays.

Figure 5.10: Cashewnuts Ready for Peeling.

17. Utilization of Cashew Apple

Cashew apple is a valuable source of sugars, minerals and vitamins especially vitamin C and B2. It can be eaten raw as a fresh fruit. Either whole apples are consumed or they can be cut into small pieces, mixed with table salt and eaten.

Figure 5.11: Cashew Apple.

Figure 5.12: Cashew Apple Pickle.

If the apples are used for making different products, the fruit has to be harvested before it falls naturally. Therefore, cashew apples for the fresh fruit market should be harvested daily. Cashew apple can be used for the preparation of various products like squash, syrup, pickle, wine, liquor and vinegar, but commercial production is possible only with the use of chemical preservatives. However, organic processing of cashew apple is possible without the use of chemical preservatives at domestic level, provided refrigerated storage is possible.

18. Environmental Service

Cashew plantations have vast potential of organic biomass available for recycling. The entire cashew biomass available in the plantation can be converted to compost or vermicompost and used, which can help in meeting nutrient requirement of cashew to a great extent. Cashew has dense green leaves with good photosynthetic capacity enabling to make use of solar energy to the fullest extent. Cashew can also be grown in high-density planting system accommodating higher number of plants per unit area. Cashew is suitable crop for carbon sequestration. Seven years old trees of cashew (genotype VTH-174) could sequester about 2.2 fold higher carbon under high density planting system (625 trees/ha) as compared to normal density planting system (156 trees/ha). The extent of carbon sequestered will depend on the amounts of C in standing biomass, age of the crop, tree density, variety *etc.* Cashew can be grown in vast degraded/wasteland existing in cashew growing regions, thereby putting such land into more productive use.

Selected References

Anonymous (2000). *Organic Farming in the Tropics and Subtropics. Cashew Nuts Naturland* e.V. - 1st edition 2000 p.12.

Anonymous (2014).Cashew handbook 2014 Global Perspective 4th edition, June 2014. Published by Foretell Business Solutions Private Limited on behalf of www.cashewinfo.com p.139

Satyagopal, K., Sushil, S.N., Jeyakumar,P., Shankar, G., Sharma, O.P., Boina, D.R., Sain, S.K., Rao, N.S., Sunanda, B.S., Ram Asre, Kapoor, K.S., Sanjay Arya, Subhash Kumar, Patni, C.S. (2014). AESA based IPM package for Cashew Nut. Department of Agriculture and Cooperation Ministry of Agriculture Government of India. pp 54.

Varanashi Krishna Moorthy, Yadukumar. N and Shankar Raj. N.S (2007). *Handbook on Organic Cashew Cultivation*, p. 71.

http://agritech.tnau.ac.in/horticulture/horti_plantation crops_cashewnut.html

Chapter 6

Organic Farming in Coffee

☆ *Y. Raghuramulu and S. Kamala Bai*

1. Introduction

In the global Coffee market, the differentiated/specialty coffees are emerging as one of the fastest growing segment with increased demand for such coffees from the consumers. Among the specialty coffees, the sustainable coffees are emerging as most important segment due to the concerns of protecting the environment, consumer health and the welfare of small growers and workers. Unlike other specialty coffees, the sustainable coffees are characterized by unique feature of certification of operations from farm to final retail outlet. The sustainable coffees broadly comprise of Organic coffee, Fair-trade coffee and Shade Grown or Eco-friendly or bird friendly coffees. Organic coffees are those produced by such management practices which help to conserve or enhance soil structure, resilience and fertility by applying cultivation practices that use only non-synthetic nutrients and plant protection methods. Further, although many producers grow coffee without use of synthetic agro chemicals, this passive approach is not sufficient to be considered organic in the absence of credible certification.

The first organic coffee cultivation was recorded at the Finca Irelanda in Chiapas, Mexico in 1967, and the first organic coffee to be imported into Europe from small farmers Cooperative came from the UCIRI cooperative in Oaxaca, Mexico in 1985. Major consumers of organic coffee are USA, Japan and European countries. From the mid 1980's onwards, a number of smallholder groups in Mexico and Central America organized themselves to produce and directly market organic coffee.

2. Production Scenario

2.1. Global Scenario

Coffee is one of the most important commodities in the world market. The unique feature of the global coffee industry is that the production is confined to around 70 developing countries, which is critical to the economies of several of them, whereas, it is largely consumed in the developed countries. Arabica and Robusta coffee are the two species that are commercially cultivated in the world and of these, nearly 58 per cent is Arabica and the balance 42 per cent is Robusta.

As per the International Coffee Organization the total production during 2015 by all exporting countries (in 60kg bags) was 143,306 (equivalent to 8.6 million metric tonnes). The world's leading coffee growing countries are Brazil (2.1million hectares), Indonesia (1.2 million hectares), Colombia (0.8 million hectares), Mexico (0.7 million hectares) and Vietnam (almost 0.6 million hectares).The top five producers (which accounts for about 72 per cent of global production) during 2015 are: Brazil (30.17 per cent), Vietnam (19.19 per cent), Colombia (9.42 per cent), Indonesia (8.59 per cent) and Ethiopia (4.68 per cent). The top five importers of coffee (accounting for 64 per cent of global production), as per 2012 statistics are: United States (24 per cent), Germany (20 per cent), Italy (8 per cent), Japan (6 per cent) and France (6 per cent).

2.2. Organic Coffee

According to FAO STAT, almost 0.76 million hectares of coffee are grown organically during 2014. This constitutes 7.7 per cent of the world's harvested area of coffee of 9.9 million hectares in 2013. The area under organic coffee was only 0.018 million hectares during 2004. Organic coffee is being produced by more than 30 countries in the world of which Bolivia, Brazil, Colombia, Congo, Ethiopia, Honduras, Indonesia, Mexico, Nicaragua, Papua New Guinea, Peru, Tanzania, Timor-Leste are the countries with more than 10,000 hectares under organic cultivation. More than 50 percent of the world's organic coffee area is in Latin America and almost 30 percent in Africa, and the largest areas are in Mexico (0.24 million hectares), Ethiopia (0.15 million hectares), and Peru (0.09 million hectares).

Nepal is having the highest share, with almost 46 percent of organic coffee (but the area is only 804 hectares), followed by Timor-Leste (45 percent), Bolivia (37 percent), and Mexico (almost 35 percent). Some of these high percentages must be attributed to the fact that coffee is grown more extensively in organic agriculture, and often in association with other crops. Recently many countries like India, Kenya, Uganda *etc.*, have taken major initiatives in promoting organic coffee production for exports. Mexico is the largest producer of organic coffee in the world with one thirds of its total production being certified as organic. In this country, smallholder groups grow majority of organic coffee.

As per the statistics available for 2011(SSI Review, 2014), organic coffee production was estimated at 2.49 lakh metric tonnes, making it the fourth-largest producer of sustainable coffee in that year. Although 67 per cent of organic coffee was sourced from Latin America, 50 per cent of organic production came from

Peru (25 per cent), Mexico (18 per cent) and Honduras (7 per cent). This trend is repeated in Africa, with a remarkable 18 per cent of organic supply coming from Ethiopia alone. In Ethiopia, Mexico and Peru, organic production accounted for 10 per cent or more of total domestic production, suggesting its economic importance in these countries.

2.3. Indian Scenario

The Indian coffee industry is characterized by predominantly small holdings. Majority of these small holdings (especially in Idukki zone of Kerala, and Bodinayakanur zone of Tamil Nadu) and all the tribal holdings (in Andhra Pradesh and the North-Eastern states) are basically organic by default. These small and tribal coffee growers do not use chemical fertilizers and plant protection chemicals due to their poor economic status and due to their belief in natural farming. Consequently, the yields are low and are only at subsistence levels. Thus, there exists a good scope for converting these small and tribal holdings into certified organic without much change in the existing cultivation practices.

The scenario of the Indian coffee growers is complex prototype *viz.*, fragmentation of the land holdings, lack of timely availability of the farm inputs and inability to get good market price for the organic produce due to lack of awareness of market. The reason is that other than corporate and large planters, small growers are unable to adopt scientific way of scaling their produce from SEED to CUP *i.e.*, from back ward linkages- soil (inputs like seeds, fertilizers, credit, insurance, knowledge and extension services) to forward linkages–CUP (collective marketing, processing, market driven agriculture production and ready to drink). India grows at least 13 unique varieties of coffee and the coffee growing regions are among the 25 biodiversity hotspots in the world. India occupies the third-largest producer and exporter of coffee in Asia and is the world's sixth-largest producer of coffee (3.9 per cent of the world's coffee). The area under coffee plantations in India is 397.147 thousand hectares in 2015-16. Most of this area is concentrated in the southern states of Karnataka (54.95 per cent), Kerala (21.33 per cent) and Tamil Nadu (8.18 per cent).Around 70 per cent of Indian production is exported to over 45 countries and is the world's fifth-largest exporter of coffee, the total exports being 277,696 MT in 2015-16.

In India, production of certified organic coffee was first initiated during 1993-94 by M/s. Bombay Burma Trading Corporation Ltd., Sidapur, Coorg district, Karnataka. They have exported around 100 tonnes per annum during the period 1994-96. Subsequently, a few groups of small growers have started organic farming in coffee mainly in Kerala and Karnataka. In India, the conditions for production of organic coffee are favourable compared to any other coffee producing country. Some of the natural opportunities in India are:

1. Coffee is mainly cultivated in deep fertile jungle soils under a two tier mixed evergreen leguminous/non- leguminous shade canopy trees. These shade trees provide a natural habitat for vast population of birds and natural enemies of insect pests/diseases, help in reducing the soil erosion and thus contribute wide biodiversity with in a small eco-system.

2. The production practices followed in Indian coffee plantations such as manual weeding, shade regulation, soil and water conservations and use of traditional farming practices like composting are considered as one of the best in the world, thus reducing excessive dependence on agro- chemical inputs.

As per the statistics of Coffee Board, Bengaluru (2008), the total area under organic coffee (both under "in-conversion" and "certified organic") in the major coffee growing states was 3550.4 hectares. The production of organic coffee (certified and in-conversion) was about 1000 MT during 2008.

The state wise area and production under organic coffee cultivation is furnished in Table 6.1.

Table 6.1: Area under Organic Coffee in India

State	Certified Area (ha)		Estimated Production (MT)	
	Arabica	Robusta	Arabica	Robusta
Tamil Nadu	736.1	22.0	125.0	2.0
Karnataka	19.3	446.0	6.2	267.4
Kerala	–	1222.0	–	454.9
Andhra Pradesh	1105.0	–	145.0	–
Total	1860.4	1690.0	276.2	724.3

Though of the total area, nearly 53 per cent was covered by arabica coffee (especially under Group ICS category), 72 per cent of total organic production was from robusta coffee.

The export of Indian organic coffee was limited in the initial years, however, it picked up considerably from 2004 onwards mainly due to entry of a few large growers during the last 3-4 years. The large and corporate growers are able to realize better premiums (up to 30 per cent) due to their better marketing strategies, while majority of small growers are not able to access the specialty market and, hence, sell their produce in the regular market. Export of organic coffee from India during 2014-15 was 88,000 kg, valued at US$240,425, while it went up to 20,160 kg, worth US$45,416 during 2015-16. Germany was the largest buyer of Indian organic coffee (57,600 kg worth US$136,880) followed by United Kingdom importing 21,200 kg, worth US $ 103,545. While Germany continued to be the largest buyer during 2015-16 also (19,200 kg, accounting for US$40,631, it was followed by Japan, which imported 960 kg worth US$ 4,786.

3. Botany and Varieties of Coffee

3.1. Botany

Botanically, coffee belongs to the genus *Coffea* of the family Rubiaceae. There are approximately 70 species under the genus *Coffea*, most of which are native of

Africa including the two species *viz., Coffea arabica* L. and *Coffea canephora Pierre ex Froehner* which are commercially cultivated in India. Another species, *C. liberica* is grown to a small extent. Arabica is the only tetraploid species of the genus *coffea* (2n=44), while the rest of the species are diploid in nature (Anon, 2011). Coffee is a perennial plant and evergreen in nature. It has a prominent vertical stem giving rise to horizontal primary branches in pairs opposite to each other. From each primary branch, several secondaries originate laterally which in turn produce several tertiary and quarternary branches. In coffee, the vegetative growth of a particular year determines the cropping wood of the succeeding year. Hence it is necessary to maintain balance between cropping and vegetative growth through bush management, which involves training, pruning, handling and desuckering as discussed in later chapters.

Generally, coffee has shallow root system, particularly robusta which has feeder roots concentrated very close to the surface of the ground. On the other hand, arabica coffee produces most of the feeder roots relatively in deeper soil. The spread of the roots depends on the type of soil and cultural practices. The leaves of coffee are opposite decussate on suckers, but in plagiotropic branches by torsion, successive nodes with the leaves lie in one plane. The leaves are shiny, wavy and dark green in colour with conspicuous veins. The shape of the leaf is usually elliptical. Coffee is a short day plant *i.e.,* floral initiation takes place during short day conditions of 8-11 hrs of day light. Flower buds are produced at the axils of mature green wood on short stalks which are known as peduncles. The group of flowers, technically called 'inflorescence' is a condensed cymose type subtended by bracts. More number of flowers per axil is produced in robusta than arabica. The fruit is a "drupe" and normally contains two seeds. Abortion of one ovule due to non-fertilization leads to the formation of a single seeded fruit, called the 'peaberry'. Seeds are elliptical or egg shaped, possessing longitudinal furrow on the plane surface. Seed coat is represented by the "silver skin". Seeds do not exhibit any dormancy. Viability is also short in coffee. Germination takes place in about 45 days.

3.2. Varieties

3.2.1. Arabica (*Coffea arabica*)

Arabica coffee under natural conditions grows like a small tree but is regulated like a bush through training. It branches profusely with dark green leaves (Figure 6.1). The flower buds are produced in clusters in the axils of leaves at each node. Under South Indian condition initiation and subsequent growth of flower buds take place from September to March. Water is essential for flowering and the blossom occurs in 9-10 days after the receipt of showers. Arabica is self fertile. The fertilized ovary grows into a fruit in about 8-9 months. Each fruit usually contains two beans / seeds. The freshly pulped beans along with their endocarp cover (parchment) are white. Sun drying and removal of parchment cover and silver skin (seed coat) gives bluish green colour to the seed which is the final produce used in preparation of the stimulating coffee beverage.

Figure 6.1: Robusta Coffee.

3.2.2. Robusta (*Coffea canephora*)

It is a bigger bush than arabica. Hence, it is popularly known as robusta coffee. The leaves are broader, large and pale green (Figure 6.2). Flowers are white, fragrant and are borne in larger clusters than in arabica. Under the conditions of South India, the buds initiate and reach maturity during November to February and precipitation in Feb-March is ideal for blossoming. The flowers open on 7th or 8th day after receipt of rain. Robusta is self-sterile, that is, its ovule cannot be fertilized with its own pollen and hence, cross pollination is necessary. The fruits mature in 10-11 months. They are generally ready for harvest two months later than Arabica.

Figure 6.2: Arabica Coffee.

4. Climatic Requirements

The climatic requirements for Arabica and Robusta coffee under South Indian conditions are as follows.

Table 6.2: climatic Requirements for Arabica and Robusta Coffee under South Indian Conditions

Factors	Arabica	Robusta
Elevation	1000-1500 m	500 – 1000m
Aspect	North, East and N-E aspects ideal	Same as Arabica
Temperature	15°C to 25°C ideal – cool, equable	20°C to 30°C ideal – hot, humid
Relative humidity	70-80 per cent	80 – 90 per cent ideal
Annual rain fall	1600 – 2500 mm	1000 – 2000 mm
Blossom showers	March – April (25-40 mm)	February–March (25–40 mm)
Backing showers	April – May (50 – 75 mm) well distributed	March – April (50 – 75 mm) well distributed

5. Soil Requirements

The coffee growing soils vary from deep volcanic soils to moderately deep soils originating from parent rocks. Under Indian conditions, deep, well-drained laterite type of soils rich in organic matter and slightly acidic (pH 6.0 – 6.5) are ideal for cultivation of coffee. Arabica can be grown in gentle to moderate slopes while Robusta can be grown in gentle slope to fairly level fields.

6. Organic Production Technologies

6.1. Establishing New Organic Plantations

Organic agriculture consists of a system of farm design and management to create an ecosystem, which can achieve sustainable productivity without the use of artificial external inputs such as chemical fertilizers and pesticides. The major aims of organic agriculture are production of quality agricultural products, which contain no chemical residues, the development of environment friendly production methods and the application of production techniques that restore and maintain soil fertility. These are achieved by suitable crop selection and rotation, recycling of plant and animal residues, proper tillage and water management. Management of weeds and pests is attained by encouraging biological control through a balanced host-predator relationship, augmentation of beneficial insect population and by mechanical removal of weeds, pests and affected plant parts. Organic farming differs in many ways from conventional farming. It is generally recognized that organic farming:

☆ Does not pollute the soil and ground water with chemical residues.

☆ Increases the biological diversity among plants and animals.

☆ Reduces leaching of minerals in the soil

☆ Depends on and makes full use of natural, local and renewable resources.

☆ Uses low energy inputs.

☆ Depends largely on natural equilibrium for crop protection.

Organic farm products are more expensive than conventional ones. Yields drop sharply during the phase of conversion as it takes some time for the soil and plants to reach equilibrium. However, yields rise again, once the management systems get established. While establishing new coffee plantations under organic production system, attention has to be paid to meet the basic requirements so that certified organic farming could be achieved. This section describes the steps involved in establishment of new plantations and their management under organic production systems.

6.2. Conversion and Isolation

Coffee being a perennial crop, claiming the produce as organic in the first year of its cultivation is not permitted. Production of organic coffee under organic management needs interim period called conversion period. The conversion period is the period required to accomplish the requirements for growing coffee under organic farming system to achieve sustainable productivity without the use of artificial external inputs. Coffee being perennial, the conversion period is three years for newly established plantations. The site selected for planting of organic coffee should be provided with appropriate isolation or buffer zone (10 meters) from the conventional estates/blocks so as to prevent drift/contamination with chemicals.

7. Field Planting of Coffee

7.1. Selection of Site

In selecting the site for coffee plantation, due consideration should be given to altitude, aspect, rainfall pattern, soil type, land slope, existence of shade trees, temperature prevailing in the area, availability of water resources, exposure to wind and transport facilities. A perennial source of water supply is an essential requirement for growing coffee. Locations with gentle to moderate slopes covered with a good canopy of evergreen trees are to be preferred. Southern and western aspects should be avoided especially at lower elevations. In case unavoidable, such areas should be provided with more shade to protect coffee from afternoon sun. In wind prone areas, wind belts consisting of tall trees like silver oak, tree coffee *etc.* should be raised.

7.2. Choice of Varieties

The varieties selected for organic coffee production must be well adapted to local conditions and tolerant/resistant to pests/diseases. In case of Arabica coffee, varieties having wider adaptability such as S.795, Sln.5-B, Sln.6 and Sln.9 may be preferred, while in case of Robusta, improved varieties like S.274 and Cx R may be selected.

7.3. Raising a Nursery

Seeds for raising nursery should be collected from organic estates/blocks only. However, if not available, seeds from conventional estates/blocks but not treated with any chemicals can be used. The organic nursery should be clearly separated from conventional nursery, if both the activities are carried out in the same estate.

7.4. Land Preparation

Clean felling of trees is not advocated when land is prepared for planting of organic coffee. Selective retention of evergreen permanent shade trees providing filter shade at a spacing of 10-12 m is desirable. The land should be divided into blocks of convenient size by laying out footpaths and roads in between. The ground level bushy growth should be cleared by uprooting. Land preparation should be completed by March/April well ahead of commencement of South-West monsoon.

7.5. Soil Conservation

The loss of top soil is negligible when the land is covered by good shade tree canopy. The soil erosion is more on steep slopes without proper shade cover. In such fields, appropriate soil conservation measures like contour planting and terracing should be practiced.

7.6. Preparations for Planting

In each block, the spots for planting of coffee and shade trees should be marked at recommended spacing soon after land preparation. The spacing recommended for tall Arabica varieties is 2.1 m x 2.1 m, while dwarf Arabica can be planted at close intervals of 1.8 m x 1.8 m. The tall Robusta variety S.274 requires a wider spacing of 3.0 m x 3.0 m while the compact robusta variety Congensis x Robusta (C x R) can be planted at relatively closer spacing of 2.5 m x 2.5 m.

Pits of size 45 cm length, 45 cm width and 45 cm depth are to be opened during the months of April - May and exposed to sun for about a fortnight to kill soil pests like cockchafers (root grubs), nematodes *etc*. Later, they should be filled with top fertile soil and well-decomposed farmyard manure or compost (1-2 kg/ pit) prepared on the estate.

7.7. Planting of Shade Trees

It is advisable to plant temporary shade trees like Dadap (*Erythrina lithosperma*) at closer spacing (3 m to 5 m) initially, for providing optimum shade to young coffee plants. Wherever large open spaces are present, evergreen permanent shade trees belonging to *Ficus* sp., *Albizzia* sp., *Artocarpus etc.* should also be planted at 10 m to 13 m intervals. For planting shade trees, pits should be taken during pre-monsoon period (April-May) and filled with topsoil after exposing to sunlight for about a fortnight. Planting of shade trees should preferably be completed before onset of S-W monsoon.

Figure 6.3: Coffee grown under Mixed Shade.

7.8. Planting of Coffee and Aftercare

Planting of coffee seedlings should be taken up during August-September towards the end of heavy monsoon rains. In areas with predominance of North-East monsoon like Tamil Nadu, planting can be done in June-July months. At the time of planting, it is advisable to add about 50g of rock phosphate to each pit, for encouraging better root growth and proper establishment of plants. In cockchafer infested fields, neem cake @ 250g per pit is advocated. After completion of planting, the coffee seedlings should be provided with cross staking and mulching to protect against wind damage and to conserve soil moisture for the ensuing dry period. Towards commencement of dry period (Nov.), the young plants in open area should be protected by erecting temporary shade huts with jungle tree twigs or bamboo thatches. Stems of young dadap plants should be coated with lime solution to prevent sun scorching.

7.9. Weed Control

Weeds pose a serious problem especially in new coffee clearings. Following measures are suggested for controlling weeds:

☆ In new clearings, cultural practices such as cover digging (30 cm deep) during second year of planting and scuffling (10 cm to 15 cm) for the next two to three years should be carried out after cessation of monsoon season (Nov-Dec). These operations would not only bring down the weed growth but also help in conservation of soil moisture. However, in sloping terrain, soil digging should be completely avoided. In such areas, only

Figure 6.4: Cow Pea as Green Manure Crop.

slash weeding and cover cropping with cow pea or horse gram should be followed to prevent soil erosion.

☆ Cultivation of green manure crops/cover crops and mulching with weed slashings and shade tree leaf litter *etc.* would also help in smothering of weeds. Once the coffee bushes cover up, the weed growth would naturally get suppressed and manual slash weeding alone would be sufficient in the subsequent years.

☆ Use of any kind of herbicides is strictly prohibited.

7.10. Nutrient Management

In newly planted fields, it is recommended to grow green manure crops like cow pea and horse gram for initial two or three years to build up soil fertility. These crops should be grown during June-September, so as to prevent them competing for soil moisture with young coffee. These green manure crops contribute around 6-10 tonnes of dry organic matter/ha and also effectively suppress weed growth in the early years. As most of these crops are leguminous in nature, they fix nitrogen from atmosphere. The green manure crops should be incorporated into soil before their flowering to improve soil fertility. The following on farm practices are important for monitoring the nutrient requirement of coffee holdings:

☆ Correction of soil pH using agricultural lime or dolomite based on soil test values at least once in 2-3 years.

☆ Application of farmyard manure or compost prepared on the farm @ 500 kg/acre per year.

☆ Deficiency in nutrient supply can be met by using other permitted products like rock phosphate, bone meal, wood ash *etc.* as per the Appendix I of the National Standards on Organic Production (NSOP).

☆ Use of bio-fertilizers may also be resorted to, in a restricted manner to improve nutrient use efficiency.

7.11. Plant Training and Pruning

The young coffee plants should be trained to provide proper shape to the bushes and to improve efficiency of operations like spraying, harvesting *etc.* at later stages. Generally, single stem system of training is recommended for coffee grown under shade. In this system, height of plants is restricted by topping (capping) at prescribed heights. The tall Arabica varieties are topped in two stages (two-tier system), while dwarf Arabicas as well as the Robustas are topped only at a single stage (single tier system). The prescribed topping heights for different coffee varieties are:

Tall Arabicas

☆ 1st topping at 75 cm

☆ 2nd topping at 135 to 150 cm (usually done after realizing 4-5 crops)

Dwarf Arabicas

☆ Single topping at 90 cm to 150 cm depending on soil fertility, wind proneness *etc.*

Robustas

☆ Single topping at 135 cm to 150 cm

In topping operation, vertically growing main stem is cut 5 cm above the node near the prescribed topping height (Arabica takes 9-12 months and Robusta takes 18-24 months). One of the top most primary branches is cut beyond the basal node to prevent splitting of main stem due to crop load. After topping, all the new suckers produced on main stem are to be removed periodically at least three to four times a year.

8. Conversion of Established Plantations and their Management

When an existing coffee plantation is proposed for conversion to organic, it is essential to know about the consequences and requirements. Traditionally cultivated estates having low to medium yield levels (*i.e.*, 100-250kg/acre in case of Arabica and 400kg/acre in case of Robusta) can be easily brought under organic farming without any significant yield reduction. However, when the intensively managed estates with higher yield levels are converted to organic, there will be a yield reduction up to 30 per cent in the first 3 to 4 years after conversion. However, if managed systematically, these plantations would attain sustainable yield levels within the next 3 years.The requirements for conversion of existing plantations into organic holdings and their subsequent management as per the 'National Standards on Organic Production (NSOP)' are described here under.

8.1. Conversion Plan

A well prepared plan of conversion would facilitate better and efficient management of organic coffee estates in the initial years. The following important criteria should be taken into consideration while planning for conversion to organic farming:

☆ Preferably, the entire farm unit should be converted to organic in a phased manner and the operator (grower) should present a conversion plan to the certification body when applying for certification.

☆ In case of conversion of a portion of existing coffee plantation into organic, adequate buffer zone should be maintained between organic plots and conventional blocks, so as to prevent drift/contamination with chemicals. The buffer zone may vary from several factors like geographic position, slope and nature of adjoining blocks *etc*. The certifying agency would be able to suggest suitable buffer zone based on site specification conditions.

☆ In case of small and marginal holdings, maintaining a buffer zone would be very difficult as it would drastically bring down the net area under organic cultivation. In view of this, a community approach is suggested for a group of contiguous farms forming a large belt or zone. The small growers within this large belt or zone may form a co-operative or a self-help group and adopt group certification to reduce cost of certification. Formation of groups would also facilitate sharing of common processing and storage facilities by all members of the group.

8.2. Conversion Period

Most of the organic production standards for agricultural crops prescribe a definite conversion period for each type of crop, so as to eliminate chemical residues in the soil and surrounding environment for production of chemical free organic products. The conversion period for organic coffee estates as per the National Standards on Organic Production is as follows:

☆ For existing plantations, a minimum period of three years is required as conversion period to qualify for organic certification. During conversion period, the yield from such blocks shall be labeled only as "In-conversion".

☆ For newly planted or replanted area, the first yield itself could be considered as 'Organic', as the coffee has a long pre-bearing period of 3-4 years.

8.3. Soil Conservation Measures

As coffee is usually grown in hilly slopes, there is a need to prevent erosion of top fertile soil. In established plantations, cradle pits dug across the slope are found to be most effective in controlling soil erosion:

☆ Cradle pits of size 50 cm wide and 25 cm to 30 cm deep should be dug across the slope in between the coffee rows in a staggered manner. The distance between these pit would vary depending upon the slope of the land.

☆ Cradle pits not only help in prevention of soil erosion but also in better harvesting of rain water and conservation of soil moisture for dry months. These pits also act as *in situ* compost pits for leaf litter, weed biomass and shade tree loppings.

☆ These cradle pits should be renovated every year just before the onset of south west monsoon.

8.4. Shade Management

Shade plays a vital role in maintaining the ecosystem and required micro-climate in coffee plantations. By maintaining optimum shade, the incidence of pests such as white stem borer and green scale and diseases such as leaf rust and black rot could be brought down substantially in Arabica coffee. Similarly, the shot hole borer incidence in Robusta could be minimized by maintaining optimum shade.

☆ A two-tier mixed shade canopy consisting of a lower canopy of temporary shade trees like Dadap (*Erythrina lithosperma*) and a top canopy of permanent shade trees belonging to *Ficus* sps., *Albizzia* sps. *etc.* is ideal for coffee.

☆ The optimum shade requirement would be 50 per cent for Arabica coffee and 30 per cent for Robusta coffee.

☆ Temporary shade plants like Dadap should be lopped just before onset of South-West monsoon to facilitate better light penetration and aeration during the monsoon season. The newly emerging branches after lopping should be thinned out towards the end of the monsoon to achieve an umbrella shaped canopy for providing shade during dry months.

☆ The canopy of permanent shade trees should be regulated once in two years by removing the excess branches so as to provide optimum shade.

8.5. Nutrition Management

Regulations on organic production stipulate that the fertility and biological activity of the soil must be maintained or increased by using natural and as much as possible local resources and organic by-products. In India, the coffee soils are fairly deep, well-drained, rich in organic matter content (1.64 to 2.81 per cent) and medium in available 'P' and 'K' status. Also, the mixed canopy of shade trees contribute substantial amounts of leaf litter (about 10 tonnes/ha) every year, which not only contributes to the humus content of soil but also help in recycling of nutrients from deeper layers. The coffee soils are also rich in the beneficial soil microflora. These favourable conditions alone can support a sustainable crop level of 150 to 200 kg clean coffee/acre. However, in order to support regular economical yields, the following measures should be adopted.

☆ Correction of soil pH using agricultural lime or dolomite based on soil test values once in two to three years.

Figure 6.5: Composting in the Coffee Plantation.

☆ Application of farmyard manure or compost prepared on the farm @ 2.0 tonnes/ha/year in two splits (pre-monsoon and post monsoon).

☆ Deficiency in nutrient supply can be met out by application of permitted inputs like rock phosphate, bone meal, wood ash *etc.* as listed in the Appendix-I of the NSOP.

☆ Bio-fertilizers may also be used for improving the use efficiency of applied nutrients.

Figure 6.6: Composting Outside the Coffee Plantation.

8.6. Weed Management

Weeds should be controlled by manual (slashing) weeding or by scuffling. However, in slopes, only hand weeding should be followed. Mulching with weed slashings and shade tree leaf litter *etc.* would be beneficial in suppressing the weed growth especially the grasses. Cultivation of cover crops like cowpea, horse gram *etc.*, in vacant spots would also help in suppression of weed growth. Use of all kinds of herbicides is prohibited.

8.7. Pruning of Coffee Bushes

Coffee, being an evergreen plant, requires regular annual pruning, in order to create a balance between the vegetative and cropping wood. Pruning helps in minimizing the biennial bearing habit especially in Arabica coffee. Pruning should be done immediately after harvesting of the crop. But if the conditions are hot and dry, delay the pruning till receipt of a few showers and recovery of plants with new vegetative growth. Pruning operation essentially involves removal of all unproductive and undesirable branches in coffee plants so as to encourage new cropping wood for the next season. The following criteria should be kept in mind during the pruning operation:

- ☆ Never cut the primary branches near to their base (main stem), as they will not regenerate into another primary branch,
- ☆ The primary branches which have exhausted all the cropping nodes should be cut back by leaving some nodes from the base,
- ☆ Remove all criss-cross branches, and those growing upwards (gormandisers), downwards and inwards towards main stem,
- ☆ Remove all the suckers,
- ☆ Remove all the diseased, damaged and lean and lanky branches,
- ☆ Remove the secondaries and tertiaries which have produced two to three crops to encourage new branches.

In case of Robusta coffee, it exhibits more determinate growth habit in which the production of secondaries and tertiaries are pre-determined. Robusta also exhibits a self-pruning habit by shedding of old laterals branches. Hence, regular pruning is not required in unirrigated fields except for periodical removal of shot hole borer affected twigs, gormandisers, suckers *etc.* However, in the irrigated Robusta fields, light pruning every year is recommended to regulate the cropping wood.The main pruning has to be followed with handling, centering and desuckering to facilitate proper growth of newly formed wood.

Handling is thinning operation for removing excess vegetative growth that arises after main pruning. Handling should be carried out at least once or twice in a year after main pruning depending on the growth characteristics of varieties. The first round of handling is done during June-July and in case required, a second round during September. In handling operation the following criteria are suggested.

☆ Retain one healthy young flush on each side of the node and remove the remaining flushes.

☆ Remove all the new growth arising on primary branches within 15 cm radius of the main stem (centering), to allow aeration of the main stem,

☆ Remove the criss-cross branches and those growing upwards (gormandisers), downwards and towards the main stem,

☆ The suckers should be removed all the time, at least three to four times a year.

9. Soil Microbial Ecology in Organic Coffee

The main characteristics of organic farming include protecting the long-term fertility of soils by maintaining high level of organic matter content and increasing soil biological activity, providing crop nutrients indirectly using relatively insoluble nutrient sources which are slowly available to plant uptake by the action of soil micro-organisms. Micro organism play dominant role in organic farming. Only limited amounts of permitted fertilizers are used in organic farming, plant production depends almost exclusively on nutrient transformations in soils which are primarily controlled by microorganisms. Soil's microorganism represents the fraction of the soil responsible for nutrient cycling and regulation of organic matter transformation. The organic residues are, in this way, converted to biomass or mineralized to CO_2, H_2O and mineral nutrients representing an important pool of nutrients (N, P and S), which are continually assimilated during the microorganism's growth. Thus, microbial biomass is considered an important source and sink of nutrients in the soil, promoting mineralization of organic matter in inorganic nutrients and consequent availability for plant growth, or immobilizing the nutrients in microbial tissues for its maintenance and growth. Consequently, soils that maintain a high content of microbial biomass are capable to accumulate and cycle nutrients in the soil system (Gregorich *et al.*, 1994).

A study was conducted (Anon 2005-06) to evaluate the microbial biomass in soil under organic and conventional farming. The results showed that the microbial biomass was significantly higher under organic than conventional management, due to greater supply of available carbon. Soil fertility status and yield data analysis in both organic and conventional block for consecutive two years was studied. In organic block soil pH, EC, OC, available nutrients are higher than conventional blocks expect available potassium. Not much variation in yield and out turn ratio was noticed in organic and conventional block (Table 6.3).

Microbiological analyses of samples from organic and conventional blocks were done. It was found that bacterial, yeast, Phosphorus Solubilising Bacteria (PSB) and azotobacter population were to be higher in organic blocks compared to conventional block (Table 6.4). Over the period, soil properties improved considerably compared to conventional block. Organic carbon was increased by 91 per cent (3.4 per cent) over conventional block (1.77 per cent). The same trend was noticed in the availability of macro and micro nutrients in organic block.

Table 6.3: Effect of Organic and Conventional Farming on Soil Nutrient Availability and Yield on Coffee

Year		2004-05		2005-06		Mean	
	Blocks	Organic Block	Conventional Block	Organic Block	Conventional Block	Organic Block	Conventional Block
a) Chemical properties							
pH		6.5	5.9	6.1	5.5	6.3	5.7
EC (dsm^{-1})		0.95	0.82	0.96	0.8	0.955	0.81
O.C (per cent)		3.35	1.75	3.46	1.8	3.405	1.775
N (per cent)		0.35	0.21	0.38	0.22	0.365	0.215
Av.P (kg/ha)		30	22	31	23	30.5	22.5
Av. K (kg/ha)		380	290	270	290	325	290
Ca (per cent)		0.15	0.9	0.16	0.1	0.155	0.5
Mg (ppm)		450	310	330	212	390	261
Mn (ppm)		60	40	65	45	62.5	42.5
Fe (ppm)		120	70	235	115	177.5	92.5
Cu (ppm)		20	11	25	14	22.5	12.5
Zn (ppm)		4.5	3	2.3	1.5	3.4	2.25
Yield of clean coffee (kg/ha)							
Yield		962.5	1005	545	650	753.75	827.5
Outturn Ratio (per cent)		18.2	18	17.5	17	17.85	17.5

Table 6.4: Study of Micro Flora in Soil Rhizosphere and Plant Parts in Organic and Conventional Farming Blocks

Microflora	Blocks	Coffee	Black pepper	Soil	Leaf	Bark	Berries
Bacteria	Organic	182.6	161.5	113.3	117.1	130.5	60.0
	Conventional	124.0	85.0	72.0	92.0	75.0	42.0
Fungi	Organic	3.17	2.83	2.33	4.67	2.17	3.83
	Conventional	13.0	9.0	4.0	16.0	10.0	3.0
Yeast	Organic	45.8	27.5	20.6	36.8	26.1	39.1
	Conventional	23.0	22.0	18.0	12.0	14.0	21.0
Actinomycetes	Organic	4.33	7.17	2.17	0.67	1.33	0.33
	Conventional	24.0	27.0	17.0	10.0	20.0	0.0
PSB	Organic	3.67	1.67	0.67	0.67	0.17	0.17
	Conventional	1.0	0.0	0.0	0.0	0.0	0.0
Azotobacter	Organic	2.33	4.33	1.33	0.33	0.17	0.00
	Conventional	2.0	1.0	0.0	0.0	0.0	0.0

The different agricultural practices can cause positive or negative effects on soil microbial and organic matter content. In the case of organic farming, studies have shown that the agricultural practices in organic farming system, such as addition of compost, straw and natural amendments, promotes positive changes in the soil microbiological and biochemical process, resulting in the increase of soil microbial biomass. In this way, the organic systems are extremely important for the increase of soil fertility and the maintenance of the environmental sustainability. An active soil microflora and a considerable pool of accessible nutrients are very important to the functioning of organic farming systems. Therefore, the maintenance of adequate status of soil microbial biomass and organic matter is crucial for success in organic farming system.

10. Plant Protection

10.1. Pest and Disease Management in New Plantations

No serious pest attacks are seen in young coffee plantations. The damage caused by sucking pests like mealy bug, green scale and foliar pests like leaf miner and grasshoppers could be controlled by spraying any plant based extracts and other permitted products as per the Appendix II of NSOP. For soil borne pests like cockchafers (root grubs) and nematodes, application of neem cake @ 250g/plant would be very effective. Young coffee plants are usually free from major diseases. However, in exposed areas, brown eye spot disease may cause defoliation. Providing adequate shade against exposure, mulching to conserve moisture and spraying with 1 per cent Bordeaux mixture can take care of this minor disease.

10.2. Pest Management in Established Plantations

The regulations on organic farming stipulate that pests and diseases should be primarily tackled by use of tolerant/resistant cultivars; manipulation of microclimate through shade regulation and pruning *etc.*; use of eco-friendly approaches like biological control. Only in exceptional cases should other permitted compounds be used as listed under the Appendix II of NSOP. The approaches for controlling major pests in organic coffee estates are given in Table 6.5.

10.3. Disease Management in Established Plantations

The details of important diseases of coffee and their management in organic production system are given in Table 6.6.

11. Cropping Systems for Organic Coffee

Cultivation of short duration vegetables and fruit crops like ginger, elephant foot yam, pineapple, banana, papaya *etc.* as intercrop can be adopted to augment income during the pre-bearing stage of coffee. Planting of compatible crops like pepper in the coffee blocks at 7-10 cm distance and associate crops like cardamom, arecanut *etc.*, in the valleys is advocated. Under such a situation, due attention should be paid for cultivation of all such mixed crops in an organic way. The nutritional requirement of coffee and its' associate crops should be met by proper recycling of all the by-products available within the farm. For this purpose, all the produce from various crops should be processed within the farm and the crop residues should be recycled after composting, so as to minimize net loss of nutrients from the farm. It

Table 6.5: Approaches for Controlling Major Pests of Coffee

Name of the Pest	*Predisposing Factors*	*Management Strategies*
White stem borer (*Xylotrechus quadripes*)	☆ The most serious pest of Arabica coffee in India ☆ Becomes epidemic scale when its population builds up reaches to high levels under favourable conditions	☆ Maintenance of optimum shade (minimum 50 per cent), preferably with mixed shade. Plant temporary shade trees like Dadap in exposed patches in the estate ☆ Trace the affected bushes, by looking for ridges on main stem and thick primaries, prior to the two main flight periods (*i.e.*, in March and September) every year. (Adopt tracing on a community approach within a village or zone for efficient control of this pest). ☆ While tracing operation, if the borer infestation has not reached root zone, such plants can be rejuvenated by collar pruning. ☆ Uproot and burn infested plant, if the borer has entered root system. Uprooted stems should not be retained in the estate beyond 2 or 3 days. They can be kept in the estate for fuel wood purpose, only after immersing them in water for about 2 weeks to kill all the pest stages. ☆ To protect the remaining healthy plants, adopt bark scrubbing using coir gloves. In this operation, the loose scaly bark on main stem as well as thick primaries is removed gently by scrubbing to minimize cracks and crevices on the stem. This operation prevents egg laying by the white stem borer beetles thereby preventing the attack. Carry out the scrubbing after a few rains when the bark is soft and easy to remove without damaging the stem. ☆ Apply 10 per cent lime solution on to the main stem and thick primaries of the healthy bushes, during April and October to prevent egg laying by beetles. ☆ Alternatively, repeated sprays (at short intervals of 10-15 days) of neem oil on to the main stem and thick primaries during the flight periods (April–May and October–December) are also effective against the pest. ☆ Gap fill the vacancies arising out of uprooting in the same season to avoid open patches and maintain optimum plant population.

Contd...

Table 6.5—Contd...

Name of the Pest	Predisposing Factors	Management Strategies
Coffee berry borer (*Hypothenemus hampei*)	☆ This pest attacks berries of both Arabica and Robusta coffee ☆ Plantation at too low altitude ☆ Abandoned or infected plantations nearby ☆ Several blossoms, coffee cherries which ripen over long period	☆ Maintain optimum shade and good drainage ☆ Practice thorough and clean harvest of the crop ☆ Use gunny bags/picking mats/polythene sheets on the ground while harvesting to minimise the gleanings ☆ Collect the gleanings and subject them to hot water treatment if infestation is noticed. Dip the infested berries in boiling water for a period of two minutes to kill all the stages of the borer inside the berry. If gleanings could not be collected, they may be swept along with the mulch and buried below a depth of 0.75m in the soil ☆ Ensure complete and timely harvest of all coffee from tree coffee species ☆ Dry the coffee to prescribed moisture level to avoid further spread of pest during storage ☆ Remove all off-season berries, infested and left over berries ☆ In the drying yard, cover the harvested coffee lot with oil smeared polythene sheet for a day to trap the beetles ☆ Maintain trap plants around the drying yard, and follow clean harvest from these trap plants and dip the berries in boiling water, to avoid further spread of pest ☆ Spray with entomopathogenic fungus *Beauveria bassiana* immediately after cessation of monsoon when beetles are waiting near the navel region ☆ Release the parasitoid *Cephalonomia stephanoderis* during post-harvest period to reduce the pest inoculum in left over fruits and gleanings ☆ Install berry borer traps during post-harvest period (from February to May-June) to trap adult beetles
Shot hole borer (*Xylosandrus compactus*)	☆ Shot hole borer is a major pest on Robusta coffee	☆ Maintain optimum thin shade in the estate ☆ Ensure good drainage, especially in flat lands by opening drainage channels ☆ Take up regular pruning and burning of affected twigs and suckers, especially during April-May and Sept.–December

Contd...

Table 6.5–*Contd...*

Name of the Pest	Predisposing Factors	Management Strategies
Mealy bugs (*Planococcus citri*; *P. lilacinus*)	☆ Mealy bug is an endemic pest on both Arabica and Robusta. ☆ The pest incidence is generally seen in exposed areas and in irrigated blocks ☆ Sometimes root mealy bugs may pose a problem in young coffee	☆ Provide optimum shade ☆ Release exotic parasitoid *Leptomastix dactylopii* as a biocontrol agent ☆ Spray neem oil (3 per cent) emulsion against mealybugs and other sucking pests on the plant ☆ In case of root mealy bugs, drench the soil around root zone with neem oil solution
Coffee leaf miner (*Leucoptera coffea*)	☆ Too much sunlight, and too dry micro-climate	☆ Improve shade

Table 6.6: Important Diseases of Coffee and their Management

Name of the Pest	Pre Disposing Causes in an Ecological System	Control Measures for Management
Coffee leaf rust (*Hemileia vastatrix*)	☆ Susceptible variety ☆ Coffee bushes planted too close together ☆ Too much or too little shade ☆ Unbalanced nutrient supply	☆ Plant resistant variety, or graft with robusta rootstocks ☆ Change plant density ☆ Maintain optimum shade ☆ Adopt judicious pruning ☆ Supply organic fertiliser to young plants ☆ Spray 0.5 per cent Bordeaux mixture as pre-monsoon and post-monsoon applications ☆ In die-back affected blocks, an additional spray of Bordeaux mixture during pre-blossom period may be necessary
Black rot (*Koleroga noxia*)	☆ Occurs only in endemic patches in valleys under thickly shaded conditions on both Arabica and Robusta during rainy season	☆ Thin out the shade ☆ Regular pruning, handling, centering (opening up the center of the bush) followed by prophylactic sprays with 1 per cent Bordeaux mixture in the black rot prone patches
Root diseases	☆ Four types of root diseases *viz.*, brown root, red root, black root and Santavery root disease occur in endemic patches only. ☆ Of these the first three diseases attack both Arabica and Robusta, while the latter is specific to Arabica coffee only. ☆ Generally, the decaying stumps of shade trees harbour the disease causing pathogens, which later spread to nearby coffee bushes	☆ Uproot shade tree stumps after timber extraction ☆ Uproot and burn affected coffee bushes ☆ Make an isolation trench around the affected plants by including a ring of surrounding healthy plants ☆ Apply 1-2 kg of agricultural lime to the uprooted pit and exposing the spot for at least six months before replanting. ☆ Apply well-decomposed farmyard manure/compost fortified with biocontrol agent *Trichoderma* (@ 5 to10 kg/plant) to the surrounding healthy plants
Brown spot (*Cercospora coffeicola*)	☆ Too dense cultivation in tree ☆ Nursery; wrong irrigation and shade ☆ Site too wet/trees to close together ☆ Too much shade	☆ Trim, produce more air circulation ☆ Regulate shade
South American leaf spot (*Mycena citricolor*)	☆ Site too cool and wet ☆ Too much shade or weeds ☆ Distance between coffee bush and tree crown too small	☆ Regulate shade and weeds ☆ Plant taller shading trees

is also advisable to go for diversification, by maintaining subsidiary activities like apiculture, dairy farming *etc.* This approach enables not only additional income but also minimizes economic losses due to crop failures and market fluctuations. Similarly, areas unsuitable for coffee planting such as scrubby, rocky jungle may be left out undisturbed as a natural habitat for birds and natural enemies of pests.

12. Harvest and Post-harvest Management

Organic coffee is a specialty coffee targeted especially for select consumers who are quality and health conscious. Despite adopting good cultivation (agricultural) practices, the quality of final produce may be adversely affected if proper care is not taken during on-farm processing stage. Hence, it is essential to adopt good agricultural and manufacturing practices (GAPs and GMPs) to maintain quality standards at all levels from estate to curing works.

12.1. Processing at Estate Level

Coffee fruits can be processed both by wet method and dry/natural method. Generally, Arabica is processed by wet method to obtain parchment/plantation coffee, while Robusta is processed by dry method to obtain cherry coffee. The plantation coffee produced by wet method of processing is preferred in the export market. But, this method requires elaborate infrastructure facilities like pulpers, washers and pollution abatement systems. While this method of processing may suit well for large holdings, the small coffee holdings may not be in a position to have all these facilities for economic reasons. In view of this, it is better for the small growers to organise themselves into co-operatives or self-help groups for establishing common processing and pollution abatement systems to save on the costs. This would also facilitate common inspection by the certifying agency.For processing of organic coffee at estate level, the following guidelines may be adopted.

   ☆ In case of holdings having both conventional and organic farming activities, the processing, drying and storage facilities should be distinctly separate for each kind of coffee.

   ☆ Only mechanical and physical processes with natural fermentation should be adopted for processing.

   ☆ The by-products like coffee pulp, cherry husk should be recycled to the field after composting.

   ☆ When wet method of processing is followed, appropriate effluent treatment measures should be implemented as per the requirements of State Pollution Control Board regulations.

12.2. Processing at Curing Factories

Once the produce is dried at estate level, the same is cured at curing factories to obtain clean marketable coffee. As far as possible, the curing factories should be separate for processing the organic coffee. In case such an exclusive arrangement is not possible, the processing of organic coffee should take place separately from the conventional coffees. There should be a provision for separate drying, handling and storage of organic coffee. In case of small organic coffee producers, it would be worthwhile to have common curing and storage facilities of a smaller capacity, if they are interested in holding their coffee for better returns. This would also facilitate common inspection by the certifying agency.

13. Storage and Transport

Not only coffee cultivation, but also all subsequent steps in the production chain, have to be certified. On-farm processing, storage, transport, export processing, shipping, export, import, roasting, packaging, distribution and retailing all have to be certified organic. The following measures have to be adopted during storage and transport of organic coffee lots.

☆ Use clean new gunny bags/IJIRA bags for packing/storing green coffee both at estate level and at curing factories.

☆ Contact of organic coffee with conventionally produced coffee must be completely avoided. In case where only part of estate is under organic, the organic coffee should be stored and handled separately from conventional coffee. In order to maintain the identity, the storage structures and transport containers should be clearly labelled for "organic' and 'conventional product".

☆ Spraying or fumigation with toxic agents is not permitted and special care should be taken to prevent contact with the areas where fumigation has taken place.

☆ The storage structures and transport containers carrying organic coffee should be cleaned using methods and materials permitted in organic processing.

☆ Adequate records should be maintained for the incoming and outgoing coffee lots so that the entire product flow can be documented and accounted for (traceability).

☆ All steps in production, processing, storage and transport chain should be documented and administered for easy traceability of the product and to ensure that no contamination with conventional coffee has taken place.

14. Value Addition with Respect to Organic Coffee

The first step in value addition is wet processing of coffee at estate level to produce washed or parchment coffee. A carefully prepared parchment coffee without major defects could fetch attractive premiums over the conventional/commercial product. Majority of certified organic coffee is exported to other countries in the form of green beans only. The main value addition takes place towards the consumer end, where coffee is converted to roast and ground product which is used for preparation of various types of coffee beverage like filter coffee, espresso, cappuccino, latte, cold coffee *etc.* To some extent some small quantities of certified organic coffee is also converted to instant or soluble coffee form which is the high end value addition. The other forms of value addition of organic coffee include decaffeinated, flavoured coffee as well as in other foods like ice-creams, yoghurt, sodas, candies and chocolate covered beans *etc.* The flavouring of roasted coffee is permitted when natural flavouring substances or preparations are used. For packaging roasted coffee, flushing with nitrogen or carbon dioxide is permitted. For the decaffeination of coffee, chemical solvents (*e.g.* methylene chloride) are not permitted, but the water method or the supercritical carbon dioxide method (the CO_2 method) may be used.

According to Organic agriculture sustainability, markets and policies (OECD, 2003), Premium is also one of the value addition obtained for organic coffee. The premiums are difficult to indicate because they are highly correlated with quality and origin of the coffee, the situation of the market at a given moment, the reputation of the producer or additional certifications such as fair-trade. Price premiums are most important for increasing net cash returns for coffee growing households.

15. Quality Control and Organic Certification Standards

Quality control (QC) is a process by which entities review the quality of all factors involved in production, where every product is examined visually, for fine detail before the product is sold into the external market. This ensures organic integrity at every link in the organic production chain and providing excellent customer service, domestically and internationally from the land on which the product is grown, to the post-harvest facilities and processing plants, to the retail store where it is available for purchase. Organic certification regulation varies from country to country that are operating throughout countries. In U.S., EU, Canada, Japan and India organic standards are formulated and overseen by the government, which means legislation is in place to ensure that only certified producers use the term "organic." When countries have no organic laws or government guidelines, certification is handled by non-profit organizations and private companies. In USA for the product to be sold as organic it follows The National Organic Standard (NOS) created by the USDA and administered by the National Organic Program. Japanese follow Japanese Agricultural Standard (JAS), European countries implement EU-regulations. India follows National Programme for Organic Production (NPOP).

15.1. Harmonization of Standards

There are many organic certification agencies that are varying from country to country with different regulations. In order to create harmonization An International Certification body, members of the International Federation of Organic Agriculture Movement (IFOAM) is working on harmonization efforts to create formal agreements between countries and to facilitate international trade. In 2011 IFOAM introduced a new program - the IFOAM Family of Standards - that attempts to simplify harmonization. The vision is to establish the use of one single global reference (the COROS) to access the quality of standards rather than focusing on bilateral agreements which is in process.

16. Developing Sustainable Market Chain

16.1. Organic Produce Organizations (OPO)

Commonly Organic Produce Organizations (OPO) or Producer Company (PC) is formed by a group of producers for either farm or allied activities. It is registered body and a legal entity where producers are share holders of the organization, who work for benefit of the members and share profit among members. They mainly deal with business activities related to primary produce/product.

Objective of OPO are:

☆ Enables producers to utilize scale of economy to procure inputs at a lower price

☆ Facilitates value addition and helps in deriving more bargaining and selling power for their produce/product.

☆ Helps to access to timely and adequate finance, technology leading to capacity building and provide sustainable market linkages.

OPOs give lot of opportunities for the members to improve their life skills, building sense of responsibility in the organic farming community. At the same time facing the crisis on capital investment at initial stage, meeting the needs of the members, lack of participation and poor perception of the organic coffee planters as service providers, these entire needs professional competency in terms of social and financial management to run OPO successfully (Singh and Singh, 2013).

16.2. Emerging Markets for Organic Coffee

Certified coffees are growing in other non-traditional markets as well. In the Republic of Korea, Australia and Singapore, they are already highly visible in retail market outlets. The same is true, but only in the largest urban areas, for China, India, Mexico, Chile and Brazil. Japan, a major consuming country accounting for approximately 6 per cent of total global coffee demand, has seen the market share of certified coffees grow faster than nearly any other segment.

17. Policy Issues with Respect to Organic Coffee

Today, consumers typically opt for organic food as healthy, safe and free from chemicals. However, there exist other sector of growers who are motivated to opt organic farming because for the concern of the environment, animal welfare and social justice. Since then, organic farming development has become more and more an instrument of state agricultural policy. With the legal definition of organic farming (Council Regulation (EEC) No. 2092/91) in the early 1990s, it became possible to specifically include organic farming as an option under the agri-environmental and other measures of the rural development programmes. Government support for organic farming now also extends into areas such as research, market development and consumer promotion.In India, Ministry of Commerce and Industry, Govt. of India launched the National Programme on Organic Production (NPOP) in the year 2000. NPOP implements National Standards on Organic Production (NSOP) through National Accreditation and Policy Programme.

More details on NPOP are available at the website www.apeda.com/organic

18. Success Stories of Organic Coffee

Surveys and pilot studies were carried out in organic coffee plantations, and economic analysis of organic coffee was made (Anon, 2000-2006). The salient outcomes are as follows:

Case Studies and Techno Economic Analysis of Organic Coffee Estates

☆ A case study on production of organic coffee carried out in private estate at Yellikodige for eight cropping season, recorded the average yield of 1178 kg/ha under organic farming, while yield before conversion was 1316 kg/ha. Analysis of soil indicated a high content of available nutrients and organic matter. The incidence of pest and disease observed was low and there was a suppression of mealy bugs by natural enemies inferring that organic coffee cultivation in Arabica is possible without much decline in the yields (Anon, 2002 -2006).

☆ A case study was conducted at Annamallis hills of Tamil Nadu region from the inception of planting in Arabica coffee to study the cost of production incurred in producing organic arabica coffee by a corporate sector. The results indicated that among the total cost of production, the cost incurred on input was highest (48 per cent) followed by labour cost (37 per cent), over head cost (12 per cent) and certification cost (3 per cent) with a unit cost of production Rs.51.14 per kg clean coffee, when compared to conventional cultivation where the labour cost is highest (62 per cent) followed by input cost (29 per cent) and over head cost (9 per cent). The break up in cost of production was analyzed for the cultural operations that were carried out under organic farming. It was noticed that the manuring/composting incurred the highest cost (44 per cent) followed by bush management (6 per cent) and weeding (5 per cent).

☆ A case study was conducted (2002-03) to analyse the economics of organic vis-à-vis conventional production of robusta coffee in Chikmagalur region, India. Field survey was carried out in the study area where self-help group consisting of fifteen growers were practicing at different phases of organic cultivation of coffee over a decade and 31 conventional robusta growers. The results indicated that the total cost of cultivation of organic robusta coffee was Rs. 43,775/ha when organic manure was applied every year and Rs. 38,375/ha when the organic manure was applied alternate year. This translated into a cost of production of Rs. 34 per kg for an average yield of 1280 kg/ha when manure was applied every year and Rs. 31 per kg for an average yield of 1225 kg/ha when manure was applied in alternate years. Among the total cost of organic robusta coffee production, labour cost accounted for 47 per cent (346 man-days per ha) followed by input cost (38 per cent), overhead expenditure (12 per cent) and certification cost (3 per cent). However, in conventional Robusta coffee cultivation, the total cost of cultivation in Chikmagalur region was Rs. 38,775/ha, which worked out to be Rs. 27/kg with an average yield realization of 1435 kg/ha. Of the total cost, labour wages accounted for 48 per cent followed by input cost (30 per cent) and overhead expenditure (22 per cent). Thus, the findings indicate that in organic coffee cultivation, input cost is more especially when the manures are met through the outside sources, hence there is a great scope to produce on farm compost by utilizing coffee farm wastes, so that the costs can be minimized.

The growers' perceived that the incidence of pest and disease was low in organic system compared to conventional system of farming. However, the premiums are low for organic coffee as the product conception is lacking among the buyers and the certification costs are more which involves cumbersome procedure and these issues need to be addressed.

19. Environmental Benefits of Organic Coffee

Organic coffee production has multiple potential environmental benefits. Organic standards require coffee farms to have a structurally and floristically diverse shade cover. Therefore, most of the organic coffee is grown the natural way – within the shade of lush forests. It provides home for wild plants, animals, and birds-which develop mutually beneficial relationships with coffee fields, enjoying the habitat while keeping insect populations under control and naturally fertilizing the soil, thus, keeping unique regional ecosystems alive. Coffee fields also store carbon from the atmosphere and protect watersheds by slowing down run-off. Organic coffee production also replaces chemical fertilizers as well as pesticides and fungicides with organic manures and less harmful alternatives and prohibits use of genetically modified organisms. Thus, these forested farms are also more resilient and better equipped to handle unusual weather patterns due to climate change, making them a safer investment for coffee farmers.

20. Conclusion and Future Strategies for Organic Coffee

World over the demand for organically produced food products is increasing rapidly due to increased awareness among the consumers, especially in the developed countries about environment protection, food safety and welfare of small farmers and workers. Coffee being a major export commodity is no exception to this trend, since it is mainly produced in the developing third world countries and consumed in developed countries. There is a good potential for production of Organic Coffee in India due to many favourable conditions, but the growth of this segment is not encouraging due to constraints like high cost of certification and lack of assured premiums. Some of the future strategies identified are as follows:

- ☆ Provide incentives and create a business atmosphere which will encourage investment by private entrepreneurs individually or in joint venture with exporters to reduce certification cost
- ☆ Access to lucrative markets (*e.g.* USA, Germany, Japan) for high value organic products
- ☆ Undertake intensive commercial organic coffee production from small growers through aggressive investment promotion program.
- ☆ Provide more attractive incentives and short-, medium- and long-term loans to small organic farmers
- ☆ Set-up agriculture data collection and market information systems and disseminate these information nation-wide on a regular basis.

Selected References

Anonymous (2000).Organic Farming in the Tropics and Subtropics. *Coffee*. Naturland e.V. – 1st edition 2000 p.19

Anonymous (2002-06) Sixty third Annual Report to Sixty sixth Annual Report Coffee Board of India, Bangalore-560 001

Anonymous (2011) *Coffee Guide*, Central Coffee Research Institute, Coffee Research Station- 577 117, Chikmagalur, Karnataka, India.

Gregorich, E. G., Carter, M.R., Angers, D. A. Monreall, C. M. and Ellerta B. H. (1994). Towards a minimum data set to assess soil organic matter quality in agricultural soils. *Canadian Journal of Soil Science* **74**(4):367-385.

Giovannucci, Daniel and Andres Villalobos (2007). *The State of Organic Coffee*: 2007 U.S. Update. Sustainable Markets Intelligence Centre (CIMS), San Jose, Costa Rica.

Giovannucci, D., Liu, P. and Byers, A. (2008). Adding Value: Certified Coffee Trade in North America. In: *Value-adding Standards in the North American Food Market - Trade Opportunities in Certified Products for Developing Countries*. Pascal Liu (Ed.) FAO. Rome 83 World Bank.

Jason Potts., Matthew Lynch., Ann Wilkings., Gabriel Huppé., Maxine Cunningham., and Vivek Voora.(2014). The State of Sustainability Initiatives Review 2014, Standards and the Green Economy. 8.*Coffee Market* pp:155-185.

Singh Sukhpal and Singh Tarunvir, (2013). *Producer companies in India: A study of organization and performance*. CMA publication No.246. Centre for Management in Agriculture, Indian Institute of Management, Ahmedabad.

Van Elzakker, B. (2001). Organic Coffee. In: '*Coffee Futures- A source book of some critical issues confronting the coffee industry*. P.S. Baker (Ed.), Published by CABIFEDERACAFE, USDA-ICO, 2001.pp.74-81.

www.apeda.org

www.ico.org

Chapter 7

Organic Farming in Tea

☆ *B. Radhakrishnan and J. Durairaj*

1. Introduction

The tea (*Camellia sinensis* (L.), O. Kuntze) is an evergreen shrub or small tree that is usually trimmed to below 1.10 m when cultivated for its leaves. It belongs to the Theaceae family, and is originated from the high regions of countries such as South west China, Myanmar and North east India. Tea is one of the oldest beverages in the world, the discovery dating back to about 2700BC. Tea, today, is cultivated across the world in tropical and subtropical regions. Fresh leaves contain about 4 per cent caffeine. The young, light green leaves are preferably harvested for tea production. Older leaves are deeper green. Different leaf ages produce differing tea qualities, since their chemical compositions are different. Usually, the tip (bud) and the first two to three leaves are harvested for processing. The leaves have been used in traditional Chinese medicine and other medical systems to treat asthma (functioning as a bronchodilator),angina pectoris, peripheral vascular disease, and coronary artery disease. Both green and black teas may protect against cardiovascular disease.

Tea is available for consumption in six main varieties, based on the oxidization and fermentation technique applied. Primarily, tea is drunk as black tea. Other sorts with less importance to the worlds market are green tea (East Asia, Arabian countries) and Oolong-tea (China, Taiwan). Recently, both Organic Green Tea as well as Instant Tea have also begun to be manufactured in increasing quantities.

2. Production Scenario

2.1. Global Scenario

World tea production (Black, Green and Instant) was 5.06 million tonnes in 2013, mainly due to increases in the major tea producing countries (Table 7.1). China was the largest tea producing country with an output of 1.9 million tonnes, accounting for more than 38 per cent of the world total, while production in India, the second largest producer, also increased to reach 1.2 million tonnes in 2013.

Table 7.1: World Tea Production (thousand tonnes)

Region	Production
Far East	3965.6
Africa	649.5
Latin America and Caribbean	95.0
Near East	253.5
Oceania	6.5
Japan	84.7
CIS	8.9
Developed	102.9 (2.03 per cent)
Developing	4961.0 (97.97 per cent)
World	5063.9

Source: FAO IGG Secretariat.

World tea consumption during 2013 was 4.84 million tonnes, mainly due to the rapid growth in per capita income levels, particularly in China, India and other emerging economies. The consumption was 4.03 million tonnes (83.1 per cent) in the developing countries, the rest of 0.82 million tonnes (16.39 per cent) in the developed countries. Growth in demand was particularly marked in China. After a spectacular rise in consumption in recent years exceeding 8 per cent annually, total consumption increased by 9 per cent in 2013, on a year to year basis, to reach 1.61 million tonnes, the largest in the world. In India, consumption expanded by 2.4 per cent in 2009 and 6.6 per cent in 2013 to reach 1 million tonnes.

2.2. Indian Scenario

In India, tea is mainly cultivated in Assam (Assam valley, Cachar), West Bengal(Dooars, Terai and Darjeeling) and Kerala, Tamil Nadu and Karnataka. According to The Indian Tea Board statistics, the total tea production during 2014-15 in India was 1197.18 M kg. The contribution from North India was 79.8 per cent, the rest from South Indian states (Table 7.2).

Table 7.2: Tea Production in different Parts of India during 2014–15 (Qty. in M kg)

Region	Production
Assam	606.80
West Bengal	324.26
Others	24.76
Total North India	**955.82**
South India	**241.36**
All India	1197.18

Source: Tea Board India.

3. Organic Tea Market Potential

The organic tea sector is still a very small part of the tea industry but the number of organic tea producers and the volume of organic tea traded in the world market have recorded high growth over the last couple of years. This development can be explained by a number of factors *viz.* awareness of pesticide residue and heavy metals in conventional teas, and other potential health hazards connected with an intensive system of tea production. Organic tea has a niche market, where the produce sells at a premium price. India leads the world in organic black tea output and Sri Lanka is quite strong as well.

The area under tea in south India is 1.20 lakh ha with the total production of 241.36 Mkg contributing around 20.4 per cent to the national production of 1197.18 mkg (2014). Organic tea area constitutes around 1,315 ha producing around 1.83 mkg which is only 0.76 per cent to the south Indian tea production. Similar trend is witnessed at the national level with the production of 11.09 mkg of organic tea and its contribution was only 0.94 per cent. In general, organic tea estates are registering around 35 to 45 per cent lower yield compared to conventional counterpart.

3.1. Domestic Market

The market potential for organic teas in India is uncertain due to the lack of awareness to the consumers about the product and regular market demands. In general, there is a belief that in areas where farmers have access to established organic markets within the country or abroad, products can achieve a higher price compared to the conventional market. This is not true in the case of tea as the cost of production of organic product is on the increase due to high labour cost, increasing input cost and low productivity compared to conventional plantations (GhoshHajra, 2005).The domestic market for organic tea is as yet not as developed as the export market. Wholesalers/traders and supermarkets can play major roles in the distribution of certified organic products at an affordable price. Large organized producers distribute their products through self-owned stalls and very few supermarkets. Considering the profile of existing consumers of organic products, supermarkets and restaurants are the major marketing channels for organic teas (GhoshHajra, 2002).

3.2. Export Market

Indian organic tea producers and exporters are well aware of the demand for organic products in developed countries. The countries with major markets of organic tea mostly do not domestically produce tea and, therefore, the demand for certified organic tea is growing at a rapid rate all over the world with Europe and the US leading the way. The channels adopted for the export of organic products, except for tea is mainly through direct sale and through export companies. Organic tea is produced by well organized tea estates which are exporting teas directly. Organic products are mainly exported to Europe (Netherlands, United Kingdom, Germany, Belgium, Sweden, Switzerland, France, Italy, and Spain), America (USA and Canada), Middle East (Saudi Arabia and UAE) and Australia (GhoshHajra, 2002).

Organic is the fastest growing segment in the US market and the EU, taken as a whole is the world's largest market followed by Japan. An average of 28 per cent growth rate and local supply unable to satisfy increasing domestic demand, EU imports of organic products account, on average, for 40 per cent of total sale offering excellent opportunities. Sweden, Austria, Denmark and the UK exhibit the highest share of supermarket organic sales; however, none, except the UK, show high annual organic food sales.

4. Botany of Tea

Under natural conditions the tea plant is an evergreen tree. It is classified under a family called Theaceae, which consists of 30 genera and 500 species. The genus *Camellia* has 82 species of which three are economically important. They are (a) *Camellia sinensis* - China variety, (b) *Camellia assamica* - Assam variety and (c) the hybrid *Camellia assamica* ssp. lasiocalyx.

4.1. China Variety

A small leaved bush with thin branches and single flowers. The China tea bush can grow to a height of 1 to 3 meters. The glossy leaves are dark green colour. A poor yielder but can withstand drought and frost.

4.2. Assam Variety

This has tree habitat and the bush grows to a height of 10 to 15 meters with thick branches, but is pruned and kept at waist level to make it easy for the tea leaves to be picked. The leaves, that droop at their ends are large and light green colour with a glossy shine. The plants produce clustered flowers, fairly high yielder, but not tolerant to drought and frost.

4.3. Cambod Variety

A hybrid of Assam and China jats having a combination of characters of both varieties, with oval, semi erect leaves. The leaves of the Cambod tea bush can be yellow or a light green color. But when autumn sets in, the leaves change to a reddish yellow or pinkish red color. The Cambod tea bush can grow to a height of 6 to 8 meters. The leaves are of an average size.

4.4. United Planters Association of Southern Indian Varieties

Tea Research Foundation (UPASI-TRF) has released 33 clones, five biclonal seed stocks and nine different graft combinations (stock-scion combinations) grafting materials with diverse characters and varying performance. Besides these, approved estate selections and clones released by other tea research institutions are also available for planting. The choice of cultivars for planting and infilling in a particular locality is based on climatic conditions such as drought, frost, wind, soil conditions *etc.*, of that area.

As there have been no studies of varieties for the organic cultivation of tea, only generalised recommendations can be offered: Organic cultivation of tea requires varieties (clones) with broad-scope resistances. Organically cultivated tea was first

Figure 7.1: General View of Tea Plantation.

produced in 1986 in Sri Lanka. Since then, it has become wide-spread mostly in India and Sri Lanka. Currently, around 5,000 ha of tea are being cultivated organically. Other producing countries are China, Japan, Seychelles, Tanzania, Kenya, Malawi and Argentina).

5. Climatic and Soil Requirements

The tea crop has very specific agro-climatic requirements that are only available in tropical and subtropical climates. The commercial cultivation of tea extends from 44°N (Georgia) to 27°S (Argentina) latitude. The plant prefers warm wet summer and dry cold winter. In south India and Darjeeling, tea is grown on slopes (5 to 30 per cent) from 700 m to 2400 m MSL. In North East India, it is grown on a flat land and above an elevation of 200 m MSL.

5.1. Temperature

A temperature regime between 13° C and 32° C is conducive to the growth of tea.

5.2. Rainfall

For effective tea cultivation, the distribution of rainfall is considered important more than the total amount. If dry spell persists for a longer period, the tea plants suffer heavily and the crop declines. Uniform and well distributed rainfall over the year (minimum 50–150 mm/month) gives better growth than seasonal rains, although seasonal stress conditions such as bright sunshine days, cold nights, and dry, desiccating winds produce chemicals in the plant which give a desirable flavor and aroma to manufactured tea. Eight to nine months of wet spell is ideal and the dry period should not extend more than three to four months. The annual average rainfall in various tea districts of southern India ranges from 1250 to 7000 mm in a year. For optimum growth, data from different tea-growing regions indicate that 23–30°C and an annual rainfall of 2500–3000 mm are necessary.

5.3. Relative Humidity

RH around 80 per cent in most of the months is ideal and RH below 40 per cent during dry weather causes damages to the plant and accentuates the drought effect. In regions with extensive dry seasons, shading trees play an important role in providing and maintaining sufficient humidity.

5.4. Wind

In all tea growing districts of South India, west facing fields are exposed to moderate to severe wind during southwest monsoon. The evaporation rate increases with wind velocity. More moisture escapes from soil at high temperature on windy days. Wind also affects the rate of transpiration by removing moist air from leaf surface (Durairaj*et al.*,2015). Therefore, tea plantations in windy regions should also be protected by wind breakers *e.g.* hedges, to reduce the intensity of evapotranspiration.

The tea growing soil should be deep, well-drained and aerated. Nutrient-rich and slightly acidic soils are best, the optimum pH-value being 4.5-5.5. China tea (*C. sinensis* var. *sinensis*) is resistant to drought, with low tolerance of shade and can tolerate short periods of frost. On the other hand, Assam tea (*C. sinensis* var. *assamica*) being a purely tropical crop, has a high tolerance of shade and reacts sensitively to drought and the cold. Tea, being a C-3 plant, is not an effective photosynthesizer, and hence, shade trees are grown in tea plantations to attain maximum productivity. An average daily amount of sunshine of 4 hours per day can be ideal. Tea production, therefore, is geographically limited to a few areas around the world and it is highly sensitive to changes in growing conditions.

6. Production of Planting Materials and Nursery Management

Vegetative propagation is carried out for clonal multiplication while biclonal seed stocks are propagated through seeds. It is recommended to establish own nurseries in the tea garden, in order to ensure a continuous supply of field worthy and required choice of healthy plants. An exclusive nursery is necessary to avoid mix up varieties of plants. The tea nursery should be in a protected site and located near a perennial water source. A site that has not been cultivated previously will be more ideal. It should have natural shade or otherwise an over head 'pandal' is to be raised on which a coir mat with 6 mm x 6 mm mesh is spread so as to allow about 33 per cent sunlight at midday into the nursery. The nursery is to have same altitude and site conditions as the tea garden is to be established. In order to grow healthy seedling plants, the fruits which provide the seeds should be mature and the seeds should be removed from the fruit within a few days after the fruits have been plucked from the tea bush.

Before sowing, the seeds are soaked in water for 3 to 4 days and then sun dried to crack the seed coat. Pre-soaking and sun-drying the seeds is said to help the seeds germinate faster. Tea seeds are first planted in sand beds or sand boxes at a depth of 1.25 cm with a spacing of 2.5 cm between each seed. The seeds germinate after two weeks and reach full germination between the 3rd to 4th weeks. As soon as the 'radicle' is about to sprout from the seed, polythene bags are filled with mixture and the seeds are transferred from the sand beds/boxes to the bags. The seeds, after germination in sand beds or boxes, or the cuttings are to be transferred to the polyethylene sleeves with a dimension of 30 cm×10 cm filled with medium. One leaf and an internodal cutting with an axillary bud prepared from the 'aperiodic shoots' arising from pruned tea bush is planted in the nursery sleeve and covered with a polythene sheet of 400 gauge. April/May and August/September are the most suitable months for planting in nursery.

Soil, sand and compost mixture of 3:1:1 for growing medium and sand and soil with the ratio of 1: 1 for rooting medium should be used in nursery bags. Liquid manure using 30 kg of fresh cow dung soaked in 200 l of water, stirred at 10 days interval for a period of 2 to 3 months is ideal for application in the nursery. The dilution rate of 1:10 and 10 l of nutrient solution for 450 plants is recommended at weekly interval for a period of 6 to 8 months in the stacking yard area. Panchagavya preparation @ 3 per cent for foliar or 1 per cent for soil application for nursery plants at 7 to 10 days interval is also beneficial for the plant growth. Alternatively spraying of vermiwash diluted with water @ 10 per cent or vermiwash, cow urine and water @ 1:1:10 (one litre of vermiwash, one litre of cow urine and 8 litres of water) at 10 days interval may be used for nursery plants. About 4000-5000 plants can be covered with 10 l spary solution while using hand operated knap sack sprayer. Growth in the nursery is faster under shade or in sealed polyethylene tunnels. The plants are allowed to grow for 6-8 months in the nursery and then transferred to the open space for hardening. Hardened nursery plants are transplanted in the field.

Although cuttings for vegetative propagation may be collected from bushes at any phase of cultivation, mother bushes that are dedicated for the purpose and

Figure 7.2: Vegetative Propagation in Tea.

which have been spaced, pruned, and manured organically, are the best source of cuttings. For organic cultivations, no genetically manipulated varieties are allowed.

7. Conversion Period

The minimum conversion period should be three years from the last usage of synthetic agrochemicals. One can start marketing the tea as "in conversion organic tea" only after the lapse of one year from the start of conversion. Tea can be marketed as "organic tea" only after the completion of conversion of three years.

8. Planting and Aftercare

8.1. Selection of Site

The area needs to be sufficiently isolated to ensure that there is no possibility of any pollutants or contaminants flowing or drifting into it from any known or unknown sources. In order to ensure this condition, a buffer zone with sufficient width is to be left on all sides of the plot depending on the topography of the area. The minimum width of buffer zone should be 100 m. The minimum depth of soil profile should be 1.5 to 2.0 m and organic matter status should be medium to high level depending on the elevation and rainfall of the area. A perennial source of water free from pollutants is required in the estate for large scale compost preparation, which is essential for organic tea cultivation.When a new tea plantation is established, all the weed plants are to be removed and later on fast growing covering plants planted to restrict the growth of unwanted species of flora. In particular, when the tea is to be in sloppy fields, the soil should be protected against erosion, which will lead to soil degradation and nutrient losses. Plantations on slopes (*e.g.* Darjeeling) should, therefore, be planted along the contour lines. Slopes and peaks that are at risk from erosion should not be put into tea cultivation, and these areas should be protected by planting permanent forests along them.

8.2. Spacing

Cultivated tea is easily tailored to the convenience and requirements of growers. The planting density should always be adapted to the site conditions such as slope, altitude, micro-climate *etc.*, as well as maintaining shade required for organic tea plantations. Different methods such as single hedge planting or double hedge planting method are adopted for planting. Tea bushes are grown at different planting densities: for the older seedling tea, from 6,000 to 14,000 plants/ha; for vegetatively propagated tea (VP tea), 12,500 to 18,000 plants/ha. Spacing of plants has to take into account such variables as soil, climate, habit, and growth pattern of the cultivars, the requirement for a continuous ground cover, access for different agricultural operations, and bush architecture suited to mechanical harvesting. About 13,000 plants could be planted in one hectare following double hedge system of planting (spacing: 135 cm 75 cm x 75 cm). One year old plants are planted in pits with a dimension of 30 cm x 45 cm. The selected plants for planting should have 14 to 16 healthy mature leaves and the root system should have reached the bottom end of the sleeves at the time of planting. The stem at the collar region should be about pencil thick and brown. Soil and water conservation measures must be adopted while new planting is taken up.

8.3. Soil and Water Conservation Measures

The hilly terrain of tea growing areas in southern India receive heavy rainfall during monsoon period and continuous dry spell from December to April aggravates drought and, therefore, water conservation measures in newly planted sites are very important. Various measures as described below could be adopted for this purpose.

☆ **Contour stonewalls:** If stones/boulders are available, construction of retaining walls serve as a good mechanical measure for preventing soil erosion. In area where stones/boulders are not available for construction of revetments, planting vegetative barriers – vetiver grass should be established. The number of contour stonewalls/contour bunds will vary depending on the percentage of the slope.

☆ **Drains:** Drains are to be designed with adequate care and caution. The main objective of drains in the fields is to carry the excess water when rainfall exceeds percolation. It helps in preventing the flow of water inside the field and cause soil erosion. The following types of drains are to be made in fields:

   ❑ **Boundary drains:** This drain is useful in diverting water coming from upper reaches and preventing its entry into the field. These drains should be 60 cm wide and 45 cm deep with slopping sides and intermittent blocks of 30 cm height in 2-3 metres interval. This should be connected to the leader drain.

   ❑ **Leader drain:** This drain should be laid out wherever there is a natural converging of the slopes. The bed of the drain is converted into saw edged pattern of steps with vertical face and gently sloping treads. Vetiver grass or weeping love grass could be planted on both the sides of boundary and leader drain to prevent soil caving in and disturbing the flow of water.

   ❑ **Contour drain:** These drains are taken once in 4 to 5 double hedges. The drain should be provided with lock and spill over arrangement to arrest the velocity of running water. The drains are 30 cm deep with intermittent block of 30 cm x 30 cm x 30 cm at every 2-3 m. distance. These drains should be connected to leader drain.

   ❑ **Staggered trenches:** The size of the trench is 180 cm long, 30 cm wide and 45 cm deep. In between, double hedges are allowed in a staggered way leaving a gap of 180 cm apart for every trench.

8.4. Weed Management

New planting and replanted areas are highly prone to invasion by weeds since soil is completely exposed and it takes three years from planting for obtaining fairly good ground cover. The weeds are to be regularly removed through approved measures under organic cultivation. Mulching with plant materials or raising cover crops will assist in keeping weeds under check and prevention of erosion. Hoeing is not recommended on those sites at risk from erosion. Manual weed control is the only option where mulch materials are not available. Having good plant population per ha without much vacancy percentage, higher pruning and high tipping are some of the cultural practices for smothering weed growth. Brush cutter is also used for control of weeds with a care of not hitting the collar zone of mature tea. The poly thread, when hits the collar zone or branches, may cause collar canker and branch canker.

8.5. Bringing up of Young Tea Plantation

Soil and water conservations measures such as boundary drains, leader drains, contour drains, staggered trenches and vegetative barriers with vetiver grass on either side of leader drain and boundary drains, along the road side, foot paths is vital in young tea area.

☆ At the time of planting, application of vermicompost @500 g or compost @ 1kg along with phosphate solubilising bacteria (PSB) and *Azospirillum* @ 10 g each per planting pit to enhance the growth and establishment of young tea are suggested. For young tea plants, application of compost @ 500g plant or 5 tonnes per ha at every six month interval till first formative pruning, first application during April/May and second application during September/October is recommended. For better frame development in young tea, apply mined SOP @ 200 kg per ha in two splits along with compost or with rock phosphate. The dosage of rock phosphate per ha is 200 kg and shall be applied in a single split (Durairaj *et al.,* 2011).

☆ Foliar application of zinc sulphate 6 to 8 kg in three to four splits (@ 2 kg per ha per split) along with same quantity of mined sulphate of potash (patent kali) is required during high cropping months of April, May, September and October/November. Bio-phos, an organic supplement for phosphorus also should be given as foliar application three to four times in a year, @ 500 ml per ha. Mined SOP @2 kg per ha during drought period beginning from December to March/April will help in alleviating the drought impact (Durairaj *et al.,*2011).

9. Nutritional Management

9.1. Nutrient Requirements

A high amount of nutrients are lost through the continual plucking of tea leaves. In conventional tea gardens, under Sri Lankan and East African conditions, the extraction of nutrients for 1000 kg tea/ha/year was found to be 45 and 42 kg for N, 8 and 6-8 for P_2O_5 and 21 and 24 kg for K_2O. The respective figures under Indian conditions were 50-65, 10-15 and 20-35 kg. The nutrient loss throughout a pruning cycle of 3 years was 785 N, 135 P_2O_5 and 570 K_2O kg/ha. Moreover, the perennial tea plant requires a considerable amount of nutrients in order to develop roots, stem and branches.

9.2. Organic Nutrition Management

During the beginning of the conversion, the tea garden is to be developed in stages from a monoculture towards a diversified crop system. Along with tea, other plants should be cultivated to improve soil fertility, provide a supply of nutrient (especially N), increase diversity (habitats for beneficial insects), supply wood (fuel and building material) and to provide feed for on-farm animal husbandry, if mixed farming is practiced. In tea gardens where integrated animal husbandry is practiced, choose those green manure plants that can also be used as fodder crops. Sufficient supply of organic matter for the tea bushes is to be ensured. Spreading the organic matter over the site should be given preference to the more work-intensive practice of composting. The foliage from green manure plants, as well as that from the other crops, should be applied as mulch material in the plantation. The loss of nutrients through pruning should not be allowed from the tea garden (*e.g.* as fuel), but should either be re-applied directly as mulch, or via composting. The compost should be applied just before the main plucking times on the site. Higher amounts of compost (average 10t/ha) are generally to be applied after deep pruning. Planting *Crotalaria agathiflora, C.anagroides* and *Glyricidia maculata* along the road side and vacant patches helps to obtain green manures from regular loppings and applying to the field. Planting vetiver grass (*Chrysopogon zizanioides*) along the edges of the roads can act as vegetative barrier and prevent soil erosion.

9.3. Pruned Fields/Fields Recovering from Pruning

Organic manures should preferably be applied in the staggered trenches. Compost shall be applied in the trenches along with burial of prunings to save cost on taking trenches. Pruned field requires more amount of potash for the sound development of frames. In addition to wood ash, mined sulphate of potash @ 200 kg per ha in two split application during April and September is suggested. Dolomite shall be applied once in a pruning cycle based on the status of soil pH. If the soil pH is between 4.7 to 4.9, the rate of dolomite required per ha is 1.0 tonne, between 4.5 to 4.6 it is 1.5 tonnes and for the pH of 4.3 to 4.4,the quantity will be 2.0 tonnes per ha.

The recovery from pruning takes a minimum of 2-3 months and foliar application of nutrients should be given when there is sufficient foliage. Foliar application of zinc sulphate @ 6 kg in three splits (@ 2 kg per ha per split) along with same quantity of mined sulphate of potash (patent kali) is required during high cropping months of April, May, September and October/November. Bio-phos, an organic supplement for phosphorus also should be given as foliar application four times in a year, @ 500 ml per ha. Mined SOP @ 2 kg per ha during drought period beginning from December to March/April will help in alleviating the drought impact.

9.4. Fields under Regular Plucking

For the fields yielding up to 1000 kg of made tea, apply six tonnes of compost and fields having the yield per ha above 1000 kg, eight tonnes of compost per ha is required to be applied in two splits. Additionally, 100 kg of mined sulphate of potash, 200 kg of rock phosphate, 25 kg each of *Azospirillum*, and phosphate solubilising bacteria (*Pseudomonas*) are to be mixed and applied along with the compost in two splits, during April-May and second application during September-October. Foliar application of zinc sulphate 8 @ kg in four splits (@ 2 kg per ha per split) along with same quantity of mined sulphate of potash (patent kali) is required during high cropping months of April, May, September and October/November. Bio-phos and mined SOP as indicated above can also be applied for fields under regular plucking.

10. Training of Young Tea Plants

To induce more laterals, apical dominance is to be arrested, three to four months after planting, by cutting off the leader stem. This operation, called 'centering' (or 'decentring'), promotes the growth of axillary buds and lateral branches are formed. All the plants should be centered as low as possible leaving 8-10 healthy mature leaves. For further lateral branch formation, good spread and establishment of plucking surface, train the growing branches two stage tipping: the first tipping at 35 cm followed by second tipping at 50 cm. 'Formative pruning' (branch formation pruning) is to be done at the end of five years after planting. The recommended pruning height for formative pruning is around 45 cm above ground level. At the time of formative pruning, remove those branches which are less than pencil size thick.

11. Pruning

Periodic pruning or removal of mature foliage is one of the major practices to be adopted to prevent the tea plant from becoming a tree, and to keep the bushes under vegetative stage and able to generate a continuous crop of fresh shoots at a convenient height for plucking. Pruning also ensures optimal utilization and productivity of the land available, as well as allowing enough space for other agricultural operations. The frequency of pruning, or the length of the pruning cycle, is determined by the particular type of tea, the cropping pattern, and elevation. A pruning cycle of four years is recommended for the fields located in low and mid elevation areas and it is five years for the fields in high elevation. In Sri Lanka, pruning cycles are 2–3 years in the low country, 4–5 years in the high country. In general, the bushes are pruned back to a comfortable plucking height every three years, and then radically cut back every 15–20 years (to a plant height of 30–40 cm). Collar pruning, reaching down to the soil, is utilised to rejuvenate the tea plants. Pruning just before the onset of the dry weather is to be avoided. A cessation of plucking (called "resting") for 2–3 months before pruning is necessary for increasing starch reserves.

The pruning material should be left in the field as mulch directly on the site, or, if applicable, used as compost material. In case the pruning material is to be used as fuel, the ashes should be used as a compost supplement (*e.g.* to replace the potassium). In order to create the conditions necessary for decomposition of such materials, the pruned material is to be sufficiently chopped (2–5 cm pieces) and evenly spread around the tea bushes (avoid creating heaps of material). In order to achieve a better carbon/nitrogen ratio for successful decomposition, the carbon-rich material is to be mixed with additional nitrogen-rich material such as neem cake, castor cake or green manure from crotalaria or glyricidia.

The different types of pruning are:

a) **Rejuvenation pruning:** The whole bush should be cut near the ground level less than 30 cm with a view to rejuvenate the bushes.

b) **Hard pruning:** Formation pruning of young tea at 30 to 45 cm for proper spread of bushes.

c) **Medium pruning:** To check the bush growing to an inconvenient height, medium pruning is done in order to stimulate new wood and to maintain the foliage at lower levels less than 60 cm.

d) **Light pruning:** Pruning depends on the previous history of the bush raising the height of medium pruning by 2.5 cm or less to manageable heights for plucking (less than 65 cm).

e) **Skiffing:** This is the lightest of all pruning methods. Remove the top 5 - 8 cm new growth to obtain a uniform level of pruning surface (more than 65 cm).

12. Shade Tree Management

Shade trees have a great role to play in the organic cultivation of tea. Some of the beneficial effects of shade trees are:

a. Nutrient supply (*e.g.* nitrogen, when legume trees are used; they retrieve nutrients from lower soil levels; reduction of nutrient losses from washing out)

b. Build-up of humus

c. Protect the tea bushes from too much sun (yield reductions are possible when the solar radiation is too intense, and there is a lack of shade)

d. Reduce erosion through wind and rain (and damage from hail)

    e. Influence the quality of the tea

    f. Positive micro-climatic effects *e.g.* during drought periods

    g. Encourage beneficial insects to settle

    h. Create a pleasant atmosphere for the harvesters

The shade trees are to be regularly thinned out to create and maintain an optimum amount of shade and the pruning material should be used for composting or mulching if possible. Thinning out will also help prevent infestations of blister blight (*Exobasidium vexans*), which thrive under too shady (and thereby moist) conditions. The shading trees should be trimmed to prevent blister blight developing directly before the rainy season (monsoon).While choosing trees to use as shade plants, it is ideal to use locally adapted varieties, enough leguminous trees, and overall, a wide variety of differing species to maintain diversity, an important characteristic of organic farming. Choose fast growing varieties of shading trees at the beginning of cultivation. As regards the number of shading trees or the intensity of shade required, the general rule that can be adopted is that the higher the tea garden is located, the less shade is necessary. The number of shading trees to be maintained varies according to site and the variety of tree (up to 500 shading trees per hectare).

Figure 7.4: Tea Field with Shade Trees.

13. Plant Protection

Tea plant is affected by a variety of pests and diseases and only organic methods of management should be adopted.

13.1. Pests and their Management

Symptoms of pest attack, their cultural and biological control measures are listed in Table 7.3.

Some of the other pests and their management are as follows:

Cockchafer Grubs

Tea plants in the new clearings are subjected to the attack of cockchafer grubs. Application of neem cake @ 250 g/pit can be very effective in controlling this pest. If noticed after planting, around 500 g per plant should be applied along the drip circle method by placement.

Phassus Borer/Red Coffee Borer

Phassus borer and red coffee borer are also important pests in the new clearings. These can be effectively controlled by the application of botanical organic insecticide-neem products. Higher concentration of neem formulation should be poured through entrance hole by ink filler and plug the hole with clay paste.

Leaf Feeders

Leaf folding caterpillars such as leaf roller, flushworm and tea tortrix can be manually removed while harvesting. Manual removal of caterpillars and pupae will go a long way in reducing the incidence of the looper caterpillar *Buzura supperssaria* and *Eterusia aedea*, the slug caterpillar. In case of severe infestation, neem formulations 0.15 per cent @1000 ml/ha or 5 per cent @100 ml/ha or 1 per cent @ 300 ml/ha in 300 to 350 l of water may be applied.

Mites

Attack of red spider mite is a serious problem in tea and it can be controlled by application of lime sulphur(dilution depending on percentage), paraffinic oil (1000 ml), wettable sulphur (1 kg) and *Paecilomyces fumosoroseus* with jaggery (1.5 kg each).

13.2. Diseases and their Management

The important diseases of tea, their symptoms and management strategies are listed in Table 7.4.

14. Harvesting and Post-harvest Treatment

14.1. Harvesting

Harvesting or plucking commences when the tea bush is three years old. It is a very vigorous process that requires hard work and perseverance in order to get the most out of the tea plant. The ideal conditions for harvesting tea are usually at high altitudes with a good amount of rainfall. Two types of plucking or harvesting is practiced in tea: coarse plucking and fine plucking. For both the techniques, harvesting tea is usually done by hand because machines are likely to damage the leaves too much for them to be of any use. Harvesting tea needs to be done in the early morning. With fine plucking, only the unopened bud, second and third leaves, popularly known as "two leaves and a bud" are to be plucked so that one gets the most from a harvest of tea. Young and tender buds that have silvery white fuzz on

Table 7.3: Pests of Tea, their Symptoms of Damage and Control Measures

Name of Pest	Symptoms	Cultural Control	Biological Control
Aphids (*Toxoptera aurantii* Boyer de Fonscolombe)	☆ Nymphs and adults suck cell sap from the plant foliage. ☆ In addition, plants may become contaminated by honeydew produced by aphids and sooty mould growing on honeydew	☆ Sturdy plants can be sprayed with a strong jet of water to knock aphids from leaves	☆ Insecticidal soaps or oils such as neem or canola oil are usually the best method of control (always check the labels of the products for specific usage guidelines prior to use)
Tea mosquito bug (*Helopeltis theivora* Waterhouse)	☆ The nymphs and adults suck the sap of the young leaves, buds and tender stems and while doing so, they injects toxic saliva which causes the break-down of tissues around the site of feeding ☆ Within 2-3 hours of sucking a circular spot is formed around the feeding point and in 24 hours it becomes translucent, light browning. ☆ Within a few days the spots appear as dark brown sunken spots which subsequently dry up. The badly affected leaves become deformed and even curl-up ☆ The tender stems develop cracks and over-callusing which lead to blockage of vascular bundle and cause stunted growth and sometimes die-back of the stems	☆ The pest lay large number of eggs on the broken ends of plucked shoots. Intensive manual removal of stalks during plucking will help to reduce its incidence ☆ Remove alternate hosts such as Guava (*Psidium guajava*), Oak (*Quercus* spp.), Melastoma (*Melastoma* sp.), Thoroughwort (*Eupatorium* sp.), Fragrant thoroughwort (*Eupatorium odoratum*), Dayflower (*Commelina* spp.), Sesbania (*Sesbania cannibina*), Jackfruit (*Artocarpus heterophylla*) etc. from in and around plantations ☆ Regulate the shade in densely shaded area areas lopping of the lower branches of shade trees. Moderate shade of 60 per cent is ideal	☆ Apply native plant crude aqueous extracts viz. *Clerodendrum viscosum*, *Cassia alata*, *Vitex negundo* and *Amphineuron* sp. @ 5 per cent concentration in case of low and moderate infestation of the pest. ☆ Entomopathogen, *Beauveria bassiana* @ 1.2 kg/acre minimizes infestation of *H. theivora* in field condition ☆ Encourage useful insects such as ladybirds
Common Thrips; mostly in Darjeeling) (*Taeniothrips setiventris*) Assam-Thrips; mostly in Assam and Dooars (*Scirtothrips dorsalis*)	☆ Feeds on tender above ground parts, creating feeding scars, distortion of leaves and discoloration of buds ☆ The infested leaves curl upward, crumble and shed ☆ Infested buds become brittle and drop down	☆ Adopt appropriate shade management ☆ Disturb the soil around the tea bush stem during the cold months to destroy the pupae	☆ Apply *Verticillium lecanii* (verelac) @ 1.5 kg per ha or paraffinic oil @ 1.5 to 2.0 l per ha in 300 to 350 l of water per ha. ☆ Use yellow pan water and yellow sticky traps @ 20 traps per ha

Contd...

Table 7.3–*Contd...*

Name of Pest	Symptoms	Cultural Control	Biological Control
	☆ The sucking marks are made one after one, forming thin pale lines on the underside of leaves parallel to the main vein		
Bunch caterpillar: (*Andraca bipunctata* Walker)	☆ The caterpillars eat the foliage of the host plant ☆ Initially, they feed upon the surface tissues only but later on the whole blade is consumed ☆ They move in groups and before going down for pupation a bunch of caterpillars may destroy several bushes of tea plantation	☆ Install light traps; collect the caterpillars from the ground, tea bushes and shading trees (in all stages of development) ☆ Apply trap bands to the shading trees	
Mites (red, pink, yellow scarlet, and purple) Red spider mite: (*Oligonychus coffeae* Nietner)	☆ **Red spider mites** usually extract the cell contents from the leaves using their long, needle-like mouth parts. This results in reduced chlorophyll content in the leaves, leading to the formation of white or yellow speckles on the leaves ☆ In severe infestations, leaves will completely desiccate and drop off. The mites also produce webbing ☆ Under high population densities, the mites move to using strands of silk to form a ball-like mass, which will be blown by winds to new leaves or plants, in a process known as "ballooning."	☆ Adopt recommended shade management to prevent the excessive build up of mites ☆ Apply mulch and incorporate organic matter into the soil to improve the water holding capacity and reduce evaporation ☆ Keep the field free of weeds ☆ Remove and burn infested crop ☆ The bushes along the motorable roads, which remain covered with dust are very often found to be severely attacked by red spider mite. Protect the roadside bushes from dust by growing hedge plants or applying	

Contd...

Table 7.3—*Contd...*

Name of Pest	Symptoms	Cultural Control	Biological Control
		water on such dusty roads at regular intervals	
	☆ **Purple mite** damaged leaves are characterized by the coppery brown discoloration; presence of numerous white cast skins of the mites along with the live mites; purple mites are prevalent on the under surface of mature leaves	☆ To prevent migration of red spider mites by restricting the pluckers from entering into un-infested areas from infested areas and cattle trespass inside the tea sections should be stopped	
	☆ **Pink mite** is the important mite pest of tea in southern India and cause considerable damage. During early stages of attack leaves turn pale and curl upwards while severe infestation leads to brownish discolouration. Pink mites attack tender crop shoots where "Assam" hybrids are more susceptible	☆ Removal of alternate hosts (*Borreria hispida, Scoparia dulcis, Melochia corchorifolia* and *Fussiala suffruticosa*) in and around plantations	
	☆ **Yellow mite** is seen on young leaves especially the top two to three leaves and the bud. Affected leaves become rough and brittle and corky lines.	☆ The bushes in ill drained or water-logged areas are subject to increased red spider damage and hence improve drainage conditions	
	☆ **Scarlet mite**: Symptoms of attack first appear on either side of the midrib and gradually spread to the entire leaf, feeding leads to brown discolouration of leaves and severe infestation leads to defoliation.	☆ Red spider mite affected fields should get a new tier of maintenance foliage since the infested bushes are very week due to defoliation of maintenance leaves	
Root knot nematode (*Meloidogyne* ssp.)	☆ Infect plants in patches in the field ☆ Formation of galls on host root system is the primary symptom	☆ Prepare nursery bed by harrowing and ploughing to expose and dry the un-decomposed weeds and roots of	

Contd...

Table 7.3–Contd...

Name of Pest	Symptoms	Cultural Control	Biological Control
	☆ Roots branch profusely starting from the gall tissue causing a '**beard root**' symptom ☆ Infected roots become knobby and knotty ☆ In severely infected plants the root system is reduced and the rootlets are almost completely absent. The roots are seriously hampered in their function of uptake and transport of water and nutrients ☆ Plants wilt during the hot part of day, especially under dry conditions and are often stunted ☆ Nematode infection predisposes plants to fungal and bacterial root pathogens	the plants. All sorts of mulching materials should be kept away from the seed nursery to avoid nematode infestation ☆ Kill the plant parasitic nematodes by uniform heating (after sieving) of the soil up to 60°–70°C for 4–5 minutes on plain tin sheets. The soil can be used after heat treatment ☆ Remove weed hosts from nursery beds ☆ Remove infested tea bushes, and remove and replace large amounts of the soil ☆ Prevention *e.g.* using plant-bags in the seed bed; use shading tree *Indigofera teismanii* as a trap plant	

Table 7.4: Diseases of Tea and their Management

Disease	Symptoms	Cultural Control	Biological Control
Blister blight (*Exobasidium vexans* Massee) Endemic to Southeast Asia, does not occur in East Africa; *Poria hypolateritia* (Red root rot)	☆ Small, pinhole-size spots are initially seen on young leaves less than a month old. As the leaves develop, the spots become transparent, larger, and light brown ☆ After about 7 days, the lower leaf surface develops blister-like symptoms, with dark green, water-soaked zones surrounding the blisters ☆ Following release of the fungal spores, the blister becomes white and velvety. ☆ Subsequently the blister turns brown, and young infected stems become bent and distorted and may break off or die ☆ The pathogens survive on leaves or stems and in fallen plant host debris. ☆ Disease is readily spread by the dispersal of spore by wind. ☆ Cloudy and wet weather favors infection	☆ Use spore trap/regular field assessment ☆ Maintain the plucking interval ☆ Pruning during November/December is effective to reduce the disease incidence for new clearing ☆ Avoid broad leaved Assam jats	☆ Spray 2-3 rounds of 5-10 per cent aqueous extracts of *Cassia alata/ Polygonum hamiltonii/Acorus calamus/ Adhatoda vasica/Equisetum arvense/ Polygonum hydropiper/Tagetis petula* at 1:9 dilutions once in 15 days interval
Red, brown and black rot disease Black rot (*Corticium theae; C. inivisum*)		☆ Prune or skiff the severely affected sections. Improve aeration by lopping side branches and 'matidals'. Thin out dense shade and improve drainage ☆ Give alkaline wash after pruning. ☆ Shorter pruning cycle helps in minimizing infestation ☆ Uproot the infected bushes and burn it ☆ Insulation of diseases patches by making trenches of 120 cm deep and 45 cm width surrounding the diseased plants for preventing the spread	

Contd...

Table 7.4—Contd...

Disease	Symptoms	Cultural Control	Biological Control
Red rust (*Cephaleuros parasiticus* Scot Nelson)	☆ Leaves develop lesions that are roughly circular, raised, and purple to reddish-brown. The alga may spread from leaves to branches and fruit ☆ Most algal spots develop on the upper leaf surface ☆ Older infections become greenish-gray and look like lichen. *Cephaleuros* usually does not harm the plant ☆ The pathogens reproduce and survive in spots on leaves or stems and in fallen plant host debris. ☆ Frequent rains and warm weather are favorable conditions for these pathogens. For hosts, poor plant nutrition, poor soil drainage, and stagnant air are predisposing factors to infection by the algae	☆ If vigour of plant is maintained by balanced nutrients, the disease is less ☆ Avoid plant stress. Avoid poorly drained sites. Promote good air circulation in the plant canopy to reduce humidity and duration of leaf wetness ☆ Identify and correct predisposing factors such as- poor drainage, low soil fertility, particularly potash, improper soil acidity, inadequate shade and continuous use of green crops like *Tephrosia candida*, *T. vogelli etc.* in addition to pruning of severely affected sections	☆ Spray 4-6 rounds of 5 per cent aqueous extracts of *Argimone maxicana*/ *Polygonum hamiltonii* at 15 days interval
Charcoal stump rot: (*Ustulina zonata* (Lév.) Sacc.) Collar and branch canker, Die back, Twig die back, stem canker Brown and Grey blight: *Colletotrichum* sp. and *Pestalotiopsis theae*	☆ The first symptoms include browning and drooping of affected leaves ☆ As the disease spreads into the shoots, they become dry and die ☆ The entire branch can die from the tip downward ☆ Dying branches often have cankers- shallow, slowly spreading lesions surrounded by a thick area of bark ☆ The fungus usually requires wounded plant tissue to gain entry and initiate infection	☆ Avoid intensive harvesting using flat shears ☆ Ensure proper nutrition through regular organic manuring ☆ Avoid predisposing factors ☆ Avoid mulching close to the stem collar and planting in gravelly soil ☆ Avoid plant stress. Grow tea bushes with adequate spacing to permit air to circulate and reduce humidity and the duration of leaf wetness ☆ Prohibit the entry of workers of the	☆ Spray 2-4 rounds of 5 per cent aqueous extracts of *Amphineuron opulentum*/ *Cassia alata*/*Polygonum sinensis* at 15 days interval.

Contd...

Table 7.4–Contd...

Disease	Symptoms	Cultural Control	Biological Control
	☆ Spores are spread when splashed by rain and can survive for several weeks on pruned branches left in the field ☆ Rainy weather favors its spread, and dry conditions promote its development	infested section into the healthy sections	

Figure 7.5: A Tea Plantation Ready for Plucking.

them should be harvested. This type of harvesting tea makes very fine and delicate flavoured tea. It is usually lighter and sweeter in taste.

Harvesting tea using the coarse plucking technique produces a lower quality of tea than fine plucking. In coarse plucking, the bud along with more than two leaves will be harvested. This is generally done at a very fast pace and this technique of harvesting tea makes a stronger flavour tea than that of fine plucking. The interval of harvesting or plucking of the tender shoots or flush is adjusted depending on environmental conditions and the need to maintain bush vigor, for obtaining the highest yield at the lowest cost and for the desired quality. General climate, daily weather conditions, ambient temperature, day length, and bush vigor and nutrient status are the main factors determining the rate of shoot growth, and this varies according to longitude and latitude.

For manual plucking, the top of the tea bushes can be trained by the plucking process to a level "plucking table". The removal of the terminal bud with the plucked shoot causes the axillary bud just below it to produce a new shoot having at least four new leaves. Later on, leaf production on the shoot ceases temporarily. The flush consists of both actively growing and temporarily dormant (or *banji*) shoots. The average time required for a bud to develop into a shoot ready for plucking varies from about six weeks in Southern Africa (high country) to eight to eleven weeks in Sri Lanka (low country). In the tropics, although shoot growth fluctuates with annual variations in temperature and moisture conditions, plucking is done uniformly over the year at intervals of 4–10 days. In Sri Lanka, plucking rounds are four days for fast-growing, vegetatively propagated teas in the warm, low country, and ten days for seedling teas under cold, high-country conditions. In Malawi (Southern Africa), harvesting rounds are one to two weeks. Shorter intervals give increased crop yield, but at higher plucking costs. In India, plucking continues throughout the year at weekly intervals during March - May and at intervals of 10-14 days during the other months. In East Africa (as in Kenya) with a high altitude and evenly distributed annual rainfall, crop is taken every month although the yields fluctuate, while in

Southern Africa, with lower altitudes, higher temperatures, and seasonal rain, 90 per cent of the annual crop is taken in four months.

Dormant buds increase in number during unfavorable dry conditions in the tropics, but when conditions become favorable, as with the onset of rains, "bud break" or a simultaneous growth of dormant buds takes place. This results in a peak or "rush" crop during the wet months of the year. During the rush period, harvest two to three leaves with a bud at 7 to 10 days interval, while during the lean period, harvest two leaves and a bud at 10 -15 days interval.

14.2. Post-harvest Management

The manufacture of "Organic tea in conversion" and "Organic tea" should be done on separate days; care should be taken to properly clean and wash the factory with water under pressure before the manufacture of "Organic tea" and after manufacture of "Organic tea in conversion". There should be a separate store for organic tea where no fumigants, insecticides or fungicides are used. Vacuum, steam or high pressure water cleaning is permitted. Organic tea should be packed in plywood chests or biodegradable packing materials on the same day of production and the organic quality grade should be clearly indicated on each chest or container along with the invoice number of dispatch. The chests of organic tea should be transported separately and there should not be any chance of it coming into contact with the conventional tea. Before shipment to the destination, it should be stored in a separate place, away from the conventional tea.

14.2.1 Types of Tea

There are three main types of tea manufactured from plucked shoots such as black, green, and oolong tea. In each case, the processing methods employed are different. On delivery to the factory, shoots are withered (for making black tea), steamed, or subjected to dry heat (for green tea).

a) **Black tea:** It is the fully fermented tea. In the manufacture of green tea, the initial heating or steaming of the shoots denatures the polyphenol oxidase enzymes in the beginning. Fermentation is suppressed by deactivating these enzymes, and therefore, green tea is rich in monomeric polyphenols (catechins) and the leaves retain their olive green colour.

b) **Oolong tea:** The fermentation process is halted at an earlier stage (partly fermented tea) and therefore, it is semi- or 50 per cent fermented and contains polyphenol dimmers and it is produced only in China (mainly in Fujian, Guangdong, and Taiwan).

c) **Instant teas:** Made either from low-quality teas (fermented and dried), or from non-dried tea in a special process directly after fermentation. Instant teas lose much of their aroma during the extraction (only hot water extraction is permitted) and subsequent freeze-drying processes.

14.2.2 Flavouring of Tea

The use of any synthetic and/or naturally identical aromas is not permitted in organic production system. This is important as flavouring of tea has a long tradition (*e.g.* the use of bergamot oil to make Earl Grey tea). However, the use of natural flavourings is allowed and laying out layers of plant blossoms (*e.g.* jasmine) could be used but the blossoms must be organically cultivated. In each case, the aroma substances used are to be approved by the certification body.

15. Economics of Organic Tea Production

The productivity of the organic tea garden, in general, was found lesser by 35 to 45 per cent when compared to conventional farming. Cost of production in organic garden is higher by around 60 to 65 per cent mainly due to reduced plucking average, organic inputs, application cost and manual weed control. Organic garden consumes additional man days of 80 to 100 per ha per year than that of conventional farming. Both variable and unit fixed cost in organic garden are much higher by 96 and 85 per cent, respectively due to lower productivity in the field. The survey undertaken in organic garden shows that there is a drop in plucking average by 15 to 20 kg which makes an increase in plucking cost by 52 to 68 per cent. Similarly cost of production is higher both in soil application of organic manure by 260 per cent and weed control by 360 per cent than conventional garden. It is essential that organic tea should fetch a premium price by around 50 to 75 per cent higher when compared the price level of normal tea to achieve sustainability of organic gardens. Introduction of support schemes for conversion and continuing organic farming could make significant impact on the profitability and also attract many clients. (Radhakrishnan and Durairaj, 2012).

Selected References

Durairaj, J., Radhakrishnan, B., Hudson, J.B., and Muraleedharan, N. (2015). *Guidelines on tea culture in South India.*(9th Ed.) pp. 231.

Durairaj, J.,Hudson,J.B.,Muraleedharan,N. and Perumalsamy,K.(2011). Establishment of young tea in an organic way. *Planters' Chronicle.* **107**:5:17.

Ghosh Hajra, N. (2002).Organic tea growing in Darjeeling hills. A case study. Proc. *PLACROSYM XV.* 10-13 December, 2002. Mysore, India, 437-447.

Ghosh Hajra, N. (2005). Organic tea growing in India: A case study in Darjeeling hills. *Proc. National Workshop on Vision and Strategies for Tea Development in Hilly States.* May 28-30, 2005, Uttaranchal Tea Dev. Board, Almora, Uttaranchal, in press.

Radhakrishnan, B.(2004). Package of practices for organic tea. *Planters' Chronicle.* **100** : 19-31

Radhakrishnan, B. (2005).Indigenous preparations useful for pest and disease management.*Planters' Chronicle.* **101**:4-16.

Radhakrishnan, B. and Durairaj, J. (2012). Economic analysis of organic tea cultivation. p. 136. In: *Abstract of papers.* PLACROSYM XX. Coimbatore, Tamil Nadu.

Satyagopal, K., Sushil, S.N., Jeyakumar, P., Shankar, G., Sharma, O.P., Sain, S.K., Boina, D.R., Chatopadhyaya,D., Sunanda, B.S., Ram Asre, Kapoor, K.S., Sanjay Arya, Subhash Kumar, Patni, C.S. Somanth Roy, Hitendra Kumar Rai, Hath, T.K., Nripendra Laskar, Ayan Roy, and Surajit Khalko, (2014). AESA based IPM package for Tea. 73p.

Chapter 8

Organic Farming in Spices

☆ *C.K. Thankamani, V. Srinivasan and S. Hamza*

1. Introduction

India has been a traditional producer, consumer and exporter of spices. During 2014-15, the country's production of spices is estimated to about 61.62 lakh tonnes from an area of 33.25 lakh hectares. During last year, India exported 8, 93,920 tonnes of spices valued Rs. 14899.63 crores. In India, spices and plantation crops are mainly grown in the Western Ghats region originating from Maharashtra, extending 1,600-km range through Goa, Karnataka, Tamil Nadu and Kerala. The Western Ghats' ecological expert panel, set up by the Ministry of Environment and Forest (MoEF), recommended promoting organic agriculture especially spices and plantation crops to protect these ecologically-sensitive hot spot areas. The Ministry of Agriculture and Co-operation has also introduced new initiative under Parampragat Krishi Vikas Yojana (PKVY) for promotion of organic farming through farmers' clusters consisting 50 or more farmers with 50 acres of land in the entire country. Further, spices have tremendous importance in everyday life of human being as ingredients in foods, medicine, perfumery, cosmetics *etc*. They are used as products having anti oxidant, antimicrobial, pharmaceutical and neutraceutical properties and, therefore, their potential as organically produced functional food has got magnified scope. Since 1990, the market for organic food and other products has grown rapidly, reaching $63 billion worldwide in 2012. This demand has driven a similar increase in organically managed farmland which has grown over the years (2001-2011) at a compounding rate of 8.9 per cent per annum (Willer *et al.*, 2013). As such, with growing awareness for healthier products in the modern world, the demand for organic spices is increasing to the tune of about 9-10 per cent annually.

Organic farming is a form of agriculture that relies on techniques such as crop rotation, green manure, compost, and biological pest control. In this system only natural fertilizers and pesticides are allowed. Organic agricultural methods are internationally regulated and legally enforced by many nations, based on large part on the standards set by the International Federation of Organic Agriculture Movements. It is based on minimal use of off-farm inputs and on management practices that restore, maintain and enhance ecological harmony. It also relies heavily on crop diversity and natural sources of manure, fertilizers and herbicides. Among the spices, black pepper, small cardamom, nutmeg, clove and cinnamon, garcinia are important crops that are cultivated mainly as mixed crop along with other plantations like coconut, arecanut, coffee, tea *etc.* and their cultivation technology through organic way is described in this chapter.

2. Black Pepper

In India, black pepper occupies an important place covering an area of 1.24 lakh hectares with production of 70,000 m tonnes (2014-15). It is mostly grown as an intercrop with coconut, arecanut and various fruit trees. In the hilly areas, inter-cultivation is also done with rubber and cashew. Black pepper is commonly cultivated as "homestead cultivation" as a secondary crop interspersed with several other crops. It is more so in Kerala state, which accounts for 97.4 per cent of the total area under the crop in the country.

2.1. Conversion Period and Isolation

The methods for organic production of black pepper, when grown as an intercrop or pure crop, should conform to the standards laid down for the purpose. For certified organic production of black pepper, at least for 36 months, the crop should be under organic management *i.e.* the first harvest after three years of planting can be sold as organic. For a newly planted or replanted area raised through organic cultivation practices, the first yield itself can be considered as organic produce provided chemicals have not been used in the previous cropping and sufficient proof of history of the area is available. In the case of cultivation on virgin land also, the conversion period can be relaxed. It is desirable that organic method of production is followed in the entire farm, but in large estates the transition can be phased out for which a conversion plan is to be prepared. In order to avoid contamination of organically cultivated plots from neighbouring non-organic farms, a suitable buffer zone with definite border is to be maintained. An isolation belt of at least 25 m wide is to be left from all around the conventional plantation. The produce from this isolation belt shall not be treated as organic. In smallholder groups, where the pepper holdings are contiguous, the isolation belt is needed only be at the outer periphery of the entire group of holdings. In sloppy lands, adequate precaution should be taken to avoid the entry of runoff water and drift from the neighbouring farms.

2.2. Climatic and Soil Requirements

Black pepper is a plant of humid tropics, requiring 2000 mm to 3000 mm of rainfall, tropical temperature and high relative humidity with little variation in day length throughout the year. Black pepper does not tolerate excessive heat and dryness. It requires a moderate, well-distributed rainfall with high temperature for better performance. Studies carried out in black pepper growing areas identified specific cultivars/varieties suitable for different agro ecological regions as well as for growing under different cropping systems. The plant starts flowering during May -June with the onset of the southwest monsoon and harvesting is usually in November-January. Virgin soils are ideal for black pepper cultivation. Soil should be well-drained and rich in organic matter (humus). The crop thrives best on red, lateritic or alluvial soils that are rich in humus. The ideal pH is 4.5 to 6.0.

2.3. Planting and after Care

When black pepper is grown in a mixed cultivation system, it is essential that all the crops in the field are also subjected to organic methods of production. For new planting, varieties that are resistant or tolerant to diseases, pests and nematode infection should be used. Black pepper as a best component crop in agri-horti systems, recycling of farm waste can be effectively done when grown with coconut, arecanut, coffee *etc*. As a mixed crop, it can also be grown with green manure/ legumes crops as intercrops enabling effective nutrient build up and weed control. Proper soil and water conservation measures by making conservation pits in the interspaces across the slope have to be followed to minimize the erosion and runoff. Water stagnation has to be avoided in the low lying fields by taking deep trenches for drainage.

2.3.1. Selection of Site

The land that is proposed for planting black pepper should be cleared of weeds and undergrowth. In level and low lands, proper drainage channels should be provided to prevent water stagnation during periods of high rainfall. Areas having 1-3 per cent slope are ideal for planting black pepper. Planting in slopes facing south should be avoided so that the vines are not subjected to the scorching effect of the southern sun during summer. In sloppy lands, adequate soil and moisture conservation measures are to be adopted.

2.3.2. Standards

Black pepper vines require a support for its growth, development and yield. These supports are called standards. Providing ideal supports plays an important role in successful establishment of black pepper vines. Since the black pepper vine is productive for 15 years and more, selection of standards assumes great significance. The standards used for trailing black pepper vines are of two types, *viz.*, living and non-living. The non-living standards include reinforced concrete posts, granite pillars and teak poles. The ideal living support should establish easily and grow rapidly to provide shade and support for the establishment of the crop and it should tolerate regular and heavy pruning. A variety of trees are used as living standards for black pepper cultivation in India. In homesteads gardens in Kerala, black pepper is usually trained on arecanut and coconut and also on mango, jack, *etc*. When interplanted in cardamom and coffee plantations, it is trailed on various forest trees that provide shade. It is cultivated both as a pure and mixed crop along with arecanut in the plains of Uttar and Dakshin Kannada districts in Karnataka. In Andhra Pradesh, the vines are trailed on coconut and oil palm. In north-eastern states especially Assam, most of the homestead gardens have arecanut palms where

Figure 8.1: Black Pepper Garden.

black pepper vines can be trailed. However, such standards are not ideal to establish large plantations under organized cultivation as monocrop at a normal spacing of 2.5 m x 2.5 m. Therefore, on a plantation scale, *Erythrina indica* is the common live standard for trailing black pepper especially in Kerala. As *Erythrina* is prone for attack by wasp (*Quadristichus erythrinae*) and nematodes, it is better to avoid them as standard. Other common standards that can be used are *Garuga pinnata, Glyricidia sepium, Leucaena leucocephala, Ailanthus malabarica* and *Grevillia robusta*. Both coconut and arecanut palms are used for trailing when black pepper is used as a mixed crop. For raising living standards, seedlings are to be planted three to four years in advance so as to attain sufficient height at the time of planting of black pepper. In case stems/stem cuttings of *Glyricidia sepium* or *Garcinia pinnata* are used, they are to be cut to suitable lengths during March-April and stacked in shade. The stacked stems start sprouting in May. After the first rain in May-June, the sprouted stems are to be planted at the edge of the pits dug for planting black pepper vines. The cuttings of standards are to be planted in narrow holes of 40 to 50 cm depth. The soil should be well pressed around the standards to avoid air pockets and keep the standards firm in the soil.

2.3.3. Spacing

Spacing of black pepper vines varies considerably and depends on the spacing at which the standards are planted. Under monocropping system, the optimum spacing is 3 m x 3 m, which can accommodate 1100 standards/ha, whereas, in sloppy land, 3 m x 2 m spacing is recommended. When non-living standards are used, much closer spacing (1.5 m x 2.0 m) is recommended. Poor growth and productivity is noticed on concrete poles. Dead standards are not used in India, though trials at ICAR-IISR have shown that pepper trailed on dead standards gives higher yield.

2.3.4. Planting

Pits of 50 cm x 50 cm x 50 cm at a distance of 30 cm away from the base of supporting tree are taken with the onset of monsoon. The pits are to be filled with a mixture of topsoil, farmyard manure @ 5 kg/pit and 150 g rock phosphate. In case of large pits, alternate layers of coconut husk and the above mixtures can be filled to help the young plants to withstand hot summer. Neem cake @ 1 kg and *Trichoderma harzianum* @ 50 g may be mixed in the pit at the time of planting. In addition to organic manure application before planting, apply 50 g Azospirillum + 50 g Phosphate Solubilizing Bacteria + 200 g VAM per plant also at the time of planting. Plant two rooted cuttings individually in the pits on northern side of each standard. At least one node of the cutting should be planted below the soil for proper anchorage. The young vines should be protected from hot sun during summer, especially in open areas, by providing shade. The young vines are to be covered with dry arecanut or coconut leaves or twigs of trees during summer which should be removed with the onset of rains.

2.3.5 Training

Training is an essential step for establishment of black pepper vines. As the cuttings grow, the shoots are to be tied to the standards regularly using suitable materials for anchorage. For the production of leader shoots from the nodes, the vines are to be lowered into the soil during the first year after planting. In this method, the leaves are removed from vines after they attain a height of 1 m and they are brought down and $3/4^{th}$ of the basal portion is buried around the standard and covered with fertile top soil. This induces a good root system and also helps in production of more leader shoots from nodes. Three to five leader shoots are enough to produce sufficient laterals and to form a full canopy around the trunk of the standard. Pruning of terminal shoot increases production of spikes and number of bearing laterals.

2.3.6. Pruning of Standards

After establishment, the side branches of standards are to be pruned to enable the standard to grow erect. After complete establishment of standards, periodical pruning is important to allow sufficient light penetration into the black pepper canopy. Normally pruning of live standards has to be carried out one week prior to manure application. At the beginning of rainy season, heavy pruning is done, leaving only one or two twigs at the top of the trees, while at the end of the rainy season, a moderate pruning is carried out, leaving three or four twigs at the top of support trees. Shade regulation by pruning branches of standards in black pepper gardens during March-and July-August is an important cultural practice to allow sufficient light for crop growth and productivity and also to reduce the incidence of diseases. Pruning reduces leaf density on the live support and stimulates growth of new leaves, which reduces root activity and places the support tree in a weaker position to compete for light, water and nutrients, relative to the pepper vines.

2.3.7. Mulching

Mulching increases infiltration of water, conserves moisture, regulates temperature, decreases evaporation, suppresses weed growth, enhances microbial activity and also improves soil fertility. Mulching around the basins of black pepper vines with organic materials especially green leaves @10 kg/vine to a radius of 1 m is to be made at the end of North-East monsoon. Mulching should be repeated once the applied mulch is decomposed. Live mulch (cover crops) such as *Calopogonium mucunoides* and *Mimosa invisa* can also be grown to provide soil cover and to prevent soil erosion. These cover crops are to be cut back regularly from the base to prevent them from twining along with black pepper vines.

2.3.8. Irrigation

Moisture requirement of black pepper vines vary with stages of crop growth. Moisture stress is one of the most serious constraints affecting the productivity of black pepper. Blossoming, spiking, flowering, spike elongation and setting are the sensitive periods for black pepper. Following summer, the first 16 weeks could be considered as critical period for shoot growth, flower bud differentiation, spike emergence, flower opening and fertilization, berry formation and development. Spike development ceases if there is prolonged dry spell immediately after good summer showers. Irrigating black pepper vines @ 100 litres per vine (hose irrigation) once a week during summer is ideal. The water is to be applied in basins taken around the plants at a radius of 75 cm. In case drip irrigation is adopted, 7 litres of water per day through drip during October to May is sufficient.

2.3.9 Weeding

Weeds are a major problem in black pepper plantations that are not maintained properly. The weeds growing in the basins absorb considerable quantities of moisture and nutrients and should be removed. Weeding may be done only when necessary by slashing and the materials should be used for mulching. Raising cover crops also controls weeds. The borders of the farm and road edges should also be raised with suitable leguminous cover crops to prevent soil erosion and weed growth.

2.4. Nutrient Management

Compost or farmyard manure may be applied from second year onwards @ 5 kg/vine which can be increased to 10 kg/vine from third yielding year onwards. The optimum time of application is May-June. This can be partially or completely substituted by vermi/leaf compost @ 10 kg/vine. Based on soil test, if found necessary, application of lime/dolomite, rock phosphate/bone meal and wood ash may be done to get a dose of 50 kg P_2O_5 and 150 kg of K_2O per ha. Leaf/vermicompost may be prepared from the green loppings, crop residues, grasses, cow dung slurry, poultry droppings *etc.* at the farm level itself, fortified with wood ash and rock phosphate and should be applied @ 2 t/ha. Such compost can be further supplemented with non-edible oil cakes like neem cake @ 1 kg/vine/year and suitable microbial cultures of *Azospirillum* and phosphate solubilizing bacteria that are native to the environment. Composted coffee pulp and husk rich in potassium source can also be recycled and used. Foliar application of micronutrient mixture specific to black pepper developed by ICAR-IISR is also recommended @ 5 g/l at the start of flowering and followed at monthly interval for higher yield.

2.5. Cropping System

Multiple cropping is the practice of growing two or more crops in the same field simultaneously. A systematically planted black pepper garden would provide adequate interspaces for cultivation of other crops especially during the pre-bearing period because of the negligible shade effect by the vines. The factors that assume significance in multiple cropping are probable competition for nutrients, moisture and sunlight between black pepper and other crops and its effect on black pepper yield. The goal of intercropping is to produce a higher yield on a given piece of land, by making use of the space that would otherwise be wasted with a single crop. However, careful planning is required before selection of crops, taking into account the soil, climate and crops and it is important not to have crops competing with each other for space, nutrients, water or sunlight. Multiple cropping in black pepper gardens is a routine practice in Kerala and parts of Karnataka. A variety of crops are grown along with black pepper, such as elephant foot yam, colocasia, ginger and turmeric and perennial fodder grass, vanilla and banana. At higher altitudes, black pepper is grown along with coffee, cardamom and tea. Cereals like upland paddy, pulses like red gram, vegetables, flowers, fodders and other annuls are intercropped with black pepper. Black pepper is also trailed on coconut and arecanut palms in most of the areas. In these cases, the rooted black pepper plants are planted away from the base of palms and as and when the vines grow, they are trailed along the ground and then trailed on to the trunk of coconut or arecanut palms. The varieties Sreekara, Subhakara and Panniyur-5 perform well as intercrops in coconut and arecanut gardens.

2.6. Plant Protection

2.6.1. Pest Management

Use of biopesticides and cultural methods for the management of insect pests forms the main strategy for the development of organic spices. For control of pollu beetle (*Lanka ramakrishnae/Longitarsus nigripennis* Mots), top shoot borer (*Cydia hemidoxa* Meyr.), and leaf gall thrips (*Liothrips karnyi* Bagnall), the cultural control measures are regulation of shade in the plantation during May-June to reduce the population of the pest in the field. Clipping of severely infected branches and spraying Neemgold 0.6 per cent and fish oil rosin 3 per cent during August, September and October is effective for the management of the pests. The underside of leaves, where adults are generally seen, and spikes are to be sprayed thoroughly. For the management of mealy bugs (*Planococcus* sp./*Pseudococcus* sp.),remove weeds and alternate host plants like hibiscus, bhindi, custard apple, guava *etc*. in and around the pepper garden throughout the year. Deep ploughing in summer or raking of soil in vineyards helps to destroy its nymphal stages and minimizing the incidence. As biological control measure, release exotic predator, *Cryptolaemus montrouzieri* @ 10 beetles/vine.

2.6.2. Disease Management

The most important disease of black pepper is foot rot caused by the soil borne fungus *Phytophthora capsici*. All parts of the plants are vulnerable to attack and the symptom expression depends upon the site or plant part infected and the extent of damage. For the management of this disease, regular adoption of phytosanitary measures is very essential. They include removal and destruction of dead vines along with root systems from gardens. Planting material must be collected from disease free gardens and the nursery preferably raised in fumigated or solarized soil. Tillage operations are to be kept to the minimum at the basin to avoid soil disturbance and root damage. Injury to the root system due to cultural practices such as digging should also be avoided. Adequate drainage should be provided to reduce water stagnation. The freshly emerging runner shoots should not be allowed to trail on the ground. They must either be tied back to the standard or pruned off. The branches of support trees must be pruned at the onset of monsoon to avoid build up of humidity and for better penetration of sunlight. Reduced humidity and presence of sunlight reduces the intensity of leaf infection. Application of *Trichoderma* multiplied in suitable carrier medium such as coffee husk, compost/tea waste, well rotten cow dung or quality neem cake may be made during May-June @ 5 kg/vine (to supply 50 g/vine inoculum) and a second round application of the same quantity during August-September. To control fungal pollu and foliar infection of *Phytophthora*, restricted spraying of Bordeaux mixture @ 1 per cent may be done. Biocontrol agents like *Pochonia chlamydosporia* can be applied @ 50 g/vine twice a year (during April-May and September-October) to control nematode problems. The fungus load in the substrate should be 10^8 cfu/g. Yellowing and wilting symptoms are noticed to the extent of more than 30 per cent when the root system of the vines is damaged beyond recovery. Hence, those vines showing such symptoms should be uprooted and destroyed. Regular application of bio control agents may be done as a standard practice during May-June and September-October seasons. Application of quality neem cake will be useful to check the nematode population and thereby slow decline disease. Planting materials from mother vines showing symptoms of stunted disease and phyllody should not be collected for raising rooted cuttings. Severely affected vines are to be uprooted and destroyed.

2.7. Harvest and Post-harvest Operations

In India, black pepper flowers in May-June and it takes about 7-8 months after flowering to reach full maturity. The harvest season extends from November - January in plains and January-March in the high ranges of Western Ghats of India. It is very important to harvest at the right stage of maturity so as to achieve a dried product of good colour and appearance. Harvesting can commence when one or two berries in the spike turn bright orange red by handpicking the whole spike and collecting in clean gunny bags. If the berries are allowed to over ripe, there is heavy loss due to berry drop and damage by birds. The spikes which are fallen on to the ground may be collected separately, cleaned and then added to the general lot.Various products from black pepper are now being made and, therefore, the maturity of pepper for harvest should be considered depending on the product diversification aimed at. The level of maturity required at harvest for processing into different pepper products is given in Table 8.1.

Table 8.1: Level of Maturity Required at Harvest for Processing

Product	Stage of Maturity at Harvest
Canned pepper	4-5 months
Dehydrated green pepper	10-15 days before maturity
Oleoresin and essential oil	15-20 days before maturity
Black pepper	Fully mature and 1-2 berries start turning from yellow to red in each spike
Pepper powder	Fully mature
White pepper	Fully ripe

The post-harvest processing operations to be followed for black pepper are: threshing, blanching, drying, cleaning, grading and packaging. During processing, care should be taken to maintain the quality at each stage of operation. Threshers with capacities varying from 50 kg/h to 2500 kg/h are available which can thresh quickly and provide clean product. The quality of the black pepper can be improved by dipping the mature berries taken in perforated aluminium vessel or bamboo basket in boiling water for a minute and drained before drying. This blanching gives uniform coloured black pepper after drying, reduces the microbial load and the time required from 5-6 days (traditional practice) to 3-4 days and also helps to remove the extraneous impurities like dust from the berries. While harvesting, black pepper has moisture content of 65 per cent to 70 per cent. Sun drying is the conventional method followed for drying of black pepper.

The berries are to be spread on concrete floor or bamboo mats or PVC sheets and dried under sun for three to five days to bring the moisture content below 10 per cent. During sun drying, it is important to turn over the material periodically to facilitate uniform drying. Dried black pepper with moisture content >12 per cent is susceptible to fungal attack and mycotoxins produced by the fungal attack render the pepper unfit for human consumption. The average dry recovery varies between 33-37 per cent depending on the varieties and cultivars. Mechanical driers developed by various agencies are also used to dry black pepper. Models of varying capacities operated either electrically or by burning agricultural wastes are available for drying of black pepper by maintaining temperature below 55^0 C.

The threshed and dried black pepper is likely to have various extraneous matters like spent spikes, pinheads, stones, soil particles *etc.* mixed with it. Cleaning and grading are basic operations that enhance the value of the produce and help to get higher returns. Cleaning on a small scale could be done by winnowing and hand picking which removes most of the impurities. Grading of black pepper is done by using sieves and shifting black pepper into different grades based on size. The major grades of black pepper are Tellicherry Garbled Special Extra Bold (TGSEB) (4.8 mm); Tellicherry Garbled Extra Bold (TGEB) (4.2 mm); Tellicherry Garbled (TG) (4.0 mm); Malabar Garbled (MG grades 1 and 2) and Malabar Ungarbled (MUG grades 1 and 2). Organically grown black pepper should be packaged separately and labeled. Mixing different types of pepper should be avoided. Recyclable/reusable packaging materials shall be used wherever possible and eco friendly packaging materials such as clean gunny bags or paper bags may be used and the use of polythene bags may be minimized to the extent possible.

Black pepper is hygroscopic in nature and absorption of moisture from air, during rainy season when there is high humidity may result in mould and insect infestation. Therefore, before storing, it is to be dried to less than 10 per cent moisture. The graded produce may be bulk packed separately in multi layer paper bags or woven polypropylene bags provided with food grade liners or in jute bags. The bags are to be arranged one over the other on wooden pallets after laying polypropylene sheets on the floor. White pepper is generally prepared by retting (with frequently changing of water) fully ripened red berries for seven to eight days followed by removal of outer skin, washing and drying to a moisture level of 12 per cent. White pepper is also prepared by fermentation using matured green pepper and black pepper.

3. Small Cardamom

Small cardamom (*Elettaria cardamomum* Maton) the "Queen of Spices" enjoys a unique position in the international spices market, as one of the most sought after spices. It is indigenous to the evergreen forests of Western Ghats in South India. India has been the world's largest producer of cardamom until 1979-80. Later, Guatemala emerged as world's premier producer and the exporter of cardamom accounting for about 90 per cent of the global trade. The cardamom of commerce is the dried ripe fruit and is one of the most important valued spices next only to saffron. It is used for flavouring various food preparations, confectionery, beverages and liquors. It is also used for medicinal purpose, both in Allopathy and Ayurveda systems and hence there is large demand for organic produce.

Small cardamom is cultivated in an area of 69,970 hectares with production of 18,000 m tonnes (2014-15).Cultivation of small cardamom is confined to three South Indian states namely, Kerala, Karnataka and Tamil Nadu. In 2014-15, Kerala had 56.8 per cent of area and contributed 88.8 per cent production; Karnataka with 35.8 per cent of area and 5.6 per cent production, while Tamil Nadu had 7.4 per cent of area and contribution of 5.6 per cent production.

3.1. Conversion Period and Isolation

Conversion period of minimum three years and isolation of 25 m wide space should be provided between organic plots. If organically produced planting materials are used and if at least two years have elapsed without use of any inorganic inputs in the field prior to planting, the yield from such a crop shall be considered as organic. In case of cultivation in virgin lands and farms wherein no chemical inputs have been applied in the past, the conversion period can be relaxed. In the case of cardamom plants grown wild in the forest, the entire produce can be considered as organic (Anon, 1998).

3.2. Climatic and Soil Requirements

Cardamom is highly sensitive to elevation and the wrong choice of cultivar or location can severely affect the growth and productivity. It is a shade loving plant (pseophyte) thriving well in elevations up to 600-1200 m above MSL under an average annual rainfall of 1500-4000 mm and temperature range of 10-35° C. The 'Malabar', type is productive under lower elevations of 500-700m MSL. The other two cardamom types, 'Mysore' and 'Vazhukka' are suitable for cultivation in lower altitudes *i.e.* 700m MSL and below. Though vegetative growth is satisfactory, fruit production is poor at lower elevations.

Cardamom is cultivated under shade trees and the soils in general have high fertility status due to addition of leaf litter. In India, soils of major cardamom growing areas come under the order alfisols, formed under alternate wet and dry conditions and the sub order ustalfs derived from schists, granite and gnesis and lateritic in nature. Cardamom, generally, grows well in forest loamy soils with good drainage. These soils are generally acidic in nature with pH from 4.2 to 6.8, high in organic matter and nitrogen, low to medium in available phosphorus and medium to high in available potassium.

Figure 8.2: Cardamom Plantation.

3.3. Site Selection and Shade Requirement

The shade canopy provides suitable environment by maintaining humidity and evaporation at suitable level. Shade requirements vary from place to place and depend on the lay of land, soil type, rainfall pattern and crop combination *etc*. In Guatemala, which receives well distributed rainfall and has cool climate round the year, cardamom is grown practically in open areas with either no shade or having only very sparse shade.

3.4. Planting and after Care

3.4.1. Preparation of Main Field

Cardamom is generally cultivated in forest area. The initial work involves cleaning the forest and thinning out overhead canopy in order to get umbrella-shaped canopy to get filtered shade. If land is sloppy, it is advisable to start cleaning from the top and work downwards. In sloppy areas, adequate soil and water conservation measures are necessary while preparing the land for planting. Planting in trenches across the slopes in low rainfall areas, diagonal planting and mulching the soil will help in soil and water conservation.

3.4.2. Spacing

Spacing should be decided based on variety and duration of crop in the field. 'Mysore' and 'Vazhukka' types are vigorous and need wider spacing. 'Malabar' type needs closer spacing as they have comparatively smaller plants. Better growth and higher yield in cardamom could be obtained when planting is done at 2m x 1m in hill slopes and 2.1m x 2.1m in flat lands.

3.4.3. Method of Planting

Cardamom is propagated through seeds, suckers and tissue culture plantlets. Economic yield of cardamom starts from third year onwards after planting, and it continues up to 8-10 years. The total life span of cardamom plants is about 15-20 years; however, pseudostem is biannual in nature. The trench method of planting of suckers (60 cm x 30 cm) with a spacing of 2 m x 1 m is ideal for better growth and yield of cardamom. In some places, seedlings are planted in holes, just scooped out at the time of planting. In other areas, considerable care is to be taken in the preparation of planting pits and they are filled with surface soil mixed with leaf mould, compost or cattle manure. 'Mysore' and 'Vazhukka' types can be planted in the pits of 90 cm x 90 cm x 45 cm, whereas, 'Malabar' type could be planted in pits of 45 cm x 45 cm x 45 cm. Pits are to be filled with mixture of topsoil and compost or well rotten farmyard manure and 40 g of rock phosphate.

Ten or eighteen months old cardamom seedlings are to be planted in pits by taking small portion of filled soil and adding rock phosphate (40 g) in the centre of pit. Planting is to be done just above the ground level to avoid rotting in heavy rains. Seedlings are normally planted in a pit at acute angle for promoting shoot production. If older seedlings are used, light root pruning is to be taken up. In case of planting suckers, they may be placed in a slanting position and the base of rhizome covered with soil. Staking of plants is to be done immediately after planting to avoid damage due to wind. In order to reduce evapotranspiration and rain damages, mulch the plant base. Adequate care should be taken to offset the transplanting shock and to save the seedlings from heavy rains. The newly planted area should be inspected regularly and gap filling may be done, whenever needed, if the climate is favourable.

3.4.4. Mulching

Fallen leaves of shade trees can be utilized for mulching. Sufficient mulch should be applied during November-December to reduce the ill effect of drought, which prevails for nearly four to five months during summer. Removal of mulch materials is also equally important during May after the pre monsoon showers to facilitate honey bee movement to obtain better pollination and capsule setting and also provide better aeration and to minimize incidence of clump rot or rhizome rot. The weeded and trashed materials can also be used for mulching.

3.4.5. Weeding and Trashing

Clean weeding is to be limited to the plant bases (50 cm) and the inter rows are to be maintained by slash weeding. Trashing the dry leaves, leaf sheaths and removal of yielded old suckers along with rhizomes may be carried out once in a year about a month after completion of final harvest, which can be used for composting. Trashing thrice a year and cleaning the plant bases during monsoon months helps in better pollination, elimination of breeding sites of sucking pests and reduce rhizome rot incidence.

3.5. Cropping System

In the inter-cropping system, plants compete for light, water, soil and space. Competition will occur when the immediate supply of a single necessary factor falls below the combined demands of the plants. Better utilization of environmental resources is the prime objective of these planting systems. Inclusion of black pepper, coffee and arecanut are highly remunerative giving higher cost benefit ratios. Mixed cropping of cardamom in arecanut gardens gives higher income than mono cropping of arecanut.

3.6. Nutritional Management

3.6.1. Nutrient Recycling

Biomass regeneration and utilization as a source of organic input has been a feasible approach towards upgrading productivity of land and crop. Cardamom, being sciophyte, requires filtered light for its growth and yield. It is commonly cultivated beneath the shade trees, which play dominant role in recycling of nutrients through their leaves and other plants. Such tree crop based land use systems are more efficient in maintaining soil fertility than annual cropping system. Shade trees help to maintain a high fertility status and soil pH in surface soils. Even though on an average about 5.5 t/ha of organic materials, leaf litter, weeds and pruned plant parts are recycled in a cardamom plantation in a year, the nutrients are mainly in the organic form and are available to the crop only by the process of mineralization. As the rate of mineralization is always low, the nutrients that become available to the crop will be able to sustain only average growth and production.

Cardamom is generally grown in the rich fertile soils of the forest eco-system. Being a perennial crop, sucker production is noticed throughout the year, while initiation of panicles and development of capsules is spread over a period of eight to nine months in a year. A steady absorption and utilization of plant nutrients take place throughout the life cycle of cardamom and, hence, a regular application of nutrients should be followed for higher yields. Application of compost or farm yard manure @ 5 kg/plant/year in May-June is recommended. Application of nutrients as organic manures such as FYM and neem cake for N; bone meal and rock phosphate for P as well as wood ash for K is effective in increasing the yield and quality of small cardamom.

Application of organic manures such as neem cake @ 1 kg or poultry manure/ farm yard manure/compost/vermicompost @ 2kg per plant is recommended once in a year during May-June. Mussorie rock phosphate or bone meal may be applied based on soil analysis. Nearly 70 per cent cardamom roots are confined to shallow depth of 5-40 cm and 30-50 cm radius, and therefore, for the maximum efficiency of applied manures, it would be necessary to apply at a radius within 50 cm, and being a surface feeder, deep placement of manures is not advisable. In zinc deficient soil, foliar micronutrient application during panicle initiation will enhance yield.

3.7 Plant Protection

3.7.1. Pest Management

Pest problem is a major production constraint in cardamom. Hence, management of pest infestation at various stages of crop is an important factor, which contributes to the high cost of cultivation and low yield in cardamom. As many 60 insect and non-insect pests attack cardamom, but only five or six are the major problematic ones. The most important pests are thrips, shoot and capsule borers, root grubs, hairy caterpillar and shoot fly.

a) **Cardamom thrips (*Sciothrips cardamomi* Ramk):**The cardamom thrips is the most destructive pest of cardamom. This pest causes 10-90 per cent damage to the crop in various areas. The adults and larvae lacerate and feed on leaves, shoots, inflorescences and capsules. When panicles are infested, it results in shedding of flowers and immature capsules. The feeding activity on tender capsules results in the formation of corky, scab-like encrustations on them. These later become malformed and shriveled with slits on the outer skin; the seeds within them are poorly developed and lack the usual aroma. The population of thrips is generally high during summer months and declines with the onset of rains. As cultural control measures, removal of dry drooping leaves as well as dry leaf sheath (trashing) during January/February, destruction of collateral host plants, and detrashing and weeding to reduce thrips infestation are recommended. Releasing *Chrysoperla zastrowi sillemi* @ 2 larvae/ plant in early stage of the plant and four larvae/plant in later stage is recommended as biological control measure.

b) **Shoot and capsule borer (*Conogethes punctiferalis* Guen.):** The shoot and capsule borer is a serious pest in nurseries and main cardamom plantations. The larvae bore into pseudo stems and feed on the internal contents leading to formation of 'dead hearts' symptoms. The larvae also bore into the panicles resulting in drying up of the same and into the capsules feeding on the seeds, thus, making capsules empty. Yield losses of 70-80 per cent have been recorded in severely infested plantations. As cultural control measures, sowing of castor seeds @0.4-0.8 kg/acre as trap crop in open areas/boundary, rogueing and destruction of infested tillers during September-October are recommended. Collecting and destroying all castor inflorescences with capsules infested by shoots and capsule borer, using pheromones in the monitoring of the pest and correct timing of application of biorationals are the control measures recommended. Also clip the inflorescence/flower parts of alternate hosts *viz., Alpinia speciosa, A.mutica, Amomum ghaticum, A. pterocarpum, Curcuma heilyherrensis,* and *Hedydium ceranarium* during off season (December to May). Apply *Bacillus thuringiensis* when early-instar larvae are found in capsule or panicle or unopened lead buds *i.e.,* within 20 days of adult moth emergence.

c) **White grub/root grub (*Basilepta fulvicornis* Jacoby):** The grubs feed on the roots in the form of irregular scraping. In the advanced stages, entire root system is found damaged resulting in drying and rotting depending on the season of attack. In the severely infested plants, leaves turn yellow and dry. As cultural control measures, avoid planting of jack, mango, fig *etc.* as shade trees as these trees are alternate hosts of the pest. Mulch plant base with leaves of wild *Helianthus* sp. to prevent egg laying of adult beetles. Adopt earthing up and detrashing. Irrigating @15–20 l water per plant reduces root grub population. Set up light trap @ one/ acre as mechanical control measure. Application of local strain of EPN (*Heterorhabditis indica*) @ 1,00,000 nematodes (IJS)/plant is the biological measure that can be adopted.

3.7.2. Disease Management

The major fungal diseases causing considerable crop loss are 'azhukal' (capsule rot), rhizome rot (clump rot), seedling rot (damping off), leaf blight, leaf blotch and nursery leaf spots.

a) **Capsule rot (Azhukal disease):** Besides capsules (fruits), the disease can also be seen in other parts of plant such as leaf, panicle, inflorescence and rhizome. Large circular or irregular water soaked lesions on leaves or capsules are the first visible symptoms. These lesions enlarge and affected portion rot. In some instances, long lesions extending over the entire length of the leaf were observed. The exposed portions of tender, unopened leaves remain rotted and decomposed without opening within two to four days. The leaves become shredded and remain attached to the pseudostem. Infected capsules show water soaked spots, which enlarge and turn grayish brown. Such affected capsules rot and drop down prematurely during rains emitting foul smell. *Phytophthora meadii* and *P. nicotianae* var. *nicotianae* are the primary fungi associated with the disease. Spraying of 1 per cent Bordeaux mixture is recommended to control the disease. Since the occurrence and spread of 'azhukal' disease is directly correlated to weather conditions, an integrated disease management strategy involving plant sanitation and biological methods are more suitable. Plant sanitation including removal and destruction of diseased plant parts and regulation of shade in plantation are to be carried out before the onset of monsoon. The first round of prophylactic fungicidal application should be done before the onset of monsoon, usually in May. Since water stagnation aggravates disease, better drainage in the plantation should be assured. Plant sanitation coupled with timely application of Bordeaux mixture (1 per cent) two to three times per year effectively controls the disease.

Application of biocontrol agents such as *Trichoderma viride* and *T.harzianum* twice during May and July with one kg of multiplied biocontrol agent in carrier media having a cfu of 23×10^8 g is successful in the management of the disease under field conditions. Apply 100 g product/plant along with neem cake (0.5 kg/plant) and 5 kg FYM/plant.

b) **Rhizome/root rot:** This disease, referred to as clump rot also, is reported from both Karnataka and Kerala states. The disease seen in nurseries is referred to as damping off. The disease affects roots and rhizomes causing decline and death of plants. Incidence of disease up to 20 per cent has been reported. Affected plants show pale and yellow foliage leading to premature death of older leaves. Root and rhizome of affected plants rot. Rotting symptom develop at the collar region as it turns soft and brownish. Aerial stem of the affected clump breaks off or comes off with a gentle pull. In nursery, soon after emergence, seedlings wither away causing damping off. Seedlings of 2-4 leaf stage show root rot and wilting and subsequent drying of leaves.The disease is caused by *Pythium vexans, Rhizoctonia solani* and *Fusarium oxysporum*. In a few cases, nematode (*Meloidogyne incognita*) also causes damage to root and in combination with *P.vexans* and *R.solani,* the damage is more. Similarly, root grub infestation is also reported to pre-dispose the plant to infection by *P.vexans* and *R.solani.* All these pathogens are soil borne and disease appears with the onset of south-west monsoon during June.

Removal of infected plants and damage of nursery site would reduce inoculum build up. Raising nursery in solarized soil is recommended to ensure pathogen free seedlings. Soil drenching with 1 per cent Bordeaux mixture or neem oil cake @ 500g/plant is effective in managing the disease. Application of Biocontrol agents such as *Trichoderma harzianum* and *T.viride* also reduces disease incidence. In order to get effective control, integration of methods such as phytosanitation, providing good drainage and soil application of Biocontrol agents is necessary.

c) **Leaf blight:** In recent years, leaf blight caused by foliar infections by *Phytophthora meadii* and *Colletotrichum gloeosporioides* is gaining importance. The disease appears during mid-monsoon, becomes severe during late monsoon periods (October-November) and decline by March. The symptoms develop as brownish spots and patches on the leaf lamina, often extend to large continuous areas and finally these portions or the entire affected leaves shred and dry. The disease is found only in certain localities. It has been observed that intermittent rain and continued mist formation in plantations favours the incidence and spread of the disease. The disease can be controlled by one or two rounds of Bordeaux mixture spray (1 per cent).

d) **Viral diseases:** Cardamom is a perennial herbaceous rhizomatous crop and it is susceptible to attack of viral disease at any stage of the crop. Because of perennial nature of the spice, infection by viral diseases reduce crop to uneconomical level within 1-3 years of infection.

   i) **Mosaic (Katte or marble) disease:** *'Katte'* or mosaic or marble virus disease caused by *Cardamom mosaic virus* (CdMV) has been responsible for low yields and rapid decline of cardamom in Guatemala, India and Sri Lanka. In India, *katte* is widely distributed in all cardamom growing tracts with incidence ranging from 0.01-99 per cent. Loss in yield due to the disease depends on growth stage at the time of infection. In early infected plants loss will be almost total. Late infection results in gradual decline in productivity.

   ii) **Cardamom vein clearing virus (Kokke Kandu):** The disease is reported from Kodagu, Hassan, Chikmagalur, Shimoga and North Canara districts of Karnataka. Because of its characteristic symptom, it is locally referred to as *Kokke Kandu*, meaning hook-like tiller. This disease may occur either singly or mixed infections with *katte*. Unlike *katte* disease, the plants affected with this disease decline rapidly with yield reduction up to 62-84 per cent in the first year of peak crop.

Viral Disease Management

Management of viral diseases is much more difficult than that of diseases caused by other pathogens as viral diseases have a complex disease cycle, efficient vector transmission and lack of effective chemical (Viricide). Most of the procedures that can be used effectively involve measures designed to reduce sources of infections inside and outside the crop, to limit spread by vectors and to minimize the effect of infection on yield.

3.8 Harvesting and Post-harvest Operations

Cardamom plants normally start bearing capsules two or three years after planting suckers or seedlings, respectively. The capsules ripen within a period of 120-135 days after its formation. Harvesting or picking commences from June-July and continues till January-February in Kerala and Tamil Nadu while in Karnataka, harvesting begins in August and prolongs till December-January. Usually harvesting is done at an interval of 15-30 days. The capsules are to be harvested when they attain physiological maturity, which is indicated by dark green colour of rind and black coloured seeds. Harvesting of ripened capsules leads to the loss of green colour and also causes splitting during curing process, while immature capsules on processing yields uneven sized, shriveled and undesirably coloured produce, and therefore, only mature capsules are to be harvested. After harvesting, the fresh capsules are to be washed in water to remove the soil particles and other dirt adhering to it and to get good quality produce. The capsule should be processed within 24-36 hours after harvest to prevent deterioration, and hence, storage of capsules after harvest for longer duration should be avoided.

Curing of cardamom is the process by which moisture of freshly harvested capsules is reduced from 80 to 10-12 per cent through indirect heating. Maturity of capsules and curing temperature influences the colour and quality of processed cardamom. Processed capsules with parrot green colour fetch a premium price in foreign countries. Hence, emphasis is to be given on the preservation of green colour during curing and subsequent storage. During curing, a temperature range of 40-45°C is maintained during all the stages of drying which helps in good retention of green colour. Gradual increase of drying temperature to 50-60°C in the last two hours of curing enables easy removal of floral remnants during polishing. During curing, if temperature exceeds the threshold levels, capsules develop brownish streaks due to heat injury. An increase in drying temperature also results in loss of oil from the seeds. Cardamom is dried by adopting two methods:

a) **Natural (Sun drying):** Freshly harvested capsules are directly dried under sun for a period of five to six days or more depending on the availability and duration of sunlight. This method does not retain green colour of capsules and also leads to splitting of the capsules. In general, sun dried capsules are not preferred for export.

b) **Flue curing:** It is one of the best methods of drying by which high quality green cardamom can be obtained. Processing of capsules is done in specially built curing houses. The washed and cleaned capsules are to be spread on wire net trays in the curing chamber. The required heat is produced by burning firewood in the iron kiln. The heat, thus, produced is passed through pipes made of galvanized iron sheets. The process of drying takes about 18-24 hours, depending on the ambient temperature. The capsules are spread thinly in the wire net trays and stirred frequently to ensure uniform drying. They are initially heated at 50 °C for the first four hours and heat is then reduced to 45 °C by opening ventilators and operating exhaust fans till the capsules are properly dried. Finally the temperature is raised to 60 °C for an hour.

Dried capsules are to be polished either manually or with the help of machines. Polishing is carried out by rubbing the dried capsules in hot state against a hard surface. The polished capsules are graded based on the quality parameters such as colour, weight per volume, size and percentage of empties, malformed, shrivelled and immature capsules. After grading, cardamom capsules are to be stored at a moisture content of less than 10 per cent to retain the original parrot green colour, to avoid exposure to moisture and to prevent mould growth. Use of 300 gauge black polythene lined gunny bags improves efficiency of storage. It is advisable to store the dried cardamom in wooden boxes at room temperature, preferably in the curing houses.

4. Nutmeg

Among tree spices, nutmeg (*Myristica fragrans* Houtt) belonging to the family *Myristicacae* is the most important, providing two spices, *i.e.* the nutmeg (dried seed) and the mace (dried aril) covering the seed. The latter is more important. A native of the Moluccas (The Banda Islands of East Indonesia), nutmeg is presently cultivated in several countries such as Grenada, Malaysia, Papua New Guinea, Sri Lanka, Fiji, some Caribbean Islands, Africa and China including India. Of late, Brazil also started its cultivation. Seedlings start flowering from four to seven years and the population consists of males and females with a low proportion of bisexual trees. Flowers are born in leaf axils of current season shoots. Male flowers are in clusters of more than three, whereas, female flowers are in clusters of two to three or single. Fruit is a drupe, spherical or slightly ovoid or pear shaped, pale yellow in color with a longitudinal groove in the centre through which it splits open when ripe exposing the red aril. The tree exhibits orthotropic and plagiotropic growth pattern capable of growing up to 20 m or more and are long lived even more than 200 years. Only the female and the bisexual trees bear fruits, but trees are obligatory cross pollinated and hence, male trees are essential. Only female trees give good yields. A 15 years old good tree is projected to yield around 2000 fruits. The major importers are USA, Canada, and Japan and European countries. In USA, the oil is generally preferred whereas, the Europeans consume it as whole spice. The area under nutmeg in India is estimated to be 21,110 hectares with a production of 14,400 tonnes (2014-15). It is mostly grown in Karnataka, Kerala, Tamil Nadu and Andaman and Nicobar Islands.

Traditionally nutmeg has been cultivated organically especially in the river basins where the rich alluvial soils are very conducive for its growth and

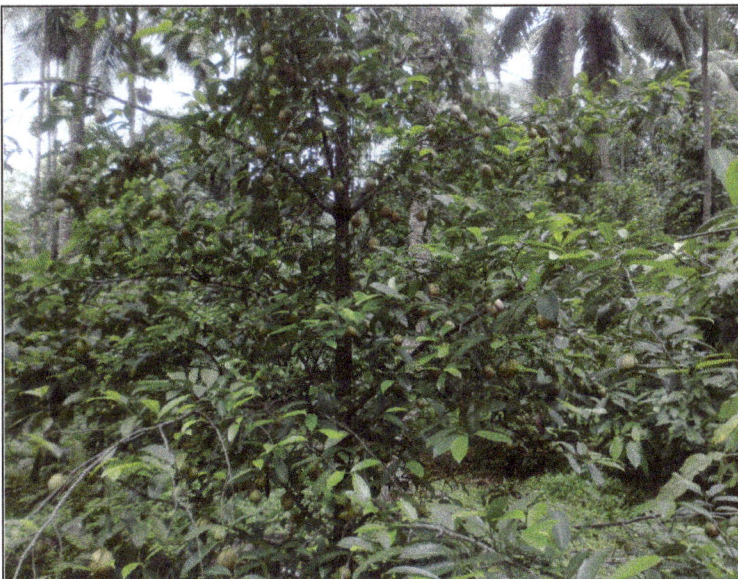

Figure 8.3: Nutmeg as Mixed Crop in Coconut.

productivity. Kalady in Ernakulam district of Kerala is the forerunner in nutmeg cultivation in India where huge trees exist even today. Nutmeg is cultivated generally as a homestead crop or intercrop in coconut/arecanut gardens. Under such situations, all the crops are to be cultivated following organic farming principles.

4.1. Conversion Period and Isolation

For an existing plantation, a minimum of three years is required as conversion period for organic cultivation. For a newly planted or replanted area raised through organic cultivation practices, the first yield itself can be considered as organic produce. In the case of cultivation on virgin land and or farms where records are available that no chemicals were used previously, the conversion period can be relaxed. It is desirable that organic method of production is followed in the entire farm, but in large scale plantations, the transition can be phased out for which a conversion plan is to be prepared. In order to avoid contamination of organically cultivated plots from neighbouring farms, a suitable buffer zone is to be maintained. An isolation distance of at least 25 m width is to be left from all around the conventional plantation. The produce from this isolation belt shall not be treated as organic. In sloppy lands adequate precaution should be taken to avoid the entry of run off water and drift from the neighbouring farms. All relevant measures should be taken to minimize the contamination from out side and within the farm. Accumulation of heavy metal and other pollutants should be avoided. For protected structure coverings, plastic mulches, fleeces, insect netting and silage wrapping, only products based on polythene and poly propylene or other polycarbonates are allowed. These shall be removed from the soil after use and shall not be burnt on the farm land. The use of polychloride based products is prohibited.

4.2. Varieties

Three improved varieties of nutmeg have been released in India for commercial cultivation, *viz.*, Konkan sugantha, a single plant selection from local seedlings population with conical tree canopy compact and adapted to the Konkan region of Maharashtra with an yield of 200-526 fruits per tree. Konkan Swad is another selection from seedlings from Ratnagiri with erect canopy conical shape recommended for the Konkan Region of Maharashtra with an average of 760 fruits per year, Viswashree, a high yielding variety from ICAR-IISR, is recommended for the entire nutmeg growing areas of Kerala.

4.3. Climate and Soil Requirements

Nutmeg thrives well in the warm humid conditions with an annual rainfall of 1500 mm to 2500 mm. It grows from sea level up to about 1300 above MSL. Clay loam, sandy loam and laterite soils are ideal for its growth. Dry climate or water logged conditions are not suitable for its cultivation. River banks and sheltered valleys are best suitable for this crop. It is a shade loving crop under warm situations whereas, it can thrive well without shade at higher elevations.

4.4. Source and Type of Planting Material

At present the existing population of nutmeg consists of seedling origin with variation in yield and other attributes. Of late, grafts of high yielding elite lines are preferred and planted. Alternately, high yielding mother plants with desirable nut and mace characters can be used for preparing grafts in view of the shortage of elite planting material. When certified organic planting materials are not available, initially conventional planting materials shall be used. Nutmeg is propagated both by seeds and vegetatively by grafting or budding. Nutmeg crop is a dioecious plant and seed propagation is not in vogue. Fully ripe, tree- burst fruits from organic farms are to be used for raising nurseries. However, initially seeds from conventional farms may be used. Farmers can also use grafted plants wherein the rootstock and scions are collected from organically cultivated area. *Myristica beddomei* and *malabarica* are related to nutmeg and also can be used as rootstocks besides nutmeg. For raising seedlings for planting or as rootstocks, naturally split healthy fruits are harvested during June-July and the seeds are extracted and sown immediately.Since the orthotropic and plagiotropic shoots can be used for grafting, the resultant plants have a different growth pattern. Orthotropic shoots give rise to erect growing plants, whereas, plagiotropic (lateral) shoots give rise to bushy spreading plants. However, these may be induced to develop orthotropic shoots by bending above 90 degrees. However, bush grafts are advantageous for high density crops.

4.4.1 Epicotyl Grafting

Grafts of elite high yielding trees can be prepared by epicotyl grafting instead of planting seedlings. For this purpose, the seed sprouts are extracted from the bed when the first leaf begins to appear and the epicotyl having a diameter of 0.5 cm or more with sufficient length to give a cut of three cm long and are used as rootstocks. Tender top shoots with two to three leaves, collected from high yielding trees can be used as scion. The stock and scion should have approximately the same diameter. A 'V' shaped cut is made in the stock and the base of the scion is tapered and is fitted carefully into the cut. Bandaging the grafted region may be done with polythene strips. The completed grafts are planted in poly bags of 25 cm x 15 cm size containing potting mixture of soil, sand and cow dung (3:3:1). The scion is covered with a polythene bag and kept in a cool shaded place protected from direct sunlight and well cared for. Those grafts showing sprouting of the scion are opened by removing the polythene bags and kept in shade for further development. The polythene bandage covering the grafted portion can be removed after three months. One year old grafts can be planted in the field.

4.5. Planting and after Care

Planting in the main field is done at the beginning of the rainy season. Pits of 0.75 m x 0.75m are dug at a spacing of 9 m x 9 m for seedlings and 5 m x 5m for bush/lateral grafts and filled with organic manure and soil 15 days before planting. If male trees are not available, one male graft has to be planted for every 10 female grafts. Alternately, one male branch can be grafted to each female graft to provide enough pollen. The plants should be shaded to protect them from sun scorch during early stages in summer. Permanent shade trees are to be planted when the site is on hilly slopes and when nutmeg is grown as a mono crop. It is grown as mixed crop in homesteads also. As a pure crop, it has been found to grow without shade in high elevations such as in Yearcaud (1000 m MSL), Tamil Nadu.

Nutmeg can best be grown as an intercrop in coconut gardens more than 15 years old where shade conditions are ideal especially along river beds and adjoining areas. Banana may be planted in between nutmeg plants to provide temporary shade and congenial micro climate. Banana and coconut should also be cultivated organically.The feeder roots are confined to the top 20-30 cm of soil and the tap root goes below two meters in nutmeg. Irrigation once in 15 days with 50 l of water per tree during summer helps build a dense root system and promote good growth.

4.5.1 Mulching

No inter-cultivation is done for nutmeg because most of the feeder roots are on the surface. Mulching with dry leaves or other organic wastes may be done to conserve moisture.

4.5.2. Soil and Water Conservation

Soil and water resources should be handled in a sustainable manner and appropriate measures are to be adopted for their conservation. In sloppy land, the soil may be disturbed to the minimum by any agricultural operations adopted. Clearing of land through the means of burning organic matter, *e.g.*, slash and burn, straw burning shall be restricted to minimum and the clearing of primary forest should be avoided.

4.5.3. Top Working

Wherever seedlings have been established and trees are poor yielders or predominantly male, top working by budding or grafting can be adopted to get desired trees. In this case, after ascertaining sex of the plants, budding can be resorted directly if they are young, whereas, for grown up trees, these have to be beheaded at the beginning of the rainy season and budding or grafting can be taken up on the coppice shoots during August-September. Instead of keeping 10 per cent male trees, one can graft a few male branches to each female tree to provide necessary pollen.

4.6. Nutrient Management

The organic manurial requirement has not been standardized in nutmeg. However, application of well rotten cattle manure or compost @ 15 kg/plant/year during May/June in the initial years is recommended. The quantity may be increased gradually so that a well grown tree of 15 years and more gets 40 to 50 kg of organic manure. The manure should be free from heavy metals. It is generally experienced that nutmeg is susceptible to potassium deficiency resulting in immature splitting of fruits. In such cases, potash deficiency may be replenished by addition of natural source such as wood ash or granite dust or sulphate of potash.

4.7. Plant Protection

Organic farming system should be carried out in such a way so as to minimize the losses from pests, diseases and weeds. Emphasis is given on the use of a balanced nutritional programme, use of crops and varieties which are tolerant to or resistant and well adapted to the environment, fertile soils of high biological activity, crop rotations, companion planting, green manures *etc*. Weeds can be slashed and applied as mulch.

4.7.1. Pest Management

The black scale (*Saissetia nigra*) infests tender stems and leaves especially in nursery and sometimes on young plants in the field. These are clustered together and are black, oval and dome shaped. They feed on plant sap and severe infestation cause the shoots to wilt and dry. Another pest is the white scale (*Pseudaulacaspis cockerelli*) which is grayish white flat and shaped like a fish scale and occurs clustered together on the lower surface of the leaves especially in nursery seedlings. The pest infestation results in yellow streaks and spots on affected leaves and in severe infestations the leaves wilt and dry. Occasionally Mealy bugs and Scale insects (*Saissetia nigra*) infect tender leaves and shoots. The shield scale (*Protopulvinaria mangiferae*) is creamy brown and oval and occurs on tender leaves and stems especially in nursery seedlings. The infestation results in wilting of leaves and shoots. These pests are to be controlled using approved botanical pesticides whenever necessary.

4.7.2. Disease Management

Die back, thread blight, fruit drop and shot hole are the major diseases reported to occur in nutmeg.

4.7.2.1. Die Back

The disease is characterized by drying up of mature and immature branches from the tip downwards. *Diploma* sp. and a few other fungi have been isolated from such trees. The infected branches should be cut and removed and the wounds are treated with Bordeaux paste.

4.7.2.2. Thread Blight

Two types of blights are noticed in nutmeg. The first is a white thread blight wherein fine whitish hypae aggregate to form fungal threads and traverse along the stem underneath the leaves in a fan shaped or irregular manner causing blighting of affected portions. The dried up leaves with the mycelium forms a major source of inoculum for the spread of the disease. The disease is caused by the fungus *Marasmius pulcherima*. The second type of blight, called as horse hair blight, is caused by *Corticium equicrinus*.Fine black silky threads of the fungus form an irregular loose network on the stems and leaves causing blighting of the leaves and stems. However, these threads hold up the detached dried leaves on the tree giving the appearance of a birds nest when viewed from a distance. The above diseases are very severe under heavy shade. They can be managed by adopting phytosanitation and shade regulation. In severe infections, Bordeaux mixture (1 per cent) may be sprayed after removal of severely infested branches.

4.7.2.3. Fruit Rot

Immature fruit split, fruit rot and fruit drop are highly prevalent in nutmeg. Immature fruit split and shedding are noticed in some trees without any apparent infection. In the case of fruit rot, the infection starts from the pedicel as dark lesions and gradually spreads to the fruit, causing brownish discolourations of the rind resulting in rotting. In advanced stages, the mace also rots emitting a foul smell. In addition to these, occasional dark sunken lesions, dark scabbing, mostly restricted to the outer layers of the pericarp without affecting the mace have also been noticed. *Phytophthora* sp. and *Diplodia natalensis* have been isolated from affected fruits. However the reasons for fruit rot could be both pathological and physiological. Bordeaux mixture 1 per cent may be sprayed when the fruits are half mature to reduce the incidence of the disease.

4.7.2.4. Shot Hole

The disease is caused by the fungus, *Colletotrichum gloeosporioides*. Necrotic spots develop on the lamina which is restricted by chlorotic halo. In advanced stages, the necrotic spots become brittle and fall off resulting in shot holes. A prophylactic spray with Bordeaux mixture (1 per cent) is effective against the disease.

4.8. Harvest and Post-harvest Operations

The female nutmeg tree starts fruiting from the sixth year, though the peak period is reached after 20 years. The fruits are ready for harvest in about nine months after flowering. The peak harvesting season is during June to August. Split opening of pericarp indicates that the fruits are ripe and ready for harvesting. Remove the outer flesh portion after harvest and separate the mace manually. Later on dry the mace and nut separately under sun. The scarlet coloured mace turns to yellowing brown and brittle when drying is over.

5. Clove

Clove is a tropical plant requiring warm humid climate. Clove plantations in India are reported to have originated from a few seedlings obtained originally from Mauritius. It is one of the ancient valuable spices of the orient, holding a unique position in the international trade. The East India Company in its 'spice garden' in Courtallam, Tamil Nadu, first introduced clove to India around 1800 AD. Induced by the success of its introduction, cultivation of clove was extended during the period after 1850 AD to Nilgiris, southern region of the erstwhile Travancore State and the slopes of Western Ghats. The important clove growing districts in India now are Nilgiris, Tirunelveli, Kanyakumari, Nagercoil and Ramanathapuram (Tamil Nadu); Kozhikode, Kottayam, Kollam, and Thiruvananthapuram (Kerala) and South Kanara (Karnataka). It is also grown in Andaman and Nicobar islands. As per the estimates for 2014-15, the total area of 2,380 hectares under clove cultivation in India is spreading over 1,070 hectares in Kerala, 1,030 hectares in Tamil Nadu, 160 hectares in Andaman and Nicobar islands and 120 hectares in Karnataka. The total production was 1260 m tonnes during 2014-15.Tanzania, Indonesia, Madagascar, Comoros and Sri Lanka are the major clove exporting countries. In recent years,

Figure 8.4: Yielding Clove Plant.

world production of clove averaged around 80,000 tonnes a year. Indonesia is the world`s biggest producer at 50,000-60,000 tonnes.

5.1. Climate and Soil Requirements

Clove is a tropical plant and requires warm humid climate. Generally it is believed that clove requires proximity to sea for the proper growth and yield, experience in India has shown that the trees do well in the hinterland conditions too. Clove thrives in all situations ranging from sea level up to an altitude of 1000 m. Deep loamy soil with high humus content found in the forest region is best suited for its cultivation. It grows satisfactorily on laterite soil, loamy and rich black soil having good drainage.

5.2. Varieties and Planting Material

Clove is propagated through 'mother of clove', that is seed obtained from ripened fruit. Fruits are taken from trees with more than 15 years of age and regular yielding nature. They are allowed to ripe on the trees and to drop down naturally. Such fruits are picked up from the ground and sown directly in the nursery. Otherwise fruits are soaked in water overnight and the seeds obtained after removal of the pericarp are sown.

5.3. Nursery Practices

Raised nursery beds are prepared on fertile soil with high percentage of organic matter. The beds normally measure one metre width and two to three metres length. Seeds should be placed flat at a depth of about 2.5 cm with a spacing of 12 to 15 cm. Care should be taken to prevent leaching of the beds in rain. Germination commences in about 10 to 15 days and completes by about 45 days. The slender and delicate seedlings grow very slowly.

5.4. Site Selection and Planting

The site for cultivation of clove should have good drainage since the crop cannot withstand water logging. It can be grown in coconut gardens of midland. At higher elevations, it can be mix cropped with pepper or coffee. Clove requires a location protected from wind. If the site is open, wind breaks must be provided. Eastern and North Eastern hill slopes, well-drained valleys and riverbanks are ideal for clove cultivation. The crop thrives well under open condition at high altitude where there is fair distribution of rainfall. The area selected for raising clove plantation is to be cleared of wild growth before monsoon.

5.5. Nutritional Management

Clove trees are to be supplied with nutrients regularly for proper growth and flowering. About 15 kg of rotten cattle manure or compost is to be applied per plant in the initial years. The quantity needs to be increased gradually so that a well grown tree of 15 years and more gets 40 to 50 kg of organic manure. Foliar spray of 0.5 per cent zinc sulphate during flower initiation will enhance yield.

5.6. Plant Protection

5.6.1. Pest Management

There are only a few pests attacking clove. Among them, stem borer, scales and mealy bugs are important.

5.6.1.1. Stem Borer (*Sahyadrassus malabaricus*)

This is the most important pest of clove. The caterpillars bore into the main stem resulting in immediate drying up of the plant above the point of attack and causing the death of the plant ultimately. Regular inspection of the plants and pouring a solution of 0.3 per cent neem based insecticide into the bore hole and plugging the opening as soon as the attack is noticed, will check the damage.

5.6.1.2. Scales (*Lecanium psidii*) and Mealy Bugs (*Planococcus* sp. *Psuedococcus* sp.)

Damages due to mealy bugs occur by sucking the sap from tender shoots. Affected portions dry up gradually. Infestation of scales is on leaves and tender shoots, and is serious in the nursery. Young seedlingsm if attacked, are killed soon. Spraying with 0.3 per cent neem based insecticide will control these pests.

5.6.2. Disease Management

Diseases are more damaging to clove than pests. The, important diseases are seedling wilt, leaf rot, leaf spot, twig blight, die back and sudden death. Bordeaux mixture (1 per cent) may be sprayed along with phytosanitation measures to reduce the incidence of diseases.

5.7. Harvesting and Post-harvest Operations

Clove tree begins to yield from the seventh year of planting and full bearing stage is attained after 15 to 20 years. The flowering season is September to October in the plains and December to February at high altitudes. Flower buds are formed on young flush. It takes about five to six months for the buds to become ready for harvest. The optimum stage for picking clove buds is when the buds are fully developed and the base of the calyx has turned from green to pink colour. Such clove buds are carefully picked by hand. Care should be taken to collect the buds at the correct stage as otherwise the quality of the produce will be poor to a considerable extent. The buds after separation from the stalks are spread evenly to dry, in-the sun on mats or cement floors. During nights, buds should be stored undercover, lest they re-absorb moisture. The period of drying depends on the prevailing climatic conditions. Normally, it is possible to dry cloves in four or five days under direct sun and in about four hours when they are heated on zinc trays over a regulated fire. Fully dried buds develop the characteristic dark brown colour and are crisp. Improperly dried and stored cloves have much darker colour and somewhat wrinkled appearance. Such a produce is considered inferior in quality.

6. Garcinia (Kudampuli)

Garcinia, the camboge (*Garcinia gummi-gutta* var. gummi-gutta) tree is a big sized glabrous and evergreen forest tree commonly seen in the Western Ghats of Kerala, Karnataka, in India and also in Sri Lanka. The tree is very much adapted to hill tops and plain lands alike. But its performance is best in river banks and valleys. It grows well in dry or occasionally waterlogged or flooded soils. The economic part of the plant is its mature fruit, which is highly acidic.

Figure 8.5: Garcinia Yielding Plant.

6.1. Production of Planting Materials

Grafts prepared through soft wood grafting or side grafting and healthy seedlings raised in the nursery are used for cultivation. If seedlings are planted, 50-60 per cent will be male; and female takes 10-12 years for bearing. Hence, planting of grafts is advocated as they ensure maternal characters including early bearing tendency.

6.1.1. Propagation by Seedlings

To select mother trees, locate trees that give a steady annual yield with a mean fruit weight of 200-275 g, high acid and low tannin content. Collect seeds from freshly harvested and fully ripe fruits and wash in running water and spread in a thin layer under roof. By the 20th day, seeds will be ready for sowing. Sow seeds @ two per bag in polybags during August-September. Usually seeds start sprouting during December, but the sprouts become visible above the soil surface only by February. In order to avoid delay in germination, simple seed treatment methods can be employed. In this method, at first the seeds are dried under shade, and later the seed coats are removed using a sharp knife without injuring the ivory coloured cotyledon. Sow these ivory coloured cotyledons afresh in polybags at a depth of 3 cm. Germination starts in 20-25 days after sowing. Keep the seedlings under shade. Irrigate them regularly on alternate days during summer months. After 3-4 months, place the seedlings under direct sunlight to trigger robust growth. At this age, apply FYM @ 50 g per bag. In six to seven months time, seedlings will be ready for planting.

6.1.2. Propagation by Soft Wood Grafting

Collect scions only from specific elite trees having regular bearing habit, which produce high yield of large and quality fruits. Select straight growing, healthy, young shoots emerging from the primary branches with whorled leaf arrangement, cut them to a length of 6-10 cm and store in polybags under humid condition. Remove leaves partly and shape the cut end to a wedge of 3-4 cm length by giving slanting cuts on two opposite sides. Stock-plants having 3-4 mm stem thickness are ideal for grafting. Behead the selected plants at two nodes below the terminal bud and remove all the leaves at the graft union. Insert the wedge of the scion into the cleft made on the rootstock and secure the graft joint firmly with a black polythene tape, 1.5-2 cm wide and 30 cm long. After grafting, cover the plants with a transparent polypropylene cover and keep under shade. Within one month grafts will establish and new leaves will start emerging. Remove the polythene cover and keep under shade. Three months after grafting the plants will be ready for planting.

6.1.3. Propagation by Approach Grafting

Here stock plants having 3-4 mm thickness are preferred and they are brought to the place where the mother tree is located. Grafting is done as in other crops and is kept intact for 45 days by which time union occurs. The main disadvantage is that only a limited number of grafts can be produced in this method. Forty-five days after grafting, they will be ready for transferring to the main nursery for hardening. One-year-old grafts can be used for field planting.

6.2. Planting and Aftercare

Prepare pits of size 1 m x 1 m at spacing of 10 m. Refill the pits with a mixture of topsoil and compost/FYM. Proper care should be given to avoid water stagnation in pits. The plants can be raised as a pure crop or as a mixed crop in coconut and arecanut gardens. Take pits of size 0.75 m x 0.75 m x 0.75 m in hard and laterite soils; 0.50 m x 0.50 m x 0.50 m in sandy and alluvial soils, at a spacing of 4 m x 4 m for grafts and 7 m x 7 m for seedlings. In slopes of 15 per cent or more, for planting grafts, rows are to be spaced at 5.0 m to 5.5 m and 3.5 m between trees in a row. For planting seedlings, rows are spaced at 8 to 12 m and at 6 to 8 m for trees in a row. Planting is generally done with the onset of monsoon showers. Under existing coconut plantation of 25 years and above, spacing shall be so adjusted that it should alternate with the palms in the rows. Under Kuttanad conditions (Kerala, India), where bunds and channels alternate, planting can be done in between two palms. The graft union shall remain just above the ground level. Provide support to the young plants. One month after planting, gently remove the polythene tape around the graft union.

6.2.1. Pruning

Grafts will grow fast from the second year onwards. Give strong support with casuarina poles at this stage. By fifth year, the tree will have 3 to 4 m height. At this stage, height of the plant may be maintained at 3.5 to 4 m and by seventh year at 4 to 4.5 m by pruning.

6.3. Nutritional Management

Apply 10 kg cattle manure or compost per seedling/graft during the first year. Gradually increase the quantity so that a well-grown tree of 15 years and above receives 50 kg of organic manure per year.

6.4. Plant Protection

Hard scales and beetles are found to infest the crop. Hard scales desap the leaves and tender shoots. Both the adult beetles and their grubs defoliate the crop inflicting heavy loss of yield. Control these pests by spraying neem based botanical pesticide. Incidence of hoppers is observed on grafts and large trees. This causes withering of leaves, drying up of branches and yield loss. Seedling blight in the nursery stage is very common. Control it by drenching nursery bed with 1 per cent Bordeaux mixture. Sometimes, fungal thread blights have been observed to cause leaf and twig blight. Adopt proper pruning and spray 1 per cent Bordeaux mixture.

6.5. Harvesting and Post-harvest Operations

Seedlings start bearing generally at the age of 10-12 years. Grafts start bearing from the third year onwards and will attain full bearing at the age of 12 to 15 years. Flowering occurs in January-March and fruits mature in July. There are reports of off-season bearers, which bear two times a year, *i.e.*, during January-July and September-February. Mature fruits, which are orange yellow in colour, drop off from the tree. Harvest mature fruits manually before they fall. Immediately after harvest, wash the fruits in running water and separate the fruit rind for processing. Separated fruit rind is first sun dried and then either smoke-dried or oven-dried at 70-80°C. In order to increase the storage life and to impart softness, mix the dried rind with common salt @ 150 g and coconut oil @ 50 ml per kg of dried rind.

7. Cinnamon

Cinnamon or 'sweet wood' (*Cinnamomum verum*) is the earliest known spice in India. It is native of Sri Lanka and Malabar Coast of India. It is grown in the Naga hills of Assam, Coastal hills of Karnataka and Western Ghats. The production is just sufficient to meet the internal demand. A few isolated trees of cinnamon are seen growing in Konkan region of Maharashtra also. Organic cinnamon is one that is grown without the use of harmful pesticides or prohibited synthetic fertilizers. Additionally, the cinnamon must be grown using farming practices that have the least amount of impact on the environment and must be harvested using fair-trade practices. The process of growing organic cinnamon starts from the seed. Cinnamon is harvested from the bark of several varieties of cinnamon trees. In order to be considered organic, the trees from which the cinnamon is harvested cannot have been grown from genetically modified seed. The cinnamon tree is then grown following specific fertilizer and pesticide criteria. Once the cinnamon tree is growing, the use of most synthetic fertilizers is prohibited for organic crops. Sewage sludge is sometimes used on nonorganic crops but is not used if the cinnamon harvest is to be considered organic. Nutrients in the soil of organic crops are maintained by farming practices such as crop rotation.

Figure 8.6: Cinnamon Plant.

7.1. Climate and Soil Requirements

Cinnamon requires hot and humid climate. Annual precipitation of 1500 mm to 2500 mm and average temperature of 27°C are ideal. It can be cultivated up to an elevation of 200 m from MSL. Prolonged spells of dry weather are not conductive for successful growth.

The quality of the bark is greatly influenced by soil and ecological factors. Well-drained soil rich in humus content is most suitable. Sandy loam soils liberally incorporated with organic manures are the best. Red dark brown soils free from rocky gravel are also good, for cinnamon cultivation.

7.2. Propagation

Cinnamon is commonly propagated through seed, though it can be propagated by cuttings and air layers. Under West Coast conditions, cinnamon flowers in January and fruits ripen during June-August. The fully ripe fruits are either picked up from the tree or fallen ones are collected from the ground. Seeds are removed from fruits, washed free of pulp and sown without much delay, as the seeds have a low viability. The seeds are sown in sand beds or polythene bags containing a mixture of sand, soil and well - powdered cow dung in a 3:3:1 ratio. The seeds germinate within 10-20 days. Frequent irrigations are required for maintaining adequate moisture level. The seedlings require artificial shading till they become 6 months old.

7.3. Planting and after Care

Pits of 50 cm x 50 cm x 50 cm cm are dug at a spacing of 3 m x 3 m. They are to be filled with compost and-topsoil before planting. Cinnamon is planted during June-July to take advantage of monsoon for the establishment of seedlings. One year old seedlings are planted. In each pit, five seedlings can be planted. In some cases, the seeds are directly dibbled in pits that are filled with compost and soil. Partial shade in the initial years is advantageous for healthy and rapid growth of plants. Watering of newly planted seedling is done profusely and periodically. In the first four years, weeding is to be done three to four times in a year. Subsequently one or two weedings are required. Seedlings grow to a height of two meters in seven years.

7.4. Coppicing

The cinnamon plants may attain a height of 10-15 m, but it is generally cut back or 'coppiced' periodically. When the plants are two years old, they are coppiced during June-July to a height of about 20 cm to 30 cm from the ground and the stump is then covered by earthing up. This helps development of side shoots from the stump. This is to be repeated for every side shoot developing from the main stem during the succeeding season so that the plant will assume the shape of low bush of about two m height and shoots suitable for peeling would develop in a period of about four years. The first coppicing can be done from the fourth or fifth year of planting.

7.5 Nutritional Management

An adult cinnamon tree should be given about 30 kg FYM and 4 kg neem cake per annum to get better growth and bark yield.

7.6. Harvest and Post-harvest Operations

The cinnamon plants will be ready for harvest in about three years after planting. Harvesting is done during two seasons, the first in May and second in November. The correct time for cutting the shoots for peeling is determined by noting the sap circulation between the wood and corky layer. This could be judged by making a test cut on the stem with a sharp knife. If the bark separates readily, the cutting could be taken immediately. Stems measuring 2.0 to 2.5 cm in diameter and 1.5 to 2.0 m length are cut early in the morning and twigs and leaves are detached. The outer brown skin is first scrapped off and the stem is rubbed briskly to loosen the bark. Two cuts are made round the stem about 30 cm apart and two longitudinal slits are made on opposite sides of the stem. The bark is separated from the wood with curved knife. The detached pieces of bark are made into compound quills. The best and longest quills are used on the outside while inside is filled with smaller pieces. The compound quills are rolled by hand to press the outside edges together and are neatly trimmed. They are dried in shade as direct exposure to sun can result in warping. The dried quills consist of mixture of coarse and fine types and are yellowish brown in colour.The quills are graded as Fine or Continental, Mexican and Hamburg or Ordinary. The Fine consists of quills of uniform thickness, colour and quality and the joints of the quills are neat. Mexican grades are intermediate in quality. The Hamburg grade consists of thicker and darker quills.

The lower grades are exported as: (a) *Quillings*: The broken lengths and fragments of quills of all grades are bulked and sold as quillings; (b) *Featherings*: This grade consists of the inner bark of twigs and twisted shoots that do not give straight quills of normal length.

Chips: This includes the trimmings of the cut shoots, shavings of outer and inner bark, which cannot be separated, or which is obtained from small twigs and odd pieces of thick outer bark.The yield varies with type of variety and age. Three to four year onwards, the tress will give 62 to 125 kg quills/ha and ranges up to 225 to 300 kg quills/ha from 10th year onwards.

8. Organic Certification for Spices

Under organic farming, processing methods also should be based on mechanized, physical and biological processes to maintain the vital quality of organic ingredient throughout each step of its processing. All the ingredients and additives used in processing should be of agriculture origin and certified organic. In cases where an ingredient of organic agriculture origin is not available in sufficient quality or quantity, the certification programme authorizes use of non organic raw materials subject to periodic re-evaluation. Labelling should clearly indicate the organic status of the product as "produce of organic agriculture" or a similar description when the standards requirements are fulfilled. Moreover, organic and non-organic products should not be stored and transported together except when labelled or physically separated. Certification and labelling is usually done by an independent body to provide a guarantee that the production standards are met. Govt. of India has taken steps to have indigenous certification system to help small and marginal growers and to issue valid organic certificates through certifying agencies accredited by APEDA and Spices Board. The inspectors appointed by the certification agencies will carry out inspection of the farm operations through records maintained and by periodic site inspections. Documentation of farm activities is must for acquiring certification especially when both conventional and organic crops are raised. Group certification programmes are also available for organized group of producers and processors with similar production systems located in geographical proximity.

Selected References

Anonymous, (1998). Concepts, Principles, Basic standards, Production guidelines, Documentation, Inspection and Certification. In: Production of Organic spices. Spices Board, Cochin.

Anonymous, (2004).Guidelines for production of organic spices in India. Spices Board, Cochin, India, pp:84 – 92.

Devasahayam, S., John Zacharaiah, T., Jayashree, E., Kandiannan, K., Prasath, D., Santhosh J. Eapen., Sasikumar, B., Srinivasan, V. and Suseela Bhai. R. (2015). *Black pepper* (Extension pamphlet) p.24.

Dinesh Kumar, M. and Babitha, J. (2006) Rates of leaf fall and nutrient recycling of shade trees in coffee (*Coffea arabica* L.), cardamom (*Elettaria cardamomum* Maton.) and black pepper (*Piper nigrum* L.) production systems of mudigere. Karnataka. *J. Spices and Aromatic Crops*, 15(2): 108-114.

Halberg, N., Alroe, H.F.,Kundsen,M.T.and Kristensen, E.S.(Eds).(2006). *Global development of organic agriculture: challenges and prospects*. CABI. p. 297. ISBN 978-1-84593-078-3.

Lotter, D.W. (2003) Organic agriculture. *Journal of Sustainable Agriculture* 21(4):59-128.

Rethinam, P. and Venugopal, K. (1994) "Cropping systems in plantation crops". In: Chadha, K.L. and Rethinam, P. (eds.). Adv. Hort. Vol.9: Plantation and Spice Crops, Part 1., Malhotra Publishing House, New Delhi. pp: 603-620.

Sadanandan, A.K. (2000) Agronomy and nutrition of black pepper. In: Black Pepper (*Piper nigrum*). Ravindran, P.N. (Ed.). Harwood Academic Publishers, Amsterdam, pp. 163-223.

Satyagopal, K., Sushil, S.N., Jeyakumar, P., Shankar, G., Sharma, O.P., Sain, S.K., Boina, D.R., Chattopadhyaya,D.M., Ram Asre., Kapoor, K.S., Sanjay Arya., Subhash Kumar., and Patni, C.S.(2014). AESA based IPM package black pepper.p.52.

Satyagopal, K., Sushil, S.N., Jeyakumar, P., Shankar, G., Sharma, O.P., Boina, D.R., Sain, S.K., Reddy, M.N., Rao, N.S., Sunanda, B.S.,Ram Asre, Kapoor, K.S.,Sanjay Arya,Subhash Kumar,Patni, C.S.,Jacob,T.K.,Santhosh J. E., Biju, C.N. Dhanapal, K., H. Ravindra, Linga Raju, Ramesh Babu, S. And Hanumanthaswamy, B.C. (2014). AESA based IPM package for Small cardamom. p: 43.

Stinner, D.H (2007). "The Science of Organic Farming". In William Lockeretz. *Organic Farming: An International History*. Oxfordshire, UK and Cambridge, Massachusetts: CAB International (CABI). pp. 40–72. ISBN 978-0-85199-833-6.

Willer, Julia Lernoud and Robert Home (2013).The World of Organic Agriculture: Statistics and Emerging Trends. Research Institute of Organic Agriculture (FiBL) and the International Federation of Organic Agriculture Movements (IFOAM, 2013).

Chapter 9

Organic Farming in Oil Palm

☆ P. Murugesan and V. Krishnakumar

1. Introduction

The *Elaeis* (Greek for "oil"), from the genus of palms, comprises two species, both called oil palms. The *Elaeis guineensis* Jacq, originally from West Coast Africa, which is the main source of palm oil; and the *Elaeis oleifera* (HBK) Cortes ("oil producing") originally from Central and South America. Though the genus *Elaeis* of the family Arecaceae contains two tropical species, only *E.guineensis* is of economic importance due to the high oil content in the mesocarp and in the kernel oil. The oil palm growing regions house vast areas of tropical rainforest rich in biodiversity on the continents of Asia, Africa and South America. The first cultivation of oil palm and use was started around 5,000 years ago and then Arab traders brought the oil palm to Egypt. It was not until the early 20th century that oil palm was planted commercially in South East Asia.

Oil palm produces two different types of oils: palm oil (from the pulp of fruit) and palm kernel oil. The oil palm provides one of the leading vegetable oils produced globally, accounting for 25 per cent of global consumption and approximately 60 per cent of international trade in vegetable oils. An estimated 74 per cent of global palm oil usage is for food products and 26 per cent is for industrial purposes. Palm oil is used in a wide variety of food products such as cooking oil, shortenings and margarine, while palm kernel oil is a raw material in the production of non-food products which include soaps, detergents, toiletries, cosmetics and candles.

For many tropical countries, oil palm is an economically important crop, fulfilling local demand for vegetable oil and generating large export incomes. It is grown by plantation companies and smallholder families, where the smallholders supply oil palm fruit to a centralised mill. As demand for vegetable oil increases, due to growing and increasingly wealthy populations, the industry is expanding rapidly onto new land and there is an increasing need for ecological intensification of production. Oil palm is among the most productive and profitable of tropical crops for bio fuel production. Production of biodiesel from oil palm has been increasing in recent years, particularly in Africa and Latin America.

2. Production Scenario

2.1. Global Scenario

Oil palms are restricted to the tropics and have mainly been cultivated in Indonesia, Malaysia and Thailand in Southeast Asia, Nigeria in Africa, Colombia and Ecuador in South America and Papua New Guinea in Oceania. Since 1980, palm oil production has increased tenfold and as per report of FAO, the global demand for palm oil will double in 2020, and triple by 2050. With global production of 62.4 million tonnes of palm oil in 2015, oil palm is grown in 43 countries of the world covering an area of 17 million ha of mature palm oil plantations across the equator. Although, it is planted on only 5 per cent of the total world vegetable oil acreage, palm oil accounts for 33 per cent of vegetable oil and 45 per cent of edible oil worldwide. Oil palm satisfies 30 per cent of the world edible oil and fat requirements with little fewer than seven per cent of the areas planted to oil crops. Indonesia and Malaysia lead the production front accounting for about 85 per cent, totaling to 53.3 million tonnes (Table 9. 1). The third largest producer of palm oil is Thailand, followed by Colombia, Nigeria, Eucador, Guatemala and Papua New Guinea. There has recently been an increase in palm oil production in South America via Colombia, Ecuador and Guatemala. The world average yield of palm oil is reported to be 12.2 tonnes fresh fruit bunch (FFB) per ha, and Guatemala, Nicaragua and Malaysia top the yield chart with 24.6, 24.3 and 21.1 tonnes per ha, respectively. Since 2006, Indonesia had exceeded Malaysia in producing oil palm, leading it to the biggest oil palm producing country in the world. Big Private Plantation dominates the total area of oil palm and Riau is the most oil palm contributor province in Indonesia.

Table 9.1: Oil Palm Production in different Countries of the World (2015)

Name of Country	Production (million tonnes)
Indonesia	33.4 (53.5)
Malaysia	19.9 (31.9)
Thailand	1.8 (2.9)
Colombia	1.2 (1.9)
Nigeria	0.94 (1.5)
Eucador	0.53 (0.9)
Guatemala	0.52 (0.8)
Papua New Guinea	0.5 (0.8)
Others	3.6 (5.8)
Total	62.39

Source: Oil World June 2016 data base. Figures in bracket is the percentage.

According to recent estimate by the United States Department of Agriculture (USDA) the estimated world palm oil production during 2016-17 will be 64.5 million tonnes (World Palm Oil production.com).

2.2. International Trade: Exports and Imports

The international palm oil exports by country during 2015 totaled US$29.1 billion. Among the continents, Asian countries accounted for the highest export of palm oil during 2015 valued at $25.2 billion (86.6 per cent). In the second place were European Union exporters at 6 per cent while 4.3 per cent of worldwide palm oil export originated from Latin American and the Caribbean. Indonesia was the leading country exporting palm oil to other consuming countries and they exported 52.9 per cent of their total production, worth US$15.4 billion and it was followed by Malaysia (32.7 per cent - $9.5 billion) and Netherlands (3.8 per cent - $1.1 billion). Exports are dominated by Indonesia and Malaysia, which account about 90 percent of the palm oil traded internationally. The three main importers, India, China and the European Union, account for slightly over half of total imports of palm oil (50.7 per cent).

2.3. Organic Palm Oil

As per the statistics available for 2011(SSI Report, 2014), the global production of organic certified palm oil fruit was 150,750 metric tonnes (Table 9.2). If 25 per cent of the palm oil fruit is considered to be composed of palm oil and that 6.5 per cent is composed of the palm kernel, organic palm oil accounted for approximately 38,000 metric tonnes, while organic palm kernel accounted for about 10,000 metric tonnes. Organic certified palm oil accounted for approximately 0.07 per cent of global palm oil production. Organic palm oil fruit production has fluctuated around the 150,000 metric tonne over the last three years, while certified organic area under cultivation has decreased considerably from 16,700 hectares in 2008 to 7,200 hectares in 2011. Ecuador and Colombia together represented 97 per cent of total organic palm oil fruit production in the world, with Colombia alone representing 89 per cent. Indonesia and Malaysia do not have any organic penetration in the palm oil sector.

Table 9.2: Organic Palm Oil Area Harvested, Production and Sale Volumes by different Countries (2011)

	Area Harvested (ha)	Production (mt)	Sales (mt)
Colombia	5,500	133,950 (88.9 per cent)	110,000
Côte d'Ivoire	100	1,100 (0.7 per cent)	400
Ecuador	1,000	13,000 (8.6 per cent)	10,000
Ghana	600	2,700 (1.8 per cent)	2,200
Total	7,200	1,50,750 (37,688 mt palm oil)	1,22,600 (30,650 mt palm oil)

Quoted from Palm Oil Market (SSI Review 2014).

In response to the urgent and pressing global call for sustainably produced palm oil, the Round Table on Sustainable Palm Oil (RSPO) was set up in 2003-04.

RSPO is multi stakeholder forum, bringing together seven sectors of palm oil, oil palm producers, palm oil processors, environmental and nature conservation NGOs and social or developmental NGOs – to develop and implement global standards for sustainable palm oil. Green palm is a certificate trading programme that allows consumers the flexibility to purchase sustainable palm oil certificates under the book and chain supply chain system. These certificates are issued to producers who are members of RSPO and certified to produce palm oil in a sustainable manner.

2.4. Indian Scenario

Oil palm was introduced into India as a small-holders' irrigated crop during 1989 to meet the growing demand for vegetable oils. India's vegetable oil economy is the world's fourth largest after USA, China and Brazil. With per capita consumption of vegetable oils @ 16 kg per year per person and for the present population of 1276 million, the total vegetable oil demand is likely to touch 20.4 million tonnes by 2017. A substantial portion of edible oil is met through import of palm oil from Indonesia and Malaysia. Palm oil has dominated Indian imports since the last two decades for its logistical advantages, contractual flexibility and consumer acceptance. India's palm oil consumption has increased from 13 per cent in 2007-2008 to 15 per cent in 2011-2012. India is also the largest importer of palm oil amounting to 44 per cent of world imports and Malaysia was the leading supplier of oil palm constituting over 20 per cent of the total imports. Other countries supplying refined and crude oil palm to India include Indonesia, Germany, Italy and China. India is the largest consumer of palm oil in the world, consuming around 17 per cent of total world consumption.

India holds only a tiny share in area and production of palm oil in the world. As per the statistics for2014-15, the area under oil palm cultivation in India was 2.3 lakh ha with production of 1.7 lakh tonnes. Andhra Pradesh is the leading palm oil producing state in India contributing approximately 86 per cent of country's production followed by Kerala (10 per cent) and Karnataka (2 per cent). Unlike in Indonesia or Malaysia, in Andhra Pradesh, palm oil cultivation is mostly irrigated. Other important oil palm producing states include Odisha, Tamil Nadu, Goa and Gujarat.

2.4.1. Schemes for Development of Oil Palm Cultivation in India

The consumption and import of oil palm has been constantly rising while there has been a negligible increase in domestic production. Creation of adverse price atmosphere due to heavy price fluctuations, availability of cheaper imported oil palm and lack of processing facilities has affected the area expansion under oilseeds particularly in the states of Odisha, Tamil Nadu, Goa and Gujarat.

The Indian Government has undertaken various schemes and programmes to increase oil palm production in India considering the heavy dependence of the country on edible oil imports. Oil Palm Development Programme (OPDP) under Technology Mission on Oilseeds and Pulses and Programme of oil palm area expansion (OPAE) under Rashtriya Krishi Vikas Yojana (RKVY) are some of the ambitious schemes of the Government. Organizations such as Oil Palm India Ltd, a joint venture between the Government of Kerala and Government of India have been promoting oil palm cultivation among the small holders. Godrej Agrovet is one of the leading private companies involved in oil palm production and has developed over 35,000 ha of oil palm in eight states. Ruchi Soya Industries Ltd. and Foods Fats and Fertilisers are the other companies engaged in oil palm business.

Oil palm development in India is undertaken recently through Mini Mission II from 2014-15 to increase the production and productivity of oil palm through area expansion in potential regions of the country. While there is a need to promote oil palm by the way of area expansion and better cultivation practices, it is equally important to focus on innovative growth strategies such as marketing of high grade derivatives and nutraceuticals, bio-mass utilization and branding of palm oil as healthy cooking medium.

3. Climatic Requirements

3.1. Geographical Position

Oil palm is a crop of the tropics, and is found abundantly in South East Asia, like Malaysia and Indonesia, on the African continent (Western and Central) and South America. From as far as 16° north in Senegal to 13° south in Malawi, and 20° south in Madagascar, isolated oil palm plantations could be found. However, it is grown commercially in more than 20 countries with most areas within 10° north and 10° south of the equator. The elevation and slope of an area intended for oil palm cultivation are also important factors that determine its suitability. In general, oil palms are not recommended for planting in areas with an elevation of more than 200 m above MSL.

3.2. Temperature

Temperature can be a limiting factor for oil palm production. The best oil palm yields are obtained in places where a maximum average temperature of 29-33°C and minimum average temperature of 22-24°C are available. Higher diurnal temperature variation causes floral abortion in regions with a dry season. If the temperature drops, the oil palm produces fewer leaves and is more often attacked by diseases. It therefore yields less.

A hot temperature enables the oil palm to make many leaves and to produce many clusters of fruit. However, the growth rate of young seedlings will be inhibited at temperatures of 15°C or lower. Night temperatures below 15 °C (sometimes experienced in Congo and Guinea) might cause "heart rot", a disease which develops in trees of five to eight years old, starting at the centre of the crown and leading often to a dying of the palm. As it is difficult to replant trees in an affected plantation because of the reduced sunlight interception, this results in a loss of production for at least 20 years.

3.3. Rainfall

Oil palm is a humid crop and requires annual evenly distributed annual rain fall of 2,500 mm to 4,000 mm or monthly rain fall of 100 to 150 mm with dry periods not exceeding 2-3 months. On average, a minimum annual rain fall of 1,800 mm is considered optimal, ranging up to 2,500 mm without harm. However, rain fall above 2,500 mm is considered unfavorable because this interferes with a lower solar radiation. In areas with dry spell, a deep soil with high water holding capacity and a shallow water table supplemented with adequate irrigation is required to meet the water requirement of the palm.

An almost continuous moisture supply is a critical factor for high oil palm yields. A water deficit greater than 30-40 cm/annum will significantly reduce fresh fruit bunch (FFB) yield. A moisture deficit affects the yield in four main ways:

1. It causes abortion of inflorescences, both male and female, about four months prior to anthesis, which results in crop reduction some 10 months after the period of stress;
2. Physiological stress at the time of sex determination of the floral initials results in the formation of higher number of male inflorescences, which adversely affects yields of mature palms about 26 months or later;
3. During periods of severe moisture stress, abortion of newly-produced inflorescences can occur, together with drying out and death of developing fruit bunches;
4. Also affects production adversely three years later.

The relative humidity should be between 75 and 100 percent throughout the year. Relative humidity of more than 80 percent is required for optimum growth. Rain fall distribution in India is not even and adequate, and hence, oil palm is to be cultivated under assured irrigation conditions by adopting recommended practices.

3.4. Solar Radiation

A high level of solar radiation is important for growth and fruit bunch production. The daily requirement of sunlight is between 5 and 7 daylight hours and at least 2,000 hours of sunshine annually. In Malaysia, the high rate of annual growth is the result of high levels of annual light interception. The sex ratio will be the highest with long periods of sunshine two years previously at the time of flower differentiation and when the dry season rainfall is at its maximum. In areas where there is a lot of sunshine and the palms are grown in soil which gives adequate water and nutrients, there will be strong photosynthesis. The leaves grow large, the fruit ripens well, and there is more oil in the fruits.

4. Soil Requirements

Oil palm can be grown in wide range of soils. The most ideal soil will be well-drained, deep, medium textured loamy alluvial soils, rich in humus content, easily penetrable with good moisture retention. The primary tap root descends deeply from the base of the trunk, but remains short when the water table is high. The finer secondary roots are in the top one meter of soil and hence, at least one-meter depth of soil is required.

Gravelly and sandy soils, particularly the coastal sands are not ideal for oil palm cultivation. Highly alkaline, highly saline, heavy clay soils with poor drainage are to be avoided. Though the crop supports water logging for short periods, areas with prolonged waterlogging are to be avoided. The soils should not be heavy with large amounts of clay, which during the monsoon season leads to water logging due to impeded drainage. Lateritic, sandy, or peat soils are problematic soils that need proper manuring and maintenance for optimum palm growth. The crop should be planted, wherever possible, on flat or undulating land: steep slopes increase both the risk of erosion and the cost of establishment and production, including the more difficult and more costly construction of access roads.

5. Botany

The oil palm tree is a member of the family *Palmae*, subfamily *Cocoideae* (which also includes the coconut), genus *Elaeis*. The genus contains two main species *viz.*, *E. guineensis* or African oil palm, and *E. melanococca* or American oil palm; the latter valuable only for hybridization purpose. The trees are unbranched with a long stout single stem, or trunk, terminating in a crown of 7–100 fronds. On an average, the fronds are produced at the rate of two per month in a regular sequence. The length of the frond is about 7 m, and each frond consists of a petiole, which is 150 cm long, and a rachis bearing 250–350 leaflets. Each leaflet may be about 130 cm long. The leaflets are arranged on two lateral planes. The root system of oil palm is relatively shallow, coarse and inefficient, with most of the active roots found in the upper 30 cm of the soil. The trees may grow to a height of 20-30 m, and a tree can live up to 50 years, but it is usually replanted at 20-25 years because of declining yields and because their height makes harvesting difficult.

The oil palm is a monoecious plant, and produces both male and female flowers separately on the same palm. The male flowers provide pollen while the female flowers develop - over a period of 5-6 months - into fruits commonly referred to as

Figure 9.1: Young Oil Palm Plantation

Figure 9.2: Yielding Oil Palm Tree.

fresh fruit bunches (FFB). The trees come to flowering 14-18 months after planting. The development of the inflorescence to the fruit regime takes 42 months, including 10 months from establishment to initial sexual differentiation, 24-26 months between sex development and flowering, and 5-6 months from flowering to yield. Hence, ecological conditions which affect earlier phases of inflorescence and flowering appear only in the yields 18 to 24 months afterwards.

Being a perennial crop, it yields continuously throughout the year. The tree produces large, spherical red fruits in bunches. Up to 200 fruits can be produced per bunch. In each productive year, an oil palm tree may produce between 8 to 12 bunches of fruit and each bunch weighs between 10 and 25 kg and contains between 1,000 and 3,000 individual fruitlets. The fruit is reddish in colour, as it contains high amounts of beta-carotene, and grows in large bunches. It consists of a hard seed (kernel) enclosed in a shell (endocarp) which is surrounded by fleshy husk (mesocarp). Palm oil is extracted from the mesocarp, while palm kernel oil is derived from the kernel after being separated from the mesocarp.

The oil palms are very efficient producers of oil. Oil content in the fruit pulp is about 50-60 percent or 20-22 percent of bunch weight; oil content in the fruit kernels is about 48-52 percent or 2-3 percent of bunch weight. Fresh fruit bunches, once harvested, must be processed in an oil mill within 24 hours to avoid deterioration of oil quality.

There are 3 oil palm varieties: *Dura, Pisifera* and *Tenera*, with the latter being mainly selected for economic production. The dura palms have kernels with a thick shell; the pisifera palms have kernels with no shell; while the fruits of the tenera palm have a lot of pulp, a thin shell and a big kernel.

6. Nursery Management

The success of a commercial oil palm plantation begins with the selection of the best planting material available from a recognized source, excellent seedling management in the nursery and subsequent sound agronomic management of the crop in the field. Adoption of the best nursery management will allow shortening the period between transplanting in the field and the first harvests (increased precocity), increasing the initial accumulated yield, and reducing initial maintenance costs in the field, particularly weed and pest control. Replacement of plants in the field will also be minimized with efficient culling/selection in the nursery. It is, therefore, important that proper nursery practices are strictly followed in order to ensure healthy growth of seedlings. The practices include: use of good top soil, shade for young seedlings, watering, weeding, manuring, pest and disease control, and culling/selection. Different systems for raising oil palm seedlings are practiced. They are as follows.

6.1. Single Stage Nursery

The single stage nursery involves planting the sprouts (germinated seeds) directly into large polybags (500 gauge and 40 cm x 45 cm size), avoiding the pre-nursery stage, and raising the nursery up to the stage of transplanting of one year. On the lower half of the bag, perforations are made at an interval of 7.5 cm for drainage. A bag can carry 15 - 18 kg of nursery soil depending on the type of potting mixture used. The water requirement for different stages of growth of seedlings is as follows: 0 - 2 months at 4 mm/day, 2 - 4 months at 5 mm/day, 4 - 6 months at 7 mm/day and 6 - 8 months at 10 mm/day. It is better to supply, if feasible, the daily requirement in two halves to prevent overflow and wastage caused by one time application. Apply 9 - 18 l of water per seedling per week according to the stage of growth and soil type.

The advantages of this system are: i) Transplanting shock is totally absent in single stage nursery system, whereas, in double stage nursery system, shock is likely to occur at the time of transfer from small poly bags to large one, and ii) Relatively lesser labour force, equipments and inputs only are needed than double stage nursery. However, some of the disadvantages are: i) It is necessary to have the full nursery infrastructure ready, large bags filled and irrigation of the full nursery area functional right from the initial seed delivery, ii) In the initial critical period, seedlings spread over large area require greater volume of water over pre-nursery system, iii). This also causes difficulties to make critical observation and supervision incurs more inputs, iv) It is definitely an unsuitable system where availability of land is a constraint, and v) Culling or removal of abnormal seedlings in the last round results in very heavy monetary loss.

6.2. Double Stage Nursery

Double stage nursery is the raising of seedlings in beds or small polybags (usually 250 gauge and 23 cm x 13 cm size) up to three to five leaf stage (pre or primary nursery) and later transplanting into bigger bags (500 gauge or more thickness and preferably 40 cm x 45 cm size) (secondary nursery). The advantages

Figure 9.3: Raising Seedlings in Primary Nursery.

of this system are: i) Less irrigation requirement for very small section of nursery area. This system allows saving of maintenance cost and also conserve reserve water supply, ii) Since all the seedlings are held in small section, it is easy to observe them critically and cultural operation takes very less time, and iii) Culling is easy and can be done quickly in the first stage resulting in less wastage. The disadvantages of this system are: i) Double operation is required which is labour intensive in the initial stages, ii) Transplanting shock is inevitable in double stage nursery, especially, if transplanted at dry period. However, with proper care and supervision, transplanting shock could be avoided.

Figure 9.4: Raising Seedlings in Secondary Nursery.

6.3. Use of Advanced Planting Material

A traditional oil palm nursery consists of seedlings developed in poly bags kept in triangular fashion, the spacing on each side being 90 cm or less. The plants are maintained in these conditions for approximately 12 months and used for planting in the field. The use of advanced planting material (APM) in oil palm allows planting of more vigorous plants in the field, which have the potential for a higher initial accumulated yield than plants coming from traditional nurseries. The most important factor in nursery plant growth is the length of the nursery stage. However, in order to achieve vigorous growth, it is necessary for the plants to receive as much sunlight as possible, which can be achieved by using a bag spacing proportional in length to plant time in the nursery. The concept of advanced stage nursery implies extending the nursery stage to 18 months or more. This requires an increase in the space between bags in order to reduce etiolation. Other practices that help production of seedlings in an advanced nursery are the use of larger bags and an increase in nutrient supply.

The potting mixture is made by mixing top soil, sand and well decomposed cattle manure in equal proportions. Use of compost prepared by mixing soaked EFB and cow dung(60:40) @ 4.8 g N/plant helps better growth of seedlings in the oil palm pre-nursery (< 3 months), while unsoaked EFB and cow dung (60:40) compost improves better in later stage of growth of seedlings (3-13 months)in the nursery.

The poly bags are to be filled with the potting mixture leaving one cm at the top of the bag. A healthy germinated sprout is placed at the centre at 2.5 cm depth. While placing the sprout, care must be taken to keep the plumule of the sprout facing upwards and the radicle downwards in the soil. It is better to plant sprouts soon after the differentiation of radicle and plumule.

6.4. Culling in the Nursery

Culling/selection need special attention in view of its great influence on palm productivity. The initial culling of inferior seedlings should be carried out just before transplanting from the pre-nursery into the main nursery, or three months after planting germinated seeds in a single stage nursery. Thereafter, culling should be done at a three-monthly interval, with final critical selection immediately prior to field planting. Regular culling should be carried out or otherwise it will be difficult to do culling if it is taken up a later stage, as the plant size increases. The most crucial culling stage is after nine months of seed planting as at this stage, the seedling will express most of the abnormal characteristics.

All deformed, diseased and elongated seedlings are to be discarded. The maximum acceptable loss rates in the pre-nursery are: seedlings that have failed to develop and dead seedlings-5 per cent ; abnormal seedlings- 10 per cent, making the total of 15 per cent at the most. Thus, if 200 germinated seeds are planted per hectare, it is acceptable to keep only 170 seedlings/ha at the end of the pre nursery.

The most common types of abnormal plants at the end of the pre-nursery are: very narrow leaves (grassy appearance); twisted, crinkled, corrugated or rolled leaves; puckered leaves (collante); exaggerated upright plants; chimeras (albines) and underdeveloped plants (dwarfs). Sick or chlorotic plants, or those that have been severely attacked by insects or by fungal diseases, together with any stunted or otherwise abnormal plants (fused leaves, leaves inserted at an acute angle, short or narrow leaves, leaves spaced too far apart), should all be discarded. Rejected seedlings should be destroyed immediately as they never yield satisfactorily.

6.5. Selection of Seedlings for Planting

Seedlings with well-developed root system, which have well bound roots with soil, with fully opened frond spread, and un-etiolated seedlings are to be selected in the nursery for the main field planting. At the time of transplanting, a normal seedling should display height of 1.0 to 1.3 m; collar girth of 18 to 22 cm and the number of functional leaves to be 10 or 12.

Avoid using seedlings which are under developed plants (dwarfs), particularly with a thin basal bulb; abnormal leaves (leaflets): rolled, twisted, too short, long or narrow, *etc*; short young leaves, giving the plant a flat appearance ; acute leaf insertion angle, given the plant a rigid and up right appearance and with juvenile character: leaflets do not differentiate for field planting.

In order to obtain high yields of crude palm oil, the most important factor at the outset is the selection of planting materials. The oil palm seeds are carefully selected and germinated under carefully controlled conditions. The seeds are then planted in polybags where they will remain for at least 12 months before planting in the field. Intensive care should be given at this period. Adequate irrigation, correct manuring, and immediate remedial treatment of any disease or pest attack are important to ensure the production of healthy, well grown seedlings for transplanting in the field.

In oil palm, being a perennial crop and with a life span of more than 25 years, it is very important for any planters to plant good quality seedlings during field planting. The planting material that was proven superior plays the most important role in ensuring a high oil yield from the plantation. Inherent genetic make-up of the planting materials determines the level of oil content in the fruitlet. Planting materials selected for planting are recommended to be obtained from proven and reputable seed suppliers or nurseries. This is to ensure only good quality planting materials with the potential to give high palm products are planted. All seedling orders should be made based on planting requirement, whereby planting policy would determine the number of seedlings to be planted. In order to obtain early high yield per hectare, high density planting is advocated unless other policies such as integration project is to be implemented. However, high yielding planting materials must be properly nurtured in the field in order to fully exploit the genetic potential of the palm as the expression of any planting material is a function of genotype and environment.

7. Land Preparation and Planting

Oil palm is mainly cultivated as an industrial estate crop, and therefore, it occupies a large area of 3,000 to 5,000ha around a central oil mill, where the harvested fresh fruit bunches (FFB) are collected and processed immediately, as any delay cause rapid deterioration of oil quality. Though the crop can also be grown by smallholders, the oil quality, in general, is much lower and does not meet the quality standards for commercialization. Smallholders' oil palm plantations are usually intercropped during the first years with various food and cash crops to provide some food and income from the field and all these crops should also be grown following organic farming practices.

Land clearing is to be done by removing the initial vegetative cover to make the area suitable for oil palm planting. Prepare the land for oil palm plantings at least three months before transplanting the seedlings to the main field. In soils with low permeability, drainage channels are to be constructed to prevent water stagnation in upper layer of soil. Pits of 60 cm^3 are to be prepared prior to planting and filled with surrounding top soil and allowed to settle. Rock phosphate is to be applied @ 200 g per planting pit.

7.1. Time of Transplanting

Transplanting to the main field has to be done during the onset of rainy season. In very impermeable soils and where there is chance for the seedlings to suffer severely during rainy season, proper drainage has to be ensured. The best season for planting is June-December *i.e.,* during monsoon. In case of planting during summer, adequate irrigation, mulching and growing cover crops like sun hemp in the basin would help in avoiding hot winds during summer. 12 -14 months old healthy seedlings with 1-1.3m height and 13 functional leaves are recommended for planting. Sowing of daincha or sun hemp in the basin is recommended which will provide favourable microclimate for the growing seedlings in the field.

7.2. Palm Density

One of the contributing factors towards achieving optimum crop productivity in oil palm is the planting density. Optimum palm density varies with terrain, soil type and weather conditions, and therefore, in order to reduce severe inter-palm competition and at the same time aim high FFB yields, palms should be planted at lower density in an environment which favors a very high level of vegetative growth. Plants should be planted at higher density in areas where conditions are less favorable for vegetative growth.

It is generally accepted that oil palm should be planted in a triangular pattern, as it allows efficient light interception and utilization of land space. For efficient utilization of solar energy the rows are to be oriented in the North-South direction. Equilateral triangular system of planting with 9 m spacing between palms, which can accommodate 143 palms/ha will allow each plant to occupy the centre of a hexagon thus allowing better use of the area. Any density higher than optimum stand would result in inter-palm competition for light, water and nutrients. Severe inter-palm competition normally results in low dry matter production which leads to depressed FFB yield through low bunch production and reduced bunch weight.

7.3. Leguminous Cover Crops

The prevailing wisdom within the oil palm industry also recommends the planting of leguminous vegetation, which increases biological nitrogen fixation, stores nutrients, and then slowly release nutrients back into the soil as the legumes die following closure of the oil palm canopy. Legumes are also thought to help prevent beetle invasions, stem soil runoff, and reduce disease spread. Once the oil palm plants are established in the field, any one of the leguminous cover crop *viz.*, *Calopogonium mucunoides*, *Centrosema pubescens*, *Pueraria phaseoloides* or *Mucuna bracteata* can be grown in the interspaces. All these cover crops have the advantage that they rapidly establish, and thus, protect the soil surface. The establishment of leguminous covers during the immature growth phase of oil palm enriches soil organic matter status and provides added nutrients to the system, thereby enhance growth and subsequent yields. During the immature period, legumes fix large quantities of nutrients, particularly nitrogen from the atmosphere and return them to the soil through decomposition of the litter. Leaf litter accumulation commences after about six months of legume establishment. Nitrogen returns in the first year of establishment are relatively low, but large amounts of nitrogen are released from the second year onwards. It has been estimated that about 200 kg N/ha is released during this period.

Figure 9.5: Growing Green Manure Crop in Oil Palm Plantation.

Some of the other benefits of raising leguminous green manure crop include:

a) It provides control of soil erosion and surface wash, particularly in the undulating/hilly terrain,

b) It improves soil physical and chemical properties,

c) it encourages the buildup of soil fauna,

d) It helps to suppress weed growth, and

e) It provides control of pests, particularly Oryctes beetles by forming vegetative barrier restricting the use of a decomposed palm residue as breeding sites.

Thus, establishment of leguminous covers is to be followed as standard practice and are maintained for at least 2 – 2 ½ years during the immature period.

7.4. Inter Cropping

Oil palm, a wide spaced perennial crop with a long juvenile period of three years, leaves considerable inter and intra row space in the field, which can be put to use to generate additional income during the juvenile phase of the crop. Commercial oil palm cultivation is generally characterized by large extent of monoculturing with palms of uniform age structure, low canopy, sparse undergrowth and intensive use of fertilizers and plant protection chemicals. The oil palm tree commences giving fruits from the third year, with the yield increasing gradually over the years until it peaks at around 20 years and thereafter, oil palm plantations are destroyed and replanted at 25 to 30 year intervals.

Inter crop selected should be compatible with the main crop and should not compete with oil palm for light, water and nutrients. Any remunerative crop can be grown, but the most suitable crops are vegetables, banana, flowers, tobacco, chillies, turmeric, ginger, pineapple *etc*. While growing inter crops in mature oil palm gardens of 8- 12 years age or palms attained a height of three meters, intercrops should be able to grow under partially shaded conditions and should not compete with oil palm for water, sunlight and nutrients (eg. cocoa, pepper, heliconia and ginger lilly). While intercropping is practiced, do not cut the oil palm fronds to accommodate the crops. Similarly, oil palm fronds should not be tied close to the stem for inter-cropping, which will reduce photosynthetic activity. Care should also be taken not to plough close to the palm base, which will cut the absorbing roots and

Figure 9.6: Growing Coffee in Oil Palm Plantation.

thereby reduce intake of water and nutrients. Maximum number of green leaves should be retained on the palm. All the intercrops should also be grown following organic cultivation practices.

7.5. Irrigation

Oil palm is a fast growing crop with high productivity and biomass production, and hence, requires sufficient irrigation, Continuous availability of soil moisture encourages vigorous growth and increased yield. Water deficiency is found to adversely affect flower initiation, sex differentiation, and therefore, results in low sex ratio due to production of more male inflorescences. Hence, oil palm cultivation is to be taken up only in areas with assured and adequate irrigation facility. In areas where perennial water source is available, basin irrigation is possible. The irrigation channels are to be prepared in such a way that the individual palms are connected separately by sub-channel. For grown up yielding palms of three years age and above, a minimum of 150 to 200 l of water per day is required. However, in older plantations, during hot summer, this quantity may be increased up to 300 l. The required quantity of water is to be applied at 4-5 days interval. For light soils, frequent irrigation with less water is to be given, whereas in heavy soils, irrigation interval can be longer. In areas with undulated terrain, and water is scarce during summer months, drip or micro sprinkler irrigation can be advantageous. If drip irrigation is installed, four drippers are to be placed in the weeded palm circle. If each dripper discharges 8 l of water per hour, 5 hr. of irrigation per day is sufficient to discharge 160 l/day. In case of micro sprinklers (180° or 360°), one each on either side of the palm can be installed. Drippers/jets should be periodically checked for proper water discharge. While irrigating, the palm basins should be adequately mulched and covered with soil to conserve moisture.

7.6. Weed control

A circular area of about 1.5 m-2 m around the tree base should be kept free of weed growth by regular weeding. It is more important for young palms, roots of which are to be kept free from competition from weed. Six rounds of weeding per year may become necessary during the immature phase of growth and three rounds per year thereafter. In addition, inter-rows are also to be weeded 2-3 times per year. The weeded materials can be used for mulching.

7.7. Soil Conservation

Soil erosion can be controlled by maintaining adequate ground cover vegetation. Perennial tree cropping systems where the trees are established along with cover crops helps in reducing soil erosion than annual crops. However, heavy rainfall can still result in high erosion rates where soil is exposed and specific erosion control measures should be taken in vulnerable areas.

Appropriate legume cover crops are to be planted as early as possible after clearing (or partial clearing with under-planting). This will also help to prevent *Oryctes* breeding in the felled palm trunk as well as improving the fertility of the soil. Spreading or stacking pruned fronds along contours, even on gently sloping

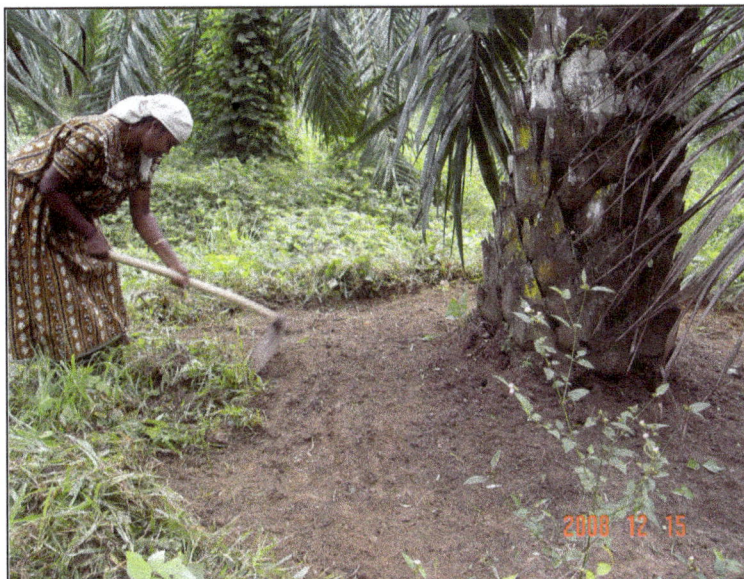

Figure 9.7: Weeding in Plant Basin.

land helps to prevent soil loss. Empty fruit bunches (EFB) should also be placed in such a way that erosion is reduced in vulnerable areas.

Constructing soil erosion prevention terraces (with or without bunds) and silt pits along roads and in fields where erosion is likely to occur, making water diversion channels, also terracing all slopes above 10 degrees help in preventing soil erosion. Use of heavy machinery where soil erosion is likely to result should be avoided. Planting in severely sloping land, where the slope is more than 25 degrees may be avoided as the cost of terracing such land and the crop obtainable is insufficient to make planting financially viable.

7.8. Pruning of Fronds

It is the removal of non-functional fronds in order to facilitate harvesting. In oil palm, two leaves are produced per month, and therefore, it becomes necessary to prune excess leaves so as to enable easy harvest of bunches. Severe pruning will adversely affect both growth and yield of palm, cause abortion of female flowers and also reduce the size of the leaves. Leaf pruning is carried out in India using chisels so that leaf base that is retained on the palm is as short as possible or otherwise it may catch loose fruits, allow growth of epiphytes and the leaf axils form a potential site for pathogens. The leaf petioles are removed by giving a clear cut at a sufficient distance from the base of the petiole using a sharp chisel for young palms and with the long sickle in taller palms.

Pruning is preferably carried out at the end of the rainy season. It is also better to carry it out during the low crop season when labourers are also available. Pruning is confined to only lower senile leaves during initial harvests but when canopy closes in later years, leaves are cut so as to retain two whorls of fronds below the ripe bunch. However, in normal practice, some of the green fronds are also removed. It is evident that FFB yield is significantly reduced if fewer than 32 fronds/palm are retained at any one time. Over-pruning influences sex differentiation which favours the formation of male inflorescence. This will lead to yield decline for about two years after heavy pruning. Increase in inflorescence abortion rate which is manifested within 9 - 11 months after heavy pruning is also a factor contributing to yield decline.

It takes more than 2 years for FFB yields to reach normal level of production. These results clearly demonstrate the need to strictly supervise pruning operations in order to avoid unnecessary yield losses. In practice, the following pruning policy should be followed.

Young palms: removal of desiccated frond only

4 - 7 years: pruning to retain 56 - 48 fronds/palm

8 -14 years: pruning to retain 48 - 40 fronds/palm

> 15 years: pruning to retain 40 - 32 fronds/palm

7.9. Pollination

Oil palm is a highly cross-pollinated crop. Though wind and insects assist pollination, wind pollination is not adequate for economical crop production. Effective pollinating insects like *Elaeidobius kamerunicus* helps in good pollination and fruit set. Release of this weevil after 2-1/2 year of planting is advisable. If the plants are not having good girth and vigour, release the weevils after three years. For introduction, cut male flowers from palms which have the weevils and transferred to a plantation where it is to be introduced. Care should be taken to see that they are not carrying any plant pathogens to other areas.

7.10. Ablation

The initial bunches produced will be very small and have low oil content. Removal of such inflorescences is called ablation. Removal of all inflorescences during the initial years is found to improve vegetative growth of young palms so that regular harvesting can commence after three and half years of planting. Ablation enables the plant to gain adequate stem girth, vigour and improves drought resistance capacity of young palms by improving shoot and root growth especially in low production areas where dry condition exists. Ablation can be extended up to 30 to 36 months depending upon the plant growth and vigour. Ablation is done at monthly interval by hand pulling out the young inflorescence using gloves or with the help of specially developed devices such as narrow bladed chisels.

7.11. Zero Burning

Open burning of felled palms during replanting is prohibited in Malaysia, and zero burning has been practiced by all plantations in the country. Zero burning involves shredding of oil palm trunks to make them 5–10 cm thick and stacking the shredded trunks in the inter rows. Decomposition of the shredded biomass takes place within two years of application, after which it does not serve as breeding

Figure 9.8: Ablation.

substrate for the rhinoceros beetles. In addition, this technique has advantages in terms of recycling large quantities of plant nutrients through decomposition and improving soil physical properties. The nutrient reserves could provide to the palms N, K, and Mg for six to seven years and P for about two years. Traditionally, empty fruit bunches (EFB) were burnt to produce bunch ash, which is a good source of K fertilizer. However, burning is also prohibited because it causes air pollution, and to overcome the problem, EFB are to be applied in the plantations as mulch within palm circles and inter rows as partial sources of nutrients.

8. Nutrient Management

Oil palm is a high nutrient demanding crop, and therefore, adequate supply of plant nutrients is essential to maintain optimum productivity. Hence, all organic materials that have nutritional value are to be applied to the field. The prehistory of the site, whether it is opened from jungle or is a replant will also determine the nutrient requirement of oil palms. The availability of soil nutrients may change appreciably with time due to removal by the crop or build up from repeated nutrient applications or through mulching of crop residues. The type of crop cover will also affect the soil nutrients available to oil palms.

In order to build up organic matter, during planting/replanting, all vegetation to be cleared should be raked together into a broad swathe, thereby leaving only the area cleared for taking up planting. In sloppy areas, place the old trunks along contours. On terraced slopes, keep the trunks on terrace edges where ever possible. Any burning of biomass should be avoided, unless serious pest and disease problems warrant it. Good legume covers should be maintained for as long as possible after planting. During the early years of planting, maintenance of a vigorous legume cover helps to supply sufficient nitrogen during 4th to 6th year. In some environments, it will be possible to maintain legume cover throughout the life of the palms.

Ground cover slows the depletion of soil organic matter from the effects of sunlight and erosion. It also adds to the organic matter content through leaf and plant litters. Thus, while clearing areas for new planting, maintain ground cover wherever needed. In case ground cover is not available, it can be provided by felled trunks and trunk chippings. It should be followed by rapid establishment of leguminous cover crops. During harvesting or pruning, the fronds should be cut into two and the petiole or frond base half stacked between palms in the palm row. Spread the upper or leafy half in the non-harvesting inter-row. Where fronds are not cut into two, the petiole base or frond stalk end should be placed squarely in the windrow with the frond tip pointing outwards. On slopes, apply the frond stacks in contours and on terraces they should be placed along the terrace edge. Loss of nitrate and phosphate by surface run-off and to ground water must be avoided. This is important on shallow soils or where heavy rainfall causes surface run-off/soil erosion.

8.1. Organic Waste Recycling

A site-specific recycling plan should be drawn up in the oil palm plantation. Besides applying organic manures from sources such as animal, fish *etc.*, all palm residues can be effectively recycled for supplying nutrients, particularly of potassium. The oil palm wastes include palm oil mill effluent(POME), empty fruit bunch(EFB), palm oil mill sludge(POMS), oil palm fronds(OPF), oil palm trunks(OPT),decanter cake, seed shells and palm pressed fibers(PPF). In order to extract one tonne of palm oil, it is estimated to produce six tonnes of old leaves, one tonne each of trunk and peel fiber, five tonnes of EFB, 0.5 tonne of palm kernel shell,0.25 tonne of pomace and three tonnes of POME. Palm oil waste management is a serious issue in most of the producing countries. Using palm oil wastes as raw materials for organic manures can not only reduce environment pollution issues, but also help in safe disposal, supply of nutrients as well as improve soil structure.

Some of the residues are:

a) Empty Fruit Bunch (EFB)

It contains a high amount of plant nutrients, and based on the nutrient composition, one tonne of EFB can supply various nutrients equivalent of 8 kg Urea, 2.9kg Rock Phosphate, 18.3kg MOP and 4.7kg Mg SO_4. It may be applied @ 250 kg/palm/year. Besides providing nutrients, it also increases soil organic matter content, improves soil structure, increases infiltration and aeration, improves soil moisture retention, and also increases cation exchange capacity. A yield of 25 t/year of FFB per hectare gives approximately six tonnes of EFB. The EFB contain about half the potassium from the FFB. Although application cost of EFB per unit of nutrient is generally higher than for inorganic fertilizers, it helps to return organic matter and conserve soil moisture. EFB should be applied preferentially in areas with low soil organic matter.

b) Palm Oil Mill Effluent

This also contains a high amount of nutrients. It may be applied @ 360 l/palm/year in the inland environment. At these application rates, each palm will receive nutrients equivalent to Urea - 2.0 to 3.0 kg, Rock Phosphate- 1.8 to 2.8 kg, MOP - 1.5 to 2.2 kg and Mg SO$_4$ - 2.3 to 3.5 kg. Application of factory effluent to the entire plantation could also be practiced. About 16 tonnes of effluent from 25 tonnes of FFB could be obtained. Efficient utilization of factory water will help to reduce costs involved in storage/pumping *etc*. Application of effluent must be avoided where contamination of streams/groundwater may occur.

c) Pruned Fronds

Fronds contribute about 70 per cent of the total dry matter through regular harvests and annual pruning and the remaining 30 per cent is contributed by rest of the palm parts. Pruned fronds are rich in plant nutrients - 2.75 per cent N, 0.223 per cent P, 1.99 per cent K and 0.45 per cent Mg. In term of fertilizer equivalent, one tonne of dried fronds contain 59.8 kg Urea, 14 kg Rock Phosphate, 39.8 kg MOP and 27.8 kg Mg SO$_4$. Placing these fronds to cover as much ground surface as possible except the palm circles and harvesting paths allows wider and uniform distribution of organic matter and plant nutrients. This placement method creates a better environment for the development of feeding roots resulting in greater efficiency in nutrient uptake from both the decaying fronds and applied manures. Proper placement of pruned fronds helps in higher moisture retention, general improvement in porosity of soil resulting in higher infiltration rates, and thus reducing soil erosion and surface wash. In the rolling/hilly terrain, placement of pruned fronds in the direction of slopes would also act as a physical barrier to soil erosion.

d) Oil Palm Trunk

Considering a planting density of 143 palms in a hectare, 163 tonnes of total dry matter can be expected at the end of the crop. The trunk contributes more than 50 per cent of the total dry matter followed by the underground bole mass. At felling stage, nutrients to the extent of 1,500 kg N, 129 kg P, 2,345 kg K, 513 kg Ca and 438 kg Mg could be expected from the available biomass. Recycling trunk helps to supply these nutrients to soil. Shredding gives a faster breakdown, with all nutrients released within 2-3 years of application. On an average, most of the oil palm residues will decompose within 12-18 months while some of the hardier materials, particularly roots, take much longer than 18 months to decompose.

Table 9.3: Nutrient Contents of Biomass from One Hectare of Plantation

Parts of Palm	Nutrient Content (kg/ha/year)				
	N	P	K	Mg	Ca
Annual pruning	108.0	10.0	139.4	17.2	25.6
Empty fruit bunches	5.4	0.4	35.3	2.7	2.3
Fibre	5.2	1.3	7.6	2.0	1.8
Shell	3.0	0.1	0.8	0.2	0.2
Factory effluent	12.9	2.1	26.6	4.7	5.4

The compost prepared mainly from fronds and empty fruit bunches contain on an average 1.8-2 per cent N, 0.2-0.3 per cent P and 0.9-1.2 per cent K. The quantity of nutrients that can be obtained from various biomass of one ha of oil palm is given in Table 9.3.

Proper residue management improves the spatial integration of nutrient release and uptake by the rooting systems of young palms. The supply of nutrient requirement that is partly provided by the recycling of biomass can help to optimize growth rates of the immature palms.

9. Plant Protection

Oil palm is very sensitive to pests and diseases, from the nursery stage to the trees in full production. Pest and disease control in an oil palm estate is, therefore, as important as the care and management for vegetative growth and production. In poorly controlled infected plantations yield losses can be as high as 50 per cent or more as compared to the potential yields.

9.1. Pests and their Management

There is growing awareness among the planters in Malaysia that palms which meet the minimum environmental standards alone be grown. A number of plantations have been accorded the ISO 14001 certification. This ensures production in a clean environment. This has important implications on the trade of palm oil in the future in the sense that developed countries would prefer importing palm oil from companies that address environmental concerns. The important pests of oil palm, their damage symptoms and control measures that are to be adopted under organic cultivation are listed in Table 9.4.

9.2. Diseases and their Management

The important diseases of oil palm, their symptoms and management practices are given in Table 9.5.

10. Harvesting

During the first three to four years, the production of the young palms is often small, of poor quality, and sometimes even not economic to be harvested. Full production starts from the sixth year onwards; it reaches its maximum four to six years later, and remains high for another 10 years. Proper and timely harvesting of fruit bunches is an important operation which determines the quality of oil to a great extent. The yield is expressed as fresh fruit bunches (FFB) in kg per hectare per year or as oil per hectare per year. The bunches usually ripen in six months after anthesis. Unripe fruits contain high water and carbohydrate and very little oil. As the fruit ripens oil content increases to 80 - 85 per cent in mesocarp. Usually the ripe fruits, attached to the bunches, contain 0.2 to 0.9 per cent FFA and when it comes out of extraction plant, the FFA content is above 3 per cent.

Over ripe fruit contains more free fatty acids (FFA) due to decomposition, and thus, increases the acidity. The fruit bunches should be handled carefully and processed as soon as possible in order to keep the FFA level low. Harvesting rounds should be made as frequent as possible to avoid over ripening of bunches. A bunch which is almost ripe but not ready for harvest for a particular harvesting round

Table 9.4: Symptoms and Management Practices of Important Pests of Oil Palm

Name of the Pest	Damage Symptoms	Management Practices
Spindle bug (*Carvalhoia arecae* (Miller))	Spindle bug - generally noticed in nursery seedlings and plantation planted young seedlings ☆ Adults and nymphs of spindle bug live in the innermost two to three leaf axils ☆ Suck sap from the spindle of leaves ☆ Necrotic lesions which later on turn into dry brown patches ☆ In severe infestation the spindle fails to open	**Cultural control** ☆ Digging and forking of the soil before and after the monsoon will help in eliminating the various developmental stages of the beetle **Biological control** ☆ Conserve predators such as wasps, green lacewings, earwigs, ground beetles, rove beetles, spiders, coccinellids, syrphids *etc.*
Mealy bug (*Dysmicoccus brevipes* (Cockerell) and Scales (*Spidiotus destructor* (Signoret)	Scales can cause considerable damage. They like aphids have proboscis, stylet or straw like mouth part which they insert into the phloem or inner cells of a plant. Upon insertion the scales draw the plants juices or sap.	**Cultural control:** ☆ Collection and destruction of infested plant parts ☆ Collect planting material from unaffected plantation ☆ Insecticidal soap is a safe and effective alternative to conventional insecticides. You can use bleach-free dishwashing liquid (1 and 1/2 teaspoons per one quart of water) in place of commercial insecticide soaps. Homemade control of plant scale can also be achieved with oil spray. Mix two tablespoons of cooking oil and two tablespoons of baby shampoo in one gallon of water. This can also be mixed with one cup of alcohol to help penetrate the shell of insect **Biological control:** ☆ Conservation and augmentation of natural enemies such as ladybird beetle *etc.* ☆ For mealybug: release coccinellid beetle, *Cryptolaemus montrouzieri* @ 10/tree
Root grub (*Leucopholis burmeisteri* (Brenske))	Root grubs or white grubs occur mostly in sandy and sandy loam soils. ☆ They are voracious feeders on roots. Adult beetles emerge during May-June few days after receipt of pre-monsoon showers, between 6.30 to 7.30 PM ☆ The early instar grubs feed on the roots of grasses and other humus. The second and third instar grubs of these	**Cultural control:** ☆ Fill the seedling bags with the soil free from root grub infestation ☆ Exposure of grubs by ploughing or digging the soil during pre and post monsoon periods **Mechanical control:** ☆ Collection and destruction of beetles during their emergence from the soil in the evening hours

Contd...

Table 9.4–Contd...

Name of the Pest	Damage Symptoms	Management Practices
	beetles feed on tender and mature roots of the palm. In severe cases, the bole of the palm is also eaten up. They feed on roots of intercrops like banana, cocoa, tapioca, yams *etc.* ☆ In oil palm seedlings, the feeding on roots results in dropping and drying of leaves ☆ Affected seedlings come off easily since the entire root system is usually eaten up. Palms with few years of infestation show a sickly appearance, with yellowing of leaves, tapering of stem, and reduction in yield ☆ The palms may topple in case of severe loss of root system	☆ Install light traps @ 1 trap/acre and operate between 6 PM and 10 PM **Biological control:** ☆ Conserve entomopathogenic nematodes such as *Heterorhabditis* spp. and *Steinernema* spp.
Red palm weevil (*Rhynchophorus ferrugineus* (Olivier))	☆ It is very difficult to detect *R. ferrugineus* in the early stages of infestation. Generally, it is detected only after the palm has been severely damaged. Careful observation may reveal the following signs which are indicative of the presence of the pest ☆ Some holes in the crown or trunk from which chewed-up fibres are ejected. This may beaccompanied by the oozing of brown viscous liquid ☆ Crunching noise produced by the feeding grubs can be heard when the ear is placed to the trunk of the palm ☆ A withered bud/crown ☆ Chewed plant tissues in and around opening of tunnels with a typical fermented odour ☆ Fallen empty pupal cases and dead adults around a heavily infested palm ☆ Breaking or toppling of the trunk	**Mechanical control:** ☆ Remove and burn all wilting or damaged palms in coconut gardens to prevent further perpetuation of the pest ☆ Avoid injuries on stems of palms as the wounds may serve as oviposition sites for the weevil. Fill all holes in the stem with cement ☆ Avoid the cutting of green leaves. If needed, they should be cut about 120 cm away from the stem. ☆ Setting up of attractant traps (mud pots) containing sugarcane molasses 2½ kg or toddy 2½ l + acetic acid 5 ml + yeast 5 g + longitudinally split tender oil palm stem/logs of green petiole of leaves of 30 numbers in one acre to trap adult red palm weevils in large numbers. ☆ Install pheromone trap @1/2 ha **Biological control:** ☆ Fill the crown and the axils of top most three leaves with a mixture of fine sand and neem seed powder or neem seed kernel powder (2:1) once in three months toprevent the attack of rhinoceros beetle damage in which the red palm weevil lays eggs

Contd...

Table 9.4–Contd...

Name of the Pest	Damage Symptoms	Management Practices
Rhinoceros beetle (*Oryctes rhinoceros* (Linneaus))	*O. rhinoceros* adults feed in the crown region of both coconut and oil palm. ☆ They bore through petiole bases into the central unopened leaves. This causes tissue maceration and the presence of a fibrous frass inside the feeding hole is an indication of its activity within ☆ Usually, a single attack is often followed by others on the same palm ☆ These attacks subsequently produce fronds which have wedge-shaped gaps or the characteristic serrated cut (fan-shaped fronds)	**Mechanical control:** ☆ Remove and burn all dead coconut trees in the garden (which are likely to serve as breeding ground) to maintain good sanitation ☆ Plant a cover crop to deter egg laying by females as they do not lay eggs in areas covered by vegetation ☆ Collect and destroy the various bio-stages of the beetle from the manure pits (breeding ground of the pest) whenever manure is lifted from the pits ☆ Examine the crowns of tree at every harvest and hook out and kill the adults ☆ Set up light traps following the first rains in summer and monsoon 200 period to attract and kill the adult beetles ☆ Set up rhinolure pheromone trap @ 1/ac to trap and kill the beetles **Biological control:** ☆ Soak castor cake at 1 kg in 5 l of water in small mud pots and keep them in the oil palm gardens to attract and kill the adults ☆ Treat the longitudinally split tender coconut stem and green petiole of fronds with fresh toddy and keep them in the garden to attract and trap the beetles ☆ For seedlings, apply 3 naphthalene balls/palm weighing 3.5 g each at the base of inter space in leaf sheath in the 3 inner most leaves of the crown once in 45 days ☆ Apply mixture of either neem seed powder + sand (1:2) @150g per palm or neem seed kernel powder + sand (1:2) @150 g per palm in the base of the 3 inner most leaves in the crown
Bag worm (*Metisa plana*, *Pteroma pendula*, and *Mahasena corbetti*)	☆ The bagworms build and live within a portable silk case (or bag), constructed by attaching fragments of leaves ☆ The larvae remain in their individual bags until the adult stage for females and the pupal stage for males ☆ All ages of palms are susceptible to bagworm attack but more damage tends to occur on matured palms of more than eight years old	**Biological control:** ☆ Plant beneficial plants (in particular, *Cassia cobanensis*) in the plantations to provide a good source of nectar to the parasitoids of the bag worm and other pests, thereby extending the life span of the natural enemies in the oil palm ecosystem ☆ Apply emulsifiable suspension of *B. thuringiensis kurstaki* to coincide with the very early instar stage

Table 9.5: Important Diseases of Oil Palm, their Symptoms and Management Practices

Name of the Disease	Damage Symptoms	Management Practices
Basal stem rot (*Ganoderma lucidum* (Karst))	The trees in the age group of 10-30 years are easily attacked by the pathogen. The fungus is soil-borne and infects the roots. The most usual symptoms are yellowing, withering and drooping of the outer fronds which remain hanging around the trunk for several months before shedding ☆ The younger leaves remain green for some time and later turn yellowish brown ☆ The new fronds produced become successively smaller and yellowish in color which do not unfold properly ☆ Soft rot occurs in the bud with a bad newly formed leaves wither away. More often the spindle is blown off leaving the decapitated stem ☆ The wilting plants also show bleeding patches near the base of the trunk ☆ A brown gummy liquid oozes out from the cracks in the tree which slowly result in the death of outer tissues ☆ As the infection advances, fresh bleeding patches appear above the old once, up to 3-5 meters height ☆ The decay of the basal portion occurs slowly and tree succumbs to the diseases in 2-3 years ☆ In the advanced stages of infection, the fungus produces fruiting body (Bracket) along the side of the basal trunk ☆ The roots of wilting trees show discoloration and severe rotting	**Mechanical control:** ☆ **Plantation sanitation:** Removal and destruction of the dead and diseased palms in order to prevent the spread of the disease ☆ **Isolation of diseased palms:** The palms in the early or middle stages of the disease should be isolated from the neighboring palms by taking trenches of 1 m deep and 30 cm wide ☆ Irrigate the palms at least once in a fortnight during summer months **Biological control:** ☆ Apply heavy doses of FYM or compost for green manure at 50 kg/tree/year along with 5 kg of neem cake
Stem wet rot/ stem bleeding (*Thielaviopsis paradoxa* (de Seynes))	The characteristic symptom is the exudation of reddish brown fluid from the cracks in the stem. ☆ The fluid trickles down to several feet on the stem and the exudates dries up forming a black crust ☆ The tissues below the cracks turn yellow and decay. As the disease progresses, more area underneath the bark gets decayed and the bleeding patch extends further up	**Mechanical control:** ☆ Improvement in agronomic practices, providing drainage, avoid flooding of the plantation *etc.* ☆ Adequate fertilization ☆ Scoop out the diseased tissue with a portion of healthy tissues, burn the exposed tissue and apply molten coal tar.

Contd...

Table 9.5–*Contd...*

Name of the Disease	Damage Symptoms	Management Practices
Bud rot **(*Phytophthora palmivora* (Butler))**	Palms of all ages are susceptible to the disease, but it is more severe in young palms of 5-20 years. The first indication of the diseases is seen on the central shoot of the tree (spindle) ☆ The heart leaf shows discolorations which become brown instead of yellowish brown. This is followed by drooping and breading off the heart leaf. With the progress of diseases, more number of leaves get affected with loss of lusture and turn pale yellow ☆ The entire base of the crown may be rotten emitting a foul smell, the central shoot comes off easily on slight pulling ☆ The leaves fall in succession starting from the top of the crown. The leaf falling and bunch shedding continue until a few outer leaves are left unaffected. But within few months the infection leads to complete shedding of leaves, within subsequent wilt and death of the tree	**Cultural control:** ☆ Remove and burn badly affected trees which are beyond recovery ☆ If diseases is detected in early stage, remove the infected tissue thoroughly by cutting the infected spindle along with two surrounding leaves
Bunch rot **(*Marasmius palmivorus* (Sharples))**	In the early stages of infection, whitish or pinkish-white mycelial threads can be seen over the bunch surface, especially at the base of the subtending frond. ☆ The fungus penetrates the mesocarp of the fruit and causes a soft, brown, wet rot which is sharply defined from healthy tissues. If affected fruits are left on the palm, the rot ultimately dries out, leaving the fibrous tissues of the mesocarp with abundant mycelia growth of the pathogen ☆ The mycelial threads spread to other bunches and grow over, and inside, the frond bases. In the later stages of infection, abundant fructifications can be seen on bunches which have been extensively colonized	**Cultural control:** ☆ **Sanitation:** Before on-set of monsoon, crown cleaning by means of removing the dead inflorescences, bunch stalks, aborted bunches *etc.* will help in reducing the inoculums buildup and harbouring of pathogen

Contd...

Table 9.5—*Contd...*

Name of the Disease	Damage Symptoms	Management Practices
Leaf spot (*Pestalotiopsis* spp.)	Tiny black spots on leaves which enlarge into 2 mm long elliptical, elongated lesions. ☆ Lesions may expand and be surrounded by black tissue and chlorosis between lesions ☆ Lesions may be present on leaf petioles and rachis	**Cultural control:** ☆ Severely diseased palm should be removed from plantation and destroyed ☆ Palms should be planted with adequate spacing to allow air to circulate between trees ☆ Remove weeds from around palms
Bacterial budrot/ Spear rot (*Erwinia* spp)	Parts of spear leaf petiole or rachi turning brown ☆ Discoloration may be associated with a wet rot ☆ Spear leaf may be wilted and/or chlorotic ☆ Leaves may be collapsing and hanging from the crown ☆ Infection of the bud results in buds becoming rotten and putrid, leading to death of the palm **Survival and spread:** ☆ Bacteria survive in crop debris and infect by water splash through damaged tissues ☆ Worse in hot wet weather. The bacteria spread in contaminated water **Favourable conditions:** ☆ Higher temperatures and high humidity are ideal growing conditions for the bacteria	**Cultural control:** ☆ Oil palm plant varieties with resistance to the bacteria ☆ Rotting tissue on spear leaves should be removed to prevent bacteria spreading to buds
Vascular wilt disease (*Fusarium oxysporum* f. sp. *Elaeidis*)	This disease is prevalent in Africa, the Ivory Coast, Ghana, Benin, Cameroon, the Congo, Zaire and Brazil. ☆ The soil born pathogen penetrates the roots and grows into the xylem where it causes blockage leading to wilting and death of the palm. In certain cases infected palm survives but its growth is greatly retarded leading to little or no yield ☆ In mature palms, the disease may either exhibit chronic or acute symptoms ☆ The chronic form is more common and characterized by having the affected fronds not usually symmetrically placed in the crown with only one to three of the phyllotactic spirals showing wilting symptom	**Control measure:** ☆ Inoculum location and disease incidence: During replanting, the proximity of the stump of felled palm of previous stand serves as a major source of inoculum. The farther the young palm is from a stump, the less chance it has of contracting the disease. It is the site of the stump that contains the inoculums. The inoculum potential decreases with distance from such site. The uprooting of stumps will not become necessary ☆ Breeding for resistance

Contd...

Table 9.5—Contd...

Name of the Disease	Damage Symptoms	Management Practices
	☆ The older fronds gradually become less turgid, wilt and eventually changing to a brown colour and becoming desiccated	
	☆ The dead fronds fracture at some point forming a cloak surrounding the palm	
	☆ As the disease progresses the remaining green fronds become shorter and may ultimately become less than one-half of their normal size	
	☆ This phenomenon occurs 3 to 12 months after the first symptom of wilting and may persist for several years before the whole crown dies and falls off	
	☆ In the acute form death usually occurs within 2 to 3 months after the appearance of the foliar symptom	
	☆ In this form, the first frond to be affected is towards the centre of the crown. The outer fronds then die swiftly, and crown dieback progresses rapidly inwards and upwards. The fronds die quickly and eventually fall off	
	☆ The disease can also occur on nursery seedlings and young field palms	
	☆ Growth of infected seedling is retarded with the inner leaves being shorter and narrower than those produced previously. This gives the seedling a flat-topped appearance, or even has a depressed centre	
	☆ In young field palm under six years old, the infected palm is characterized by the presence of bright lemon-orange fronds with a few pinnae on one side become chlorotic. This is followed by desiccation of the entire frond	

Figure 9.9: Harvested FFB for Processing.

should not be over-ripe by next round. In lean period of production, harvesting can be made less frequent and it should be more frequent in peak periods.

Harvesting should be done at 10-12 days interval. During rainy season, harvesting should be done at closer interval of 6-7 days as ripening is hastened after hot summer. In young plantations, one can get more bunches with less bunch weight and in adult plantations, the bunch weight is more but the bunch number is less. Other factors that determine frequency are: extraction capacity of the mill, transportation facilities, labour availability and skill of the workers.

Ripeness of the fruit is determined by the degree of detachment of the fruit from bunches, change in colour and change in texture of the fruit. Ripening of fruits start from top downwards, nigrescens fruits turning reddish orange and the virescens (green) to reddish brown. Fruits also get detached from tip downward in 11 - 20 days time. Ripeness is faster in young palms than in older palms for the bunches of equal weight. The criteria used in determining the degree of ripeness based on the fruit detachment are as follows:

a. Fallen fruits: 10 detached or easily removable fruits for young palms and five for adult palms,

b. Number of fruits detached after the bunch is cut; five or more fruits/kg of bunch weight,

c. Quantity of detachment per bunch; fruit detachment on 25 per cent of visible surface of bunch.

Bunches are to be cut without damaging the petiole leaf that supports it. Narrow chisel is usually used for harvesting till the palm reaches two meters above the ground. For taller palms up to four meters, a wider chisel of 14 cm is to be used. Harvesting could be carried out with a curved knife of 6 - 9 cm width attached to a wooden pole or light hollow aluminium pipe. In uneven stands, an adjustable, telescopic type of pole can be used. A man can harvest 100-150 bunches per day, provided the palms are not very tall and that he is assisted by somebody who carries the bunches to the field collection points.

A fresh fruit bunch (FFB) weights on average 20-30kg depending on the age of the tree. Oil content of fruit pulp is 50-60 per cent or 20-22 per cent of the bunch weight. Oil content of kernels is 48-52 per cent of the kernel weight, or 2-2.5 per cent of the bunch weight. Under optimal conditions yields may reach 25-30 tonnes FFB/ha/year, and with an average extraction rate of 21-23 per cent, this corresponds with an approximate yield of 6 tonnes of oil per hectare. Under sub-optimal conditions, the average yield drops to 4-5 tonnes of oil per hectare.

11. Utilization and Use

Oil palm gives the highest yield of oil per unit area of any crop and is, worldwide, a major supplier of vegetable oil. Oil palm gives five to seven times more oil per ha than other traditional oil crops like groundnuts or soybeans. It is used in food production and in industrial applications. Yields from commercial oil palm estates differ from those obtained by smallholders. Average yields on industrial plantations range from 12 to 18 tonnes FFB/ha/year, with a yield potential of 18 to 22 tonnes.

Palm trees produce two distinct vegetable oils, palm oil and palm kernel oil, both of which are important in world trade. Other mainly local uses include: palm wine from the tree sap, leaves for thatching, soap production. Empty fruit bunches are used as soil manure and amendment.

11.1. Palm Oil

It is obtained from the fleshy mesocarp of the fruit, which contains about 50-60 per cent oil. It is light yellow to orange red in color, the depth of color depending on the amount of carotene present, the amount of oxidation by lipoxidases before processing, and oxidation catalyzed by iron during processing and bulking. The oil melts over a range of temperatures up to 50° C.

Palm oil contains a high proportion of saturated palmitic acid, as well as considerable quantities of oleic and linoleic acids which give it a higher unsaturated acid content than coconut and palm kernel oils. High free fatty acid content will be there in poorly prepared palm oil, which renders it unsuitable for edible purposes in importing countries and, therefore, requires additional treatment before proper commercialization. Palm oil is widely used in the manufacture of soap and candles, but this use tends to decline. With the improvement in quality, now a days, it is being increasingly used for edible purposes, including the manufacture of margarine and composed cooking fats.

11.2. Palm Kernel Oil

It is obtained from the kernel or endosperm which contains about 50 per cent oil, after the shell or endocarp is removed. It is hard oil, closely resembling coconut oil with which it is readily interchangeable. It has a high proportion of saturated, predominantly lauric acids. It is solid at ambient temperatures in temperate countries, and is nearly colorless. It is used in edible fats, in the confectionery and bakery trades, in the preparation of ice-cream and mayonnaise, and in the manufacture of toilet soaps, soap powders and detergents. The press cake, after the extraction of oil from the kernels, can be used as an important livestock feed.

11.3. Palm Wine

It is produced from the sap obtained by tapping the male inflorescence, after incising it once or twice a day and collecting the sap that is funneled by a piece of bamboo into a bottle. The fresh sap is sweet and contains about 40g per liter of sucrose and 30g per liter of glucose, but it ferments quickly by the action of bacteria and naturally yeasts into a milky palm wine with a slight sulfurous odor. Palm wine can be further distilled into a local brandy.

11.4. Biodiesel

Oil palm was also cultivated for the production of *biodiesel* in Indonesia and Malaysia, in particular. Though the Malaysian Government appeared to consider biodiesel as an alternative outlet to its oil palm production and a potential medium for future economic growth, however, when it became clear that this high demand for biodiesel was disturbing the complete oil palm industry and market, the plan was suspended, mainly because of deleterious impact on the environment as well as the pressure it was creating to open more estates by clearing virgin forest lands.

12. Site Specific Management Practices

Some of the site-specific management practices for improving oil palm productivity are given below:

1. Maintain sufficient fronds to help attain optimum leaf area for maximum yield. Remove old, dead and damaged fronds from palms. Light competition may affect yields in older palms or when planted too close together, and therefore, proper spacing is to be adopted while planting.

2. Palms which are unproductive for six years after planting should be removed to limit competition with productive palms.

3. Fill the gaps in vacant spots and in-filling unplanted areas to maximize productivity of plantation.

4. Adopt selective thinning in dense areas to reduce competition for light and thereby improve yields.

5. Pest and disease outbreaks are to be detected before they become problematic and large-scale control measures are necessary. The fungus *Ganoderma* is responsible for basal stem rot in oil palms and it is a major threat to oil palm production in South East Asia. Therefore effective monitoring and management of pests (*e.g.* leaf eaters) and disease (*e.g* Ganoderma).

6. Spread the pruned fronds widely in inter-row area and between palms within rows and mulch palm basins with pruned fronds to provide

nutrients to the soil, conserve moisture, prevent weed growth and reduce soil erosion. Also use EFB for mulching in areas adjacent to the mill. Using EFB as mulch can replenish soil organic matter and provide nutrients.

7. Adopt regular weeding and eradicate woody perennial weeds to reduce competition for nutrients and water. Legume cover crops can help reduce soil erosion, fix atmospheric N2, supply litter to replenish organic matter and provide habitat for predators of insect pests.

Some of the site-specific management practices for improving crop recovery are given below:

1. Regular harvesting ensures efficient collection of ripe fruits, which delivers maximum oil yield, and therefore adopt harvest interval of seven days.

2. When ripe, fruits start to detach from bunches and the minimum ripeness standard is that one loose fruit before harvest. Follow the correct stage of harvesting in order to extract maximum oil yield.

3. Transport the harvested crop to palm oil mill on the same day to reduce the amount of free fatty acid (FFA) in the crude palm oil. High FFA levels make palm oil unfit for human consumption, and create problems for refining oil for biodiesel.

4. Maintain proper harvest audits to ensure maximum efficiency, crop recovery and oil yield.

5. Develop good in-field accessibility (clear paths, bridges wherever needed) for easy harvesting and transportation of harvested produces, organic manures and its application *etc*.

6. Keep the circle around palms weed free to allow efficient collection of loose fruit below palms.

7. Construct and maintain palm platforms wherever needed, which improves nutrient use efficiency, prevent soil erosion and improves harvesting efficiency.

13. Environmental Issues

Mono cropping of oil palm has been criticized for its negative impacts on the natural environment, including large scale deforestation, loss of natural habitats, and increased greenhouse gas emissions. Environmental groups also find use of palm oil biofuels objectionable, due to the fact that the deforestation caused by oil palm plantations is more damaging for the climate than the benefits gained by switching to biofuel and utilizing the palms as carbon sinks.

In Indonesia, increasing demand for palm oil and timber has led to the clearing of tropical forest land in national parks. According to *United Nations Environment Programme* report during 2007, at the rate of deforestation at that time, an estimated 98 percent of Indonesian forest would be destroyed by 2022 due to legal and illegal felling, forest fires and the development of oil palm plantations. On the other hand, Malaysian government has pledged to conserve a minimum of 50 percent of its total land area as forests.

Selected References

Anonymous.(2003). *Sustainable palm oil: Good Agricultural Practice Guidelines*, Part of the Unilever Sustainable Agriculture Initiative. www.growingforthefuture. com, www.unilever.com p.17.

Corley, R.H.V and Tinker, P.B.(2016).*The oil palm*. 5th Edition. Wiley Blackwell. p.665.

Jason Potts., Matthew Lynch., Ann Wilkings., Gabriel Huppé., Maxine Cunningham., and Vivek Voora.(2014). The State of Sustainability Initiatives Review 2014, Standards and the Green Economy.11. *Oil Palm Market* pp:235-282.

Prasad, M.V,. Arulraj, S an. Mounika B. (Eds.) *Oil palm cultivation practices*. Directorate of Oil Palm Research. ISBN No. : 81-87561-35-1.

Salmiyatia., Arien Heryansyahb., Ida Idayuc. and Eko Supriyantod (2014). Oil Palm Plantations Management Effects on Productivity Fresh Fruit Bunch (FFB). *APCBEE Procedia* 8:282-286.

Satyagopal, K., Sushil, S.N., Jeyakumar, P., Shankar, G., Sharma, O.P., Boina, D.R., Sain, S.K., Reddy, M.N., Ram Asre., Murali, R., Sanjay Arya, and Subhash Kumar(2015). AESA based IPM package for Oil palm. 47 p.

Sunitha,S.,Shareef,M.V.M. and Kochu Babu,M.(2010).Potential for biomass recycling in oil palm. In: *Organic Horticulture-Principles, Practices and Technologies*. H.P.Singh and George V.Thomas (Eds.).Westville Publishing House, New Delhi.pp:200-204.

Verheye, W. (2010). *Growth and Production of Oil Palm*. In: Verheye, W. (ed.), *Land Use, Land Cover and Soil Sciences*. Encyclopedia of Life Support Systems (EOLSS), UNESCO-EOLSS Publishers, Oxford, UK. http://www.eolss.net

ICAR-Indian Institute of Oil Palm Research, Research Centre, Palode, Pacha, Thiruvananthapuram-695 562

Chapter 10

Organic Plant Protection Technologies

☆ Chandrika Mohan and A. Josephrajkumar

1. Introduction

To meet the nutritive demand for ever growing population, agriculture technologies were boosted to intensive approach characterized by innovations designed to increase yield. Techniques included planting multiple crops per year, high yielding varieties, increased use of fertilizers, plant growth regulators, chemical pesticides and farm mechanization. Even though modem agriculture has definitely helped us to satisfy the demand to certain extent, heavy dependence on these inputs causes several disruptions to the ecological balance and it has been damaging the wealth of this planet largely. When farming is viewed in this direction, the concept of organic farming gains practical significance as it is a production system respecting the sustainability of nature. Organic agriculture aims to maintain the farming to be a live production system with sustainable nature. It supports us to live in harmony with nature. In our attempt to save the crops, pest control strategy in modem agriculture has a heavy dependency on pesticides. Pests and diseases along with weeds are known to cause an estimated crop loss of about 35 per cent of the potential food production worldwide (Sharma, 2002). In spite of our heavy dependence on insecticides for providing protection to the varieties of crops grown, their deleterious effect on the beneficial natural enemies and pollinators and the residues they leave in the palm and the environment limit the use of these poisons. An Integrated Pest Management (IPM) schedule comprising combination of different technologies – mechanical, sanitation, cultural, prophylactic, chemical and biological methods are the existing strategy globally recognized to manage pests of coconut and has proved to be quite feasible.

2. Pest Management Concept in Organic Farming

The basis of pest and disease management in organic farming systems is the reliance on the inherent equilibrium in nature. Use of bio control agents plays a crucial role in this aspect. The natural enemies are insect predators (insects that consume part or all of pest insects), parasites (insects that use other insects to produce their offspring, thereby killing the pest insect in the process), and pathogens (diseases that kill or decrease the growth rate of insect pests). Predatory insects on organic farms include lady beetles, lacewings, and spiders. Parasitic insects include wasps and flies that lay their eggs in/on pest insects, such as larvae or caterpillars. The emphasis on organic plantations should ideally be on the use of varieties resistant to pest and diseases.

Other methods that can be generally employed for the management of pests and diseases are: clean cultivation, improving soil health to resist soil pathogens and promote plant growth; crop rotation; encouraging natural biological agents for control of diseases, insects and weeds; using physical barriers for protection from insects, birds and animals; modifying habitat to encourage pollinators and natural enemies of pests; and using semio-chemicals such as pheromone attractants and trap pests. Biopesticides including micro organisms, parasites, predators and natural plant based pesticides from neem, tobacco and garlic are effective in managing pests of coconut and other intercrops. There are several examples of use of effective bio control agents for suppression of pest and diseases of coconut and other component crops.Pest management requires an overall rescheduling of various components of IPM in organic farming to maintain ecosystem sustainability. In organic agriculture, pest control strategies that compromise with the nature are suitably blended and employed to realize desirable results. Pest control strategies of a preventive rather than reactive nature are advisable in organic farming. The non-chemical and bio-rational methods of pest management that constitute the major components of IPM approach are ideal, eco-friendly and feasible in a sustainable crop production system like organic farming.

2.1. Biological Pest Suppression

Among the various components of IPM, biological pest suppression that utilizes the natural enemies of pests *viz.*, Entomophaga and Entomopathogens is the most effective tool for organic plant protection. One of the major steps involved in the biological control of pests is conservation of natural enemies which helps in regulating a particular species of pest in the ecosystem conserving the biodiversity of the ecosystem. A comparatively low level of pest incidence that occurs due to natural disturbances in the ecosystem can be corrected more easily and feasibly by intervention of an appropriate biocontrol agent. When a situation arises where natural enemies present in the ecosystem exert a low level of pest suppression, the need for augmentation through release of recommended natural enemies becomes more imperative. Biocontrol method is an advanced approach in the management of some of the major pests of coconut, oil palm, coffee and black pepper besides many annual crops. When an exotic pest is to be tackled, the need for deliberate introduction and colonization of its natural enemy to suppress the introduced pest becomes essential.

In our attempts to enrich the biodiversity by organic farming practices, several methods of conservation of natural enemies like providing nesting places, perching sites, water pan, nectar and pollen rich plants can be practiced (Sharma, 2002). Attempts to restore natural control mechanisms especially the natural enemies by selectivity of biodiversity within the ecosystem become more relevant in organic farming. Manipulation of the ecosystem inducing added activity of the natural enemies will be beneficial to reduce the pest population resulting in increased yield. Biocontrol agents have a range of attractive properties that include host specificity, lack of toxic residue, no phytotoxic effects, eco-friendly, human safety and the potential for pest management to be self sustaining. Biocontrol agents can also be produced locally which is in terms of choosing and matching natural enemies to small scale needs. Successful use requires fundamental knowledge of the ecology of both the pest and natural enemy. Biological control agents broadly involve entomophaga and entomopathogens.

2.1.1. Entomophaga

Entomophaga comprising mainly parasitoids and predators constitutes one of the major components in the natural control of crop pests.

2.1.1.1. Parasitoids

In the IPM of coconut pests, parasitoids are extensively utilized for the management of the black headed caterpillar *O. arenosella*, which is one of the dominant caterpillar pests of coconut palm. The larvae of *O. arenosella* feed on the parenchymatous tissues on the under surface of leaflets and construct galleries of silken webs reinforced with excreta and scrapes of leaf bits. Diagnostic symptoms of infestation are the presence of galleries on the lower surface of the leaflets with different stages of the pest and the upper epidermis intact. Pest infestation usually starts from the outer and middle whorl of leaves. On an average 40 per cent yield

Figure 10.1: *Goniozus nephantidis*, **Larval Parasitoid of Coconut Black Headed Caterpillar.**

decline is noticed in severely affected palms (Chandrika *et al.*, 2010) and palms of all ages are susceptible to its infestation. Concealed habitat of the pest well covered by the larval galleries makes pesticide spraying practically very difficult. Hence, biological suppression with the release of parasitoids provides the best solution for the management of *O. arenosella* (Chandrika and Sujatha, 2006).

Among the various parasitoids of *O. arenosella* (Table 10.1) hymenopterans predominate. Based on the stage of the pest on which the parasitoids develop they are treated as egg, larval, pre-pupal, larval-pupal and pupal parasitoids. Among the 40 species of parasitoids reported from black headed caterpillar of coconut in India, the gregarious larval ectoparasitoids *Goniozus nephantidis* Muesebeck and *Bracon brevicornis* Wesmael, the pre-pupal parasitoid *Elasmus nephantidis* Rohwer and the pupal parasitoid *Brachymeria nosatoi* Habu are the most promising ones (Pillai and Nair, 1993). The major desirable attributes of these parasitoids are their greater searching ability, capacity to withstand high temperature, production of higher proportion of female progeny, occurrence throughout the year, and abundance during peak period of the pest and their distribution in all pest infested areas.

Table 10.1: Important Parasitoids of *Opisina arenosella*

Name of Parasite	Family	Target Pest Stage	Nature of Parasitoid
Apanteles taragamae Vier.	Braconidae	Early larva	Solitary
Bracon brevicornis Wesmael	Braconidae	Late larva	Gregarious
Goniozus nephantidis Mues.	Bethylidae	Do	Do
Elasmus nephantidis Rohw.	Elasmidae	Pre pupa	Do
Goryphus nursei Lam.	Ichneumonidae	Larva-pupa	Solitary
Meteoridea hutsonii Nixon	Braconidae	Do	Do
Antrocephalus hakonensis Ashm.	Chalcididae	Pupa	Do
Brachymeria nosatoi Habu	Chalcididae	Do	Do
B. nephantidis Gahan	Chalcididae	Do	Do
Xanthopimpla punctuate F.	Ichneumonidae	Do	Do
X. nana nana Schlz.	Ichneumonidae	Do	Do
Trichospilus pupivorus Ferr.	Eulophidae	Do	Gregarious

Techniques have been developed for mass production of the promising parasitoids. The pest infested area should be monitored regularly and parasitoids releases should be initiated at the post monsoon period during November – December if there is any pest incidence. Parasitoids are to be released at the fixed dosages depending on the target stage of the pest at fortnightly intervals till the pest population is suppressed. As the release should synchronize with the stage of pest in the field, a sample of leaflets should be collected from infested palms at random and examined for live pest stages.

The parasitoid *Goniozus nephantidis* is released if the pest is at 3rd instar larval stage or above @ 20 parasitoid/palm and *Bracon brevicornis* @30 parasitoid/palm. The pre-pupal parasitoid *Elasmus nephantidis* and pupal parasitoid *Brachymeria nosatoi* are also very effective in managing the pest. They are released @ 49 and 32 per cent, respectively for every 100 prepupa and pupae estimated to be present on the palm (Sathaimma *et al.*, 1987). Before releasing in field, the parasitoids should be fed with honey and newly emerged parasitoid can be released in the field after three days of emergence. *G. nephantidis* and *B. brevicornis* could easily be mass multiplied on larvae of the rice moth *Corcyra cephalonica*. The pre-pupal parasitoid, *Elasmus nephantidis* is a highly host and stage specific parasitoid and always requires a steady supply of pre-pupa of *O. arenosella* for mass multiplication. This is the major constraint for mass production of *E. nephantidis* in the parasite breeding laboratories. Pupal parasitisms by many species of *Brachymeria* are observed in nature. Among them *B. nosatoi* is most potential for effective biocontrol since it possesses all the desirable attributes with high percentage parasitism, long life span, good searching ability and tolerance to high temperature. *B. nephantidis*, another major *Brachymeria* species in nature sometimes act as hyper parasitoid on primary parasitoids of *O. arenosella*.

Field Performance of Parasitoids

Using IPM technologies or exclusive release of promising biocontrol agents many demonstrations were laid out in the last two decades by ICAR-CPCRI in coastal Kerala and Karnataka. Studies conducted in an endemic area during 1990-1993 at Thodiyoor (Kollam District, Kerala) with the field release of the three stage specific parasitoids at fixed norms and intervals in *O. arenosella* infested coconut garden (2.8 ha) resulted in highly significant reduction (94 per cent) in pest population (Sathiamma *et al.*, 1996). Regular monitoring and release of stage specific parasitoids induced 52.6 and 94.7 per cent reduction in pest population after one and two years, respectively of parasite release in a heavily infested tract at Neendakara, Kollam Dist., Kerala. Large scale (1,400 ha) field validation of the biosuppression technology of coconut black headed caterpillar done during 1999-2002 in different geographic locations in coastal Karnataka (Ullal and Jeppinamogru) and coastal Kerala (Purakkad and Ayiramthengu) could achieve 93-100 per cent reduction in *O. arenosella* population in a period of two years with regular monitoring and release of stage specific parasitoids (Chandrika *et al.*, 2010) (Figure 10.2). Reports of biological suppression of *O. arenosella* in coastal districts of Odisha by the release

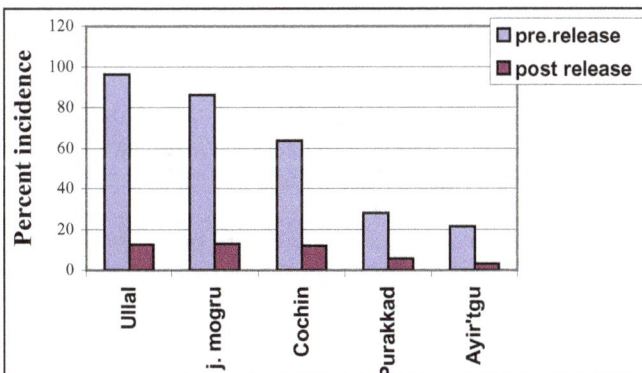

Figure 10.2: Incidence of *O. arenosella* in the IPM Demonstration Plots (1999-2002).

of parasitoids are also available. Successful biocontrol of *O. arenosella* was reported in Andhra Pradesh and timely augmentation of *G. nephantidis* @ 10 adults/palm at fifteen days interval suppressed the pest population in Karnataka. The higher frequency of occurrence of *G. nephantidis* in all the locations during post release period indicated the suitability of this larval parasitoid to adapt in various locations in the coastal belts of Kerala and Karnataka.

Parasitoids Reported from other Pests

Although parasitoids are reported from other pests of coconut, they were not successfully introduced as effective biocontrol agents. The scoliid wasp *Campsomeriella collaris* Fabr. parasitizes larvae of white grubs in the field. *Scolia cyanipennis* Fabr. and ectoparasitic mites are also observed in association with the different stages of the rhinoceros beetle. The egg parasitoid *Chrysochalcisea indica* was recorded from the coreid bug *Paradasynus rostratus* and the mymarid parasite *Parallelaptera* sp. on the eggs of *Stephanitis typica*. The parasites reported on the scale insect of coconut, *Aoniediella orientalis*, include *Aphytis chrysomaphali* Marcat and *Cauca parvipennis* Gaham, and *Cocobius reticulatus* (Aphelinidae). Further study in this line is required to collect and catalogue more parasitic species and to standardize the techniques for field application and evaluation.

2.1.1.2. Predators

Many species of spiders, coccinellids, reduviids, predatory ants, mites and thrips are well documented as effective predators. Spiders are important predators of insect pests. They suppress pest species to low densities at all growth stages. They are generalist predators and polyphagous in habit. Spider fauna contributes a major group of natural enemies in cropping systems (Hazarika and Chakraborti, 2007). Habitat manipulations like conserving ground litter, using safer pesticides *etc.* contribute to their conservation. Insect and spider predators are abundant in the coconut ecosystem. Insect predators are frequently observed in the breeding grounds of the rhinoceros beetle. They feed on the eggs and early instar larvae of the beetle. The important predators are *Santalus parallelus* Payk. (Coleoptera: Histeridae) (Antony and Kurien, 1966), *Pheropsophus occipitalis* Macleay, *P. lissoderus* Chaudior, *Chelisoches morio* (Fabricius) (Coleoptera: Chelisochidae) (Sathiamma *et al.*, 1982) and species of *Scarites sp.* (Coleoptera: Carabidae), *Harpalus* and *Agrypnus* sp. near *bifoveatus* Candeze (Coleoptera: Elateridae) (Kurien *et al.*, 1983).

An exotic predator *Platymeris laevicollis* Distant (Coleoptera: Reduviidae), was imported from Zanzibar to India for the control of rhinoceros beetle. As compared to the indigenous predators, *P. laevicollis* feeds on adult beetles. The predator is long lived (170-240 days) and fecundity high (110-170 eggs/female) and could easily be mass multiplied on ground roaches. Field release of predator was done in coconut plantations in Kerala and Karnataka @ 6 bugs/palm and could achieve significant reduction in beetle population and the damage to the palm. Leaf damage was reduced to 13.1 per cent, nil spathe damage and 1 per cent spindle damage as compared to 59.2 per cent, 2.5 per cent and 37.0 per cent, respectively, recorded during pre release observations. But the predators failed to establish under field conditions (Antony *et.al.*1979; Kurien *et al.*, 1983).

A variety of predatory fauna is also available in nature in association with *O. arenosella* (Table 10.2). The most important insect predators are the carabid beetle, *Parena nigrolineata* Chaud, anthocorid bug *Cardiastethus* spp, chrysopids, *Ankylopteryx octopunctata candida* Fab., coccinellids, ants and spiders. Spiders, though generalist predators and polyphagous, contribute a major group of natural enemies in coconut ecosystem. Spiders are the most dominant group of predators on *O. arenosella*. Species of *Cheiracanthium, Rhene* and *Sparassus* are the important spider predators and contributed nearly 21 per cent of the total spider fauna available on coconut foliage. They consume the immature and adult stages of *O. arenosella*. *Cheiracanthium* sp. consumed the prey @1.19 caterpillars per day. Male *Cheiracanthium* took 204-224 days and females 194-206 days to reach adulthood. Longevity varied from 35-122 days (male) and 51-127 days (female). Maximum population was recorded in the field during July. *Rhene indicus* is one of the common spiders feeding on *O. arenosella* caterpillars. Per day consumption was 0.7 caterpillars per predator. It reached maturity in 71-98 days (male) and 70-98 days (female). Longevity was 25-77 days (male) and 71-296 days (female). August was the peak period of the predator activity in the field (Pillai and Nair, 1990a, 1993; Sathiamma *et al.*, 1985 a, b; 1987, Nasser and Abudurahiman,1990, 1998).

Table 10.2: Important Predators of *Opisina arenosella*

Name of Predator	Predator Stage	Prey Stage
Ankylopteryx octopunctata candida Fab. (Chrysopidae)	Nymph	Egg/Early larva
Calleida splendidula F. (Carabidae)	Larva/adult	All stages
Parena nigrolineata Chaud. (Carabidae)	Do	Do
Cheiracanthium melanostoma Thor. (Clubionidae)	Do	Do
Rhene indicus (Salticidae)	Do	Do
Sparassus sp (Sparassidae)	Do	Do
Cardiastethus sp. (Anthocoridae)	Nymph/adult	Egg/Early larva

Predators of Scale Insects, Mealy Bugs, Lace Bug, White Flies

Scale insects, mealy bugs, coreid bug, lace bug and whiteflies are observed in most of the coconut plantations as minor pests. Their population is kept under check by the natural predators which are very active in the coconut ecosystem. Predators play a predominant role in limiting coconut scale population. Reports of a range of predators *viz.*, *Scymnus saverini* Wsc., *Scymnus* sp., *S. apiciflavus* Mots. *Pharellus minutissimus* Sic. *Azyatrinitatis* Hshl, *Pentilia insidiosa* Muls. and *Scymnus aeinipennis* on scale *Aspidiotus destructor* are available (Mohandas and Remamony,1993). The coccinellids *Chilocorus nigritus* Fab. and *Pseudoscymnus dwipakalpa* Ghorpade, *Cybocephalus* sp are predacious on *A. destructor* and *Aonidiella orientalis*. Sadakathulla (1993) developed techniques for the mass production of *C.nigrita* on *A. destructor* reared on pumpkin fruit (*Cucurbita maxima* Wall.). *C. nigrita* consumed 11 adults and 120 crawlers of *A. orientalis* in 24 h. The mite *Saniosulus nudus* Summers preyed on an average 15 crawlers of *A. orientalis* in 24h (CPCRI, 1990, 93). Fourteen insects and twenty-three spiders are observed as predators on *Stephanitis typica*. The identified insect predators are the mirid *Stethoconus praefectus* D., chrysopid *Ankylopteryx octopunctata octopunctata* Fabr., reduviids *Endochus inornatus* Stal., *Rhinocoris fuscipes* Fab., *Euagoras plagiatus* Burm. and earwig *Chelisoches morio*. Fab. The mealy bugs associated with coconut in India are *Palmicultor palmarum, Pseudococcus longispinus, Pseudococcus cocotis, Dysmicoccus* sp. and *Rhizoecus* sp. *Spalgis epius* (Lycaenidae) and species of *Pullus* and *Scymnus* (Coccinellidae) are the natural enemies recorded from mealy bug colonies. They exert limited check of the pest population. The predators of coreid bug include the ant *Oecophylla smaragdina* and the reduviid *Endochus* sp. Two types of whiteflies *viz.*, areca whitefly, *Aleurocanthus arecae* and spiralling whitefly *Aleurodicus dispersus* have been recorded from coconut in India. Their population is kept under check by the activity of predators *viz.*, *Seragium parcesetosum, Jauravia pallidula, Cybocephalus* sp. *etc.* Two species of lady bird beetles namely *Chilocorus subindicus* and *Scymnomorphus* sp. were found predatory on spiralling whitefly.

Predators of Coconut Eriophyid Mite

The nut infesting eriophyid mite *A. guerreronis* Keifer has emerged as one of the serious pests of coconut in India in 1998 and has become the major pest of coconut palm in a very short spell. Mites live in colonies on the tender portion of the buttons covered by the perianth and suck sap from the meristematic tissues. The symptom of attack is the appearance of elongated white patches below the perianth, later turning to pale yellow and brown. As the nut grows, the injury forms wartings and longitudinal fissures on the nut surface. Severe infestation results in button shedding, reduced kernel weight and reduction in fiber content. Currently botanical pesticides are recommended for mite management. A variety of predatory mites and smaller insects are associated with *A. guerreronis* in different parts of the

Figure 10.3: Eriophyid Mite Infested Coconuts.

world. Predatory mites belonging to Phytoseiidae, Bdellidae and Tarsonemidae are encountered in various collections. In India, the phytoseiid mite *Neoseiulus baraki* is the most dominant predator in the field. Other predatory mites include *Neoseiulus paspalivorus* and *Bdella* species. The insect predators encountered with coconut mite population in the field are thrips, coccinellids and syrphid maggots. But these are found only occasionally and in very few numbers. An increasing trend of incidence and better establishment of predatory mites in the field over the years are observed. The activity of the predators is high during June to December in the field. Compared to the young developing nuts below three months, more predators are encountered in 4-6 months old nuts. The predatory mites are larger in size compared to the coconut mite and they gain entry only later into the nuts. This is one of the limiting factors for the wider use of the predators. However, conservation of the predatory fauna in the ecosystem is beneficial to regulate the coconut mite in nature. Marimuthu *et al.* (2003) had reported three species of predatory mites *viz. Amblyseius paspalivorus, Bdella* sp and a tarsonemid attacking mite colonies in the field in Tamil Nadu. Predatory mites belonging to phytoseiidae, Tarsonemidae and Bdellidae attacking eriophyid mite colonies in the field was reported by Naseema Beevi *et al.* (2003). Mallik *et al.* (2003) reported that among the Phytoseiid mites, *A. paspalivorus* and the tarsonemid *Lupotarsonemus* sp are the predatory mites more encountered with mite colonies in Karnataka. The tarsonemid is found attacking on egg and the phytoseid attacking on all stages of mite. Predatory insects are very seldom reported on eriophyid mite. Nair *et al.* (2003) reported one species each of syrphid, thrips and coccinellid associated with mite colonies in the field.

The efficiency of the predators is neither fully assessed nor widely used in the biological control programmes of *O. arenosella*. Habitat manipulations like conserving ground litter, using safer pesticides *etc.* contribute to their conservation. The earwig *Chelisoches morio* (Fab.) feeds on the eggs and early instar grubs of red palm weevil in the field. Venkatesan *et al.* (2003) reported that *C. exiguus* could be reared using artificial diet. Lyla *et al.* (2006) evaluated *C. exiguus* against *O. arenosella* and found that the predator proved to be very efficient in suppressing the pest population. *Bergirus maindroni* Grou. (Mycetophazidae), *Dicrodiplois* sp. (Cecidomyiidae) *Spalgis epius* (Westwood) (Lycaenidae) and species of *Pullus* and *Scymnus* (Coccinellidae) were the natural enemies recorded from mealy bug colonies. They exerted limited check of the population (CPCRI, 1995).

Mites form important pests of arecanut palm. *Tetranichus fijiensis* Hirst, *Oligonychus indicus* Hirst, *O. biharensis* Hirst (Tetranichidae) and *Raoiella indica* (Tenuipalpidae) are the major species recorded (Nair, 1986; Nair and Daniel, 1982). Puttarudraiah and ChannaBasavanna (1956) had recorded some coleopteran predators on both *O. indicus* and *R. indica*. They included *Aspectes indicus* Arrow (Dermestidae), *Cybocephalus semipictis* Champ (Nitidulidae), *Stethorus parcepunctatus* Kapur, *S. tetranychi* Kapur, *Jauravia sorar* Wsc. and *Spilocarea bisselecta* Muls. (Coccinellidae). These predators particularly *Stethorus* kept the mite population in check during summer months. Kapur (1961) had described a new species, *Stethorus keralicus* from Kerala on *R. indica* and Daniel (1976) had studied the biology and predatory habits. Daniel (1979) recorded a number of indigenous predators and among them, two species of *Stethorus* and a staphylinid beetle were the major predators of *O. indicus*. The coccinellid, *S. keralicus* Kapur and the phytoseiid, *Amblyseius channabasavanni* Gupta and Daniel were the key predators of the palm mite *R. indica*.

Tea mosquito bug, *Helopeltis* sp. is the most serious pest of cashew and cocoa in India. Sundararaju and Sundarababu, (1999) reviewed tea mosquito bug's pest status, hosts and pest management practices. *Crematogaster wroughtonii* Forel (Formicidae) has been reported as a predator of nymphs of the pest (Ambika and Abraham, 1979). Spiders, *Hyllus* sp., *Oxyopes schireta*, *Phidippus patch* and *Matidia* sp. have been reported as predators of *H. antonii* (Sundararaju, 1984; Devasahayam and Radhakrishnan Nair, 1986). Reduviid bugs *viz.*, *Sycanus collaris* (Fab.), *Sphedanolestes signatus* Dist., *S. minisculus* Bergar, *Irantha armipes* Stal., *Endochus inornatus* Stal., *E. cingaensis* Stal., *Occamus typicus*, Dist. and *Alcmena* sp. have been recorded as predators of *H. antonii* (Sundararaju, 1984).

2.2. Entomopathogens

Biological control using entomopathogens is an important component of IPM. It is possible to use specific micro-organisms that kill arthropods. These include entomopathogenic fungi, nematodes, bacteria and viruses. The microbial agents which play a vital role in the bio suppression of various pests of coconut have been examined and in the case of black beetle (*Oryctes rhinoceros* L.) this group of natural enemies has been thoroughly studied.

2.2.1 Fungi

Entomopathogenic fungi are of considerable importance in crop pest control because of their ability to infect a wide range of insect pests through *non per os* infection. In coconut pest management, entomopathogenic fungi such as *Beauveria bassiana*, *Metarhizium anisopliae*, *Hirsutella thompsonii* are the best utilized globally. Fungi require relatively high humid microclimate (>70 per cent RH).

a) Green Muscardine Fungus as a Potential Enemy of Rhinoceros Beetle

The rhinoceros beetles cause damage to palms of all age groups by boring into the unopened spindles and inflorescence. As the pest bores deeper into the host, it pushes out the chewed up tissues, which are seen extruding from the entry points. Once these injured spindles open up, the green leaves present a geometric 'V' shaped cut pattern. The damage to inflorescence is seen as round oblong holes on the spathes, which soon dry up. The pest occurs throughout the year and breeds in cattle dung, compost, dead and decaying organic debris like coconut and other palm trunks, cocoa pod shells, oil palm bunch waste, coir dust, rotting paddy straw and sugar cane waste *etc.* The life cycle is completed in 6 months on an average.

Among the entomopathogens, the green muscardine fungus *Metarhizium anisopliae* (Metsch.) Sorokin (Deuteromycotina: Hypomycetes) is one of the most effective and successful biocontrol agents in coconut ecosystem as a potential pathogen of *O. rhinoceros*. *M. anisopliae* var. *major* (spore size 10-14 μm) is highly infective variety used widely for the control of this pest. All the stages of the host excepting the eggs are mycosed. The fungus is very active during monsoon when the relative humidity is 70-90 per cent and the temperature is 26-28°C. The infected grub becomes sluggish and mortality occurs within 10-15 days. The body of the infected grub becomes hardened and white powdery fungal colonies appear in the joints of the integument. Within a week green coloured spores are produced and finally the cadavers become black and mummified. The mass production of the fungus has been developed at ICAR-CPCRI using solid (cassava chips and rice bran mixture supplemented

Figure 10.4: *Metarhizium anisopliae* **Infected Grub of Rhinoceros Beetle.**

with nitrogen source) and liquid (coconut water) media. Different substrates like broken rice/wheat grains, millets *etc*. also are found to be cheaper substrates for multiplication of the fungus. For the field application of the fungus, the fungal spores are mixed with sterile water and used to drench the breeding materials of the rhinoceros beetle @ of 5×10^{11} spores/m^3. The fungus survives in the breeding material for long periods. Use of this fungus for biocontrol of rhinoceros beetle has been popularized as a women friendly technology and ICAR-CPCRI is presently facilitating farm level production of this fungus to cater local needs.

b) Hirsutella thompsonii: A Promising Fungal Pathogen against Coconut Mite

The fungus *Hirsutella thompsonii* has been reported as a predominant pathogen among microbial pathogens of eriophyid mite. In India the incidence of *H. thompsonii* was recorded from Kerala, Karnataka, Tamil Nadu, Andhra Pradesh, Pondicherry and Lakshadweep Islands. In Kerala, local strain of *H. thompsonii var. synnemetosa* could be isolated from field samples. ICAR-CPCRI could collect virulent native isolates of this fungus from different locations of India. Talc based formulations of the virulent strains of this fungus are being evaluated for the suppression of coconut eriophyid mite in the field and preliminary results indicated 70-80 per

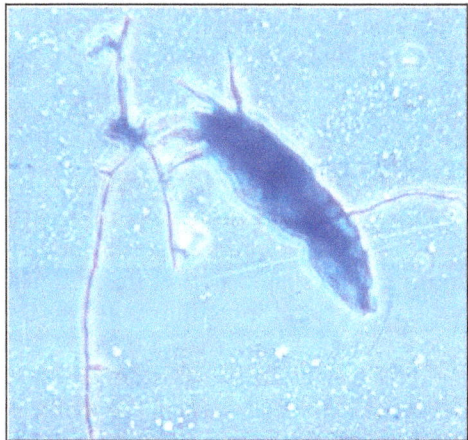

Figure 10.5: Coconut Eriophyid Mite Infected with *Hirsutella thompsonii.*

cent suppression in pest population. Other fungal species associated with eriophyid mite include species of *Paecilomyces, Beauveria, Metarhizium, Sporothrix, Verticillium,*

Acremonium, Aspergillus, Penicillium and *Fusarium*. However, the bio-efficacy of these fungi as biocontrol agents of mite in field conditions is not fully studied.

2.2.2. Virus

a) Oryctes rhinoceros Nudi Virus

The Oryctes rhinoceros nudi virus (OrNV) was discovered in 1960 in Malaysia and has been effectively used to control the rhinoceros beetle in coconut and oil palm in Southeast Asia and the Pacific. It is a classical example of successful inoculation and long term control of an insect pest. The virus consists of rod shaped virions and replicates in the nuclei of infected cells. On the basis of its (ultra) structure, OrNV was previously considered to be a so called non–occluded baculovirus (NOB). Due to the lack of occlusion bodies, it was later removed from the family Baculoviridae. OrNV contains a double stranded DNA genome of about 130 kilobase pairs. The

virus gains entry into the host through contaminated food and it multiplies in the mid gut epithelial cells and fat bodies of grubs and adults. Infected grubs become lethargic, stop feeding and move to the surface of the feeding media. With the multiplication of the virus in the mid gut epithelium of the grub the fat body disintegrates and the haemolymph content increases giving a translucent appearance for the grub. The diseased grub dies within 15-20 days. The infection by OrNV results in reduction of the longevity of the beetle by 45 per cent and fecundity by 95 per cent relative to the healthy beetles. Diseased adults also become inactive.

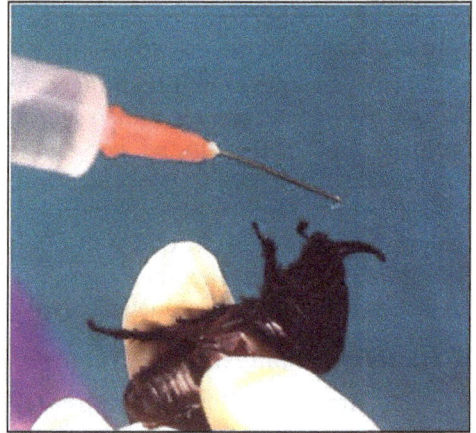

Figure 10.6: Inoculating Rhinoceros Beetle with Oryctes Rhinoceros Nudi Virus.

Mass Production of the Virus

The Oryctes virus is mass multiplied and maintained in the laboratory on live grubs. The OrNV infected grubs are dissected, the swollen midgut is taken out and the viral suspension of the midgut is prepared by grinding with phosphate buffer. This viral inoculum is fed to the healthy grubs to induce infection. The inoculated grubs are reared on sterilized food (cow dung, coir pith or saw dust) moistened sufficiently. The development of OrNV infection is to be monitored in the inoculated grubs and the whole procedure of inoculation of healthy grubs is repeated to maintain the viral culture in the host. The grubs showing external symptoms such as translucent gut or extroversion of rectum has to be dissected immediately; otherwise secondary infection with other pathogens like *Pseudomonas etc.* takes over and destroys the OrNV infection.

Field Release

For field release, either the adult beetles are inoculated orally using a syringe with the OrNV inoculum or they are allowed to wade through the virus inoculum (2g of infected larval gut tissues in 1 litre of phosphate buffer) contained in a basin for 30 minutes. After the inoculation, the beetles are confined together in a box containing rotten coconut wood powder mixed with the virus inoculum for 24 hours. The treated beetles are removed the following day and confined for a week in a box containing fresh coconut petioles provided as food, since the infected beetles begin to excrete the virus only a week after infection. The infected beetles are liberated after dusk in the field. 12-15 infected beetles per hectare are recommended to disseminate the virus in nature.

Impact of the Release of OrNV

In India, the population of *O. rhinoceros* and its damage on coconut palm was checked substantially when the OrNV was released in the pest infested areas of Lakshadweep Islands. In Minicoy, there was a reduction of 76.9 per cent leaf damage, 93.8 per cent spathe damage in a period of 30 months. In Androth islands leaf, spathe and fresh spindle damage showed reduction of 75.4, 56.1 and 74.8 per cent, respectively in a period of two years and the OrNV population showed an increase from nil to 60.6 per cent showing the establishment of the pathogen in the pest infested area (Mohan *et al.,* 1989). Similar results were observed in the mainland in Chittilappilly, Trichur, Kerala, where pest incidence reduced from 100 to 23 per cent and there was 80.6 per cent reduction in leaf and 100 per cent reduction both in spathe and fresh spindle damage when OrNV was re-released in a period of 3 years during 1989 to 1992. Establishment of OrNV in native population of *O. rhinoceros* at Sipighat, Andamans by release of the virus is also reported (Jacob, 1996). From nil infection during the pre-treatment period in 1989 the infection has increased to 61 per cent in 1991. By augmenting the natural population of virus infected beetles, sustained reduction in beetle incidence and crop damage can be achieved even in already infected contiguous areas (Biju *et al.,* 1995). The biggest singular advantage OrNV offers over other microbial agents is in being an auto transmissible pathogen capable of passing from generation to generation. The method of its propagation involves the release of diseased beetle and no spraying of the virus to the crop is required. Significant reduction in rhinoceros beetle damage by the combined use of two potential biocontrol agents (*M. anisopliae* and OrNV) was reported from large scale field studies (2,400 ha) done in Alappuzha (75.1 and 79.4 per cent reduction in leaf and spindle damage) and Kasaragod (66.6 and 95.8 per cent reduction in leaf and spindle damage) districts during 1999-2002 (Nair *et al.,* 2010).

2.2.3. Bacteria

Bacterial pathogens *Acinetobacter calcoaceticus* and *Pseudomonas alcaligenes* have been identified as pathogens of *O. rhinoceros* grubs. *Bacillus thuringiensis* Berliner and *Serratia marcescens* B. are observed to be pathogenic to *O. arenosella* in the field. So far no effective microbial agent could be identified on red palm weevil which is a fatal enemy to coconut palm. *Pseudomonas aeruginosa* was identified as facultative pathogen of red palm weevil. A yeast isolate has also proved to be a transitional pathogen producing mortality in weevils/grubs. Reports on the infectivity of *Bacillus thuringiensis* subspecies kurstaki (Btk.) and a polyhedrosis virus to larvae of red palm weevil are available. These pathogens were isolated from the field in Egypt and stored for 4 years as air dried smears on glass slides, before being tested. Record of eugregarine protozoan pathogen *Pseudomonocystis* sp., which infect 22.7 per cent of the third instar white grubs in the field is available.

2.2.4. Entomopathogenic Nematodes

Entomopathogenic nematodes (EPN) in the families Heterorhabditidae (represented by the genus Heterorhabditis) and Steinernematidae (represented by the genera Steinernema and Neosteinernema) have been used to suppress populations of soil and cryptic insect pests in a variety of agro-ecosystems. EPN are obligate parasites of insects and kill their hosts with the aid of bacteria carried in the nematode's alimentary canal which provide nutrients to the nematodes, produce antibiotics and inhibit competing microbes, and kill the host through septicemia. They are associated with mutualistic bacteria in the genus Xenorhabdus for Steinernematidae and Photorhabdus for Heterorhabditidae. The positive attributes of these nematodes as biological control agents are that they have a broad host range, are safe to invertebrates, plants and other non-target organisms, have no known negative effect on the environment, are easy to mass produce *in vivo* and *in vitro*, are easily applied using standard spray equipment, can search for their host, kill rapidly (within 48 h), have the potential to recycle in the environment, are compatible with biological pesticides, are amenable for genetic selection for desirable traits and are exempt from registration in many countries.

Field Efficacy

Moisture conditions have been recognized as one of the most important factors in the soil environment affecting survival, virulence and persistence of nematodes. Entomopathogenic nematodes need high relative humidity to survive and a film of free water for movement. They may become dormant at very low soil moistures.

In vivo Multiplication of Entomopathogenic Nematodes (EPN)

In vivo mass production of EPN has been standardized using *Galleria mellonella* larvae in the laboratory by filter paper inoculation technique. *Steinernema* sp. infected cadavers turn creamish-white in colour characteristic of the bacterium (*Xenorhabdus poinarii*) housed in it whereas, *Heterorhabditis* sp. infected insects become reddish-brown due to *Photorhabdus luminescens*. After required incubation, the cadavers are placed in White's trap for the emergence of nematodes. From the White's trap, the nematodes wriggle out and migrate all over the Petri dishes under minimum moisture content. Nematodes, thus, emerged out are harvested by adding water and later filtered for concentrating the nematodes. On an average 2-3 lakh nematodes are produced from each insect.

Use of EPN against Coconut Insect Pests

EPN has been found promising in the management of red palm weevil and white grubs infesting coconut. An entomopathogenic heterorhabditid nematode was isolated from Egypt and United Arab Emirates planted with date palm trees and has shown potential for the control of red palm weevil. Among the species of EPN evaluated against the grubs of red palm weevil, *R. ferrugineus*, in filter paper based bioassay, *Heterorhabditis* sp. was found to be more virulent than *Steinernema* sp. and the local isolate *H. indicus* was found to be more virulent inducing 92.5 per cent mortality @ 1500 IJ/grub. Talc based EPN formulation of *H. indicus* elicited higher mortality than water suspension based formulation on coconut petiole based bioassay. Among the four species of EPN evaluated against the coconut white grub in soil-column bioassay, *Steinernema abassi* was found to be virulent inducing 37-45 per cent mortality of white grubs @ 5000 IJ/grub in a period of 96-120h. Species specificity against various pests, concentration of IJ and moisture content hold the key in the field success of EPN on a large scale. Field evaluation of the promising EPNs against red palm weevil and white grubs are to be undertaken on priority as these two major pests are not having any effective biocontrol agents.

2.3. Limitations of Biocontrol Methods

Under the present market system, many biological control products have not competed well with less expensive and more effective synthetic pesticides. The down sides of biocontrol agents are that most are niche products, pest control is not immediate, there can be lack of environmental persistence, and efficacy can be low and unpredictable particularly in outdoor environments. Timely availability of quality bioagents, quantity, farmers perception about biological control, short shelf life of bioagents, formulation, storage and registration *etc.* are some of the limiting factors for popularization of biocontrol technology. Biosystematics of insect groups especially parasitoids and predators is of prime importance in biocontrol. Illustrated easy to use identification guides should be made available for research and extension workers. Identification service is lacking for major groups of crop pests and their natural enemies. Regional approaches to catalogue biodiversity would be ideal.

2.4. Botanical Pesticides

Botanical pesticides are having a prominent position in pest management under organic agriculture. As botanical pesticides are safe for the non-target organisms, they are preferred in organic pest control strategies. Botanical pesticides are used either as repellents or as primary insecticides having adverse effect on the physiological functioning of insect pests. Derivatives from various species of plants have been in use for plant protection from ancient days onwards in agriculture. More than 600 plant species are reported to have pesticidal properties. Tulsi, Mahuva, Lantana, Tobacco, Marigold, *Clerodendron infortunatum etc.*, are some important plants used for pest management. Insecticidal properties of neem were proved beyond doubt in ancient times and in modern agriculture, neem based pesticides are getting importance because of their versatile ability to control many pests of agricultural importance. *Azadirachtin* is the active principle having insecticidal property present in neem tree. Decoction of neem leaf, neem oil cake, powdered neem seed and seed oil are used in various dilutions mixed with washing soap to effect satisfactory control of various groups of foliage pests in many field crops. In coconut mite management, use of neem pesticides is recommended either as spray

Figure 10.7: *Clerodendron infortunatum* **Plant.**

or as root feeding (Nair *et al.*, 2005). Neem cake/pongamia cake mixed with sand is effectively utilized for prophylactic leaf axil filling to repel rhinoceros beetle from coconut palm (Rajan *et al,.* 2009).

2.5. Semiochemicals

Semiochemicals which are considered to be a modern tool in IPM concept has a pivotal role in eco-friendly pest management programmes. Semiochemicals

Figure 10.8: *Clerodendron infortunatum* **Induced Malformation in Rhinoceros Beetle.**

are compounds released or emitted by plants and animals for specific communication purposes. If the communication is between the same species (Instraspecific), they are called pheromones, and on the other hand if the communication is between two different species (Interspecific), they are called kairomones. Insect pheromones fall into several categories usually related to functions *viz.*, sex pheromones, aggregation pheromones, alarm pheromones, trail pheromones, *etc*. Pheromones that have been used most successfully in pest monitoring/control are sex pheromones of Lepidoptera and aggregation pheromones of Coleoptera. The application of pheromones in crop protection may be indirect as population monitoring agents or direct as tools in mass trapping, lure and kill and mating disruption techniques (Yadav *et al.*, 2004). In plantation crops, successful attempts have been recorded for the control of coconut red palm weevil, *Rhynchophorus ferrugineus* (Nair and Saritha, 2003).

2.6. Light Traps or Attractant Traps

Red palm weevil of coconut are attracted to food lure traps employing the natural food substances. White grubs are major pests of coconut, arecanut, tuber crops, sugarcane and ground nut and they are successfully managed using light traps. Trapping of adult beetles during their emergence period coinciding with onset of monsoon is found to be one of the effective strategies in IPM of white grubs.

2.7. Agronomic Practices

Figure 10.9: Pheromone Trap for Rhinoceros Beetle.

Historically, agronomic practices or cultural manipulations of plants were the most important methods for controlling and preventing crop losses. Farmers continuously refined them with experience. Among the oldest techniques for managing pests are sanitation, destruction of alternate hosts, tillage, avoidance and planting of crop cultivars that resist pest attacks. The worldwide awareness of safe environment provided impetus to foster non-chemical pest management strategies. Of these strategies, cultural control constitutes the most farmer-oriented approach, where weak points in the biology and behaviour of insects are exploited and pressure is exerted on the population by manipulating the environment. An understanding of natural defense mechanisms present in the host against the pest is useful to prepare a pest control programme where agronomic methods are to be utilized. Also, a thorough knowledge of pest, host, host-plant interaction and optimum environment for both are necessary in this method. In an agricultural country like India, agronomic management practices are of immense importance because of low cost of inputs, easy in their adoption and acceptance by farmers. Nutritional management of the plant is important in managing pest problems and is very important in sucking pests like coconut eriophyid mite.

2.7.1. Sanitation

Removal and destruction of breeding materials, manipulating over-wintering sites, pruning of infested plant parts *etc.* are sanitation methods that can be employed in organic agriculture. Sanitation is important to prevent introduction of insects, pathogens, nematodes and weeds into pest free fields and to reduce losses in infested fields. Sanitation practices such as use of healthy seed material, burning and destruction of crop refuse, clean storage, *etc.*, reduce pest population, discourage breeding and hibernating sites and prevent carry over to the next crop season. These methods are economical, effective and easy to adopt. Depending upon the crops and types of pests that affect the crops, sanitation methods can be practiced by farmers.

Crown cleaning in coconut reduces the coreid bug damage and also removes the nesting sites of rodents. Phyto-sanitation is the key factor in the management of major pests of plantation crops like rhinoceros beetle and red palm weevil of coconut and stem borers affecting crops like cashew, coffee, cocoa, *etc.* Avoiding breeding sites of black beetle in the immediate vicinity controls its attack in coconut plantations. Timely pruning and pod harvest at ripening time is important in reducing pest menace in cocoa.

2.7.2. Tillage

Tillage is an integral factor in destroying food sources and habitat of the pests within the field. In addition tillage plays a major role in rhizosphere microflora

Figure 10.10: Summer Ploughing for Pest Management.

and microfauna. Tillage operation changes the texture, nutritional composition and pH of the soil which becomes less favourable to weeds and soil inhabiting insects. Deep ploughing during the pre-monsoon and post-monsoon seasons helps in control of soil insects like white grubs, termites and army worms. Many times, tillage operations are also essential in managing nematodes and rodents.

2.7.3. Resistant Cultivars

Genetic resistance is one of the oldest methods of pest control. Growing resistant cultivars is the most effective and economic means of controlling plant pests. Resistant cultivars are also the first line of defense against pests. Resistant varieties should be used in concert with other pest suppression or pest control measures. Plant resistance provides a built-in ability to allow fewer pests and cut off the extra load of insecticides in the crop. Keeping in mind the diversity and intensity of pests in a particular place, selection of resistant/less susceptible varieties holds well in pest management. Tall varieties of coconut are found to be tolerant to red palm weevil damage. Orange dwarfs are tolerant to mite damage in coconut.

2.7.4. Trap Crops

Trap crops or catch crops are species of plants which are planted to attract and retain a pest species or to provide a more favourable habitat to increase natural enemies. Trap crop provides protection either by preventing the pest from reaching the main crop or by concentrating them in certain parts of the field, where they can be economically destroyed. Trap crops have been utilized in control of nematodes. Marigold as a trap crop for nematodes in vegetable fields and black pepper garden is common at many places.

2.7.5. Regulation of Alternate Hosts of Pests

Many economically important pests are known to survive on collateral hosts which constitute mainly the weeds grown in the vicinity of the agricultural land or a subsidiary crop grown in the field. The coreid bug of coconut not only infests coconut but also infests collateral hosts like guava, cashew, cocoa and tamarind. Tea mosquito bug is a common pest of cocoa and cashew. Knowledge on the feeding behaviour and biology of the pests of cultivated crops would enable the farmer in utilizing the technology effectively.

2.8. Pest Surveillance

Pest surveillance and monitoring system plays a major role in pest management in an organic farming system. The farming community has to be aware about the pest problems of the crop, the behavior and population ecology of the pests and their natural enemies for employing appropriate technology at a time of urgency. This is possible only through an effective transfer of technology programme by an interactive involvement of research- extension agencies and farming community.

3. Ecological Engineering for Pest Management

Ecological engineering for pest management has recently emerged as a paradigm for considering pest management approaches that rely on the use of cultural techniques to effect habitat manipulation and to enhance biological control. This novel approach is based on informed ecological knowledge rather than high technology approaches such as synthetic pesticides and genetically engineered crops.

Below Ground

There is a growing realization that the soil borne, seed and seedling borne diseases can be managed with microbial interventions, besides choosing appropriate plant varieties. The activities that can increase the beneficial microbial population and enhance soil fertility are:

1. Crop rotations with leguminous plants which enhance nitrogen content
2. Keeping soils covered (mulched) year-round with living vegetation and/ or crop residue
3. Adding organic matter in the form of farm yard manure (FYM), vermicompost, crop residue, which enhances the soil organic carbon as well as below ground biodiversity of beneficial microbes and insects.
4. Application of balanced dose of nutrients using biofertilizers based on soil test report
5. Application of biofertilizers with special focus on mycorrhiza and plant growth promoting rhizobia (PGPR)
6. Application of *Trichoderma harzianum/viride* and *Pseudomonas fluorescens* for treatment of seed/seedling/planting materials in the nurseries and field application (if commercial products are used, check for label claim and certification under organic cultivation). However, no registration is required for biopesticides that are produced by farmers for own consumption in their fields.

Above Ground

Natural enemies play a very significant role in control of foliar insect pests. Diversity of natural enemies contributes significantly to management of insect pests both below and above ground. They may require food in the form of pollen and nectar; shelter, overwintering sites and moderate microclimate, *etc.* and alternate hosts when primary hosts are not present.

In order to attract natural enemies, the activities to be practiced include the following:

1. Raise the flowering plants/compatible cash crops along the borders of the plantation by arranging shorter plants towards main crop and taller plants towards the border to attract natural enemies as well as to avoid immigrating pest population.
2. Grow flowering plants on the internal bunds inside the field.
3. Allow naturally growing weed plants such as *Tridax procumbens, Ageratum* sp, *Alternanthera* sp *etc.* which act as nectar source for natural enemies.
4. Reduce tillage intensity so that hibernating natural enemies can be saved.

5. Grow appropriate companion plants which could be trap crops and pest repellent crops. These crops will also allow natural enemies as their flowers provide nectar and the plants provide suitable microclimate.

6. Due to enhancement of biodiversity by the flowering plants, the number of parasitoids and predators (natural enemies) will also increase. The major predators are a wide variety of spiders, ladybird beetles, long horned grasshoppers, Chrysoperla, earwigs, *etc.*

4. Future Thrusts in Pest Management

The increasing awareness on the ecological aspects of pest management and the inclination to conserve ecosystem without disruption for maintenance of sustainability among various stake holders are appealing in recent years. Utilization of frontier areas like information technology can make tangible effects in safer pest/disease management systems. Proper planning on the adoption of apt control measures to save the crop is possible through reliable prediction of pest/disease out break based on weather forecast. Employment of modem tools like remote sensing and geographical information system (GIS) will be useful in assessing crops loss and pest/disease incidence and using the information from such sources, advance action plan can be taken up to schedule effective plant protection strategies over larger areas within short period. Exploring the potential of farmer's knowledge and finding out scientific basis for such indigenous technical knowledge of farmers are to be priority areas in our attempts to manage the plant health problems in organic agriculture. Bio informatics is a recent field by which speedy dissemination of information is achieved. This helps in proper understanding and timely action to tackle situations like accidental introduction or sudden out breaks of pest species.

5. Conclusion

All the proven cases with the promising bioagents have confirmed a meaningful way of pest management utilizing the indigenous fauna for the biological suppression of the pests of plantation crops especially coconut. The indigenous natural enemies proved to be quite useful in controlling the pests, particularly rhinoceros beetle and leaf eating caterpillar. The biosuppression of rhinoceros beetle by employing viral pathogen OrNV is documented as one of the classical examples of biocontrol of an insect pest. Proper monitoring and release of stage specific parasitoids could effectively bring down the population of coconut leaf eating caterpillar in the field. The current scenario warrants an augmentative release of these promising bioagents in areas wherever pest infestation is found. Spider fauna plays an important role in the natural suppression of pests in the field. Conservation of these biocontrol agents of the pests has become quite imperative. Limitation on use of pesticides or to identify safer and ecofriendly pesticides is required to integrate the biocontrol agents in IPM. As predators are mostly polyphagous, release of predators when parasitoids are active has to be avoided. Pollen and nectar bearing plants in the vicinity provides supplementary food for predators such as chrysopids, coccinellids and mites. More detailed studies on the bioecology of different species of parasitoids and predators affecting the key pests of plantation crops are to be undertaken with a view of utilizing them for biocontrol suppression of the pests. Farmer field schools have proved as the most effective tool in technology transfer and capacity building of farmers. Training farmers and setting up farm level production units of biocontrol agents would ensure timely distribution of quality bioagents to the stake holders. Community level approach is highly essential to curb pest/disease menace in plantation crops.

Selected References

Abraham, C.C. (1994). Pests of coconut and arecanut. pp. 709-726. In: Advances in Horticulture Vol. 10 (Part 2) K. L. Chadha and P. Rethinam (Eds.). Malhotra Publishing House, New Delhi, India.1282p

Bedford, G.O. (2013). Biology and Management of Palm Dynastid Beetles: Recent Advances. *Annual Review of Entomology*. **58**: 353-372.

Chandrika M. and Sujatha, A. (2006). The Coconut leaf caterpillar, *Opisina arenosella* Walker CORD: **22** (Special Issue), 25 -78.

Cock, M.J.W. and Perera, P.A.C.R. (1987). Biological control of Opisina arenosella Walker (Lepidoptera:Oecophoridae). *Biocontrol News and Information*. 8(4):283-310.

Faleiro, J. R. (2006). A review of the issues and management of the red palm weevil Rhynchophorus ferrugineus (Coleoptera: Rhynchophoridae) in coconut and date palm during the last one hundred years. *International Journal of Tropical Insect Science* 26(3): 135-154.

Howard, F.W., Moore, D., Giblin-Davis, R.M. and Abad, R.G. (2001). Insects on palms, CAB International, Wallingford, Oxon, UK pp: 42-70.

Koshy, P.K. and Chandrika Mohan (2006). Status and prospects of integrated pest management strategies in selected crops - coconut. pp: 522-579. In: Integrated pest management: Principles and Applications Vol.2. America Singh, O. P. Sharma and D.K. Gang (Eds), CBS Publications, New Delhi.

Nair, C.P.R., Rajan, P. and Chandrika Mohan. (2005). Coconut eriophyid mite, *Aceria guerreronis* Keifer – An overview. *Indian Journal of Plant Protection* 33(1):1-10.

Pillai, G.B. (1993). Biological control of insect pests of plantation crops. In: Organics in soil health and crop production (Ed.) P.K. Thampan, Peekay Tree Crops Development Foundation, Cochin pp: 235-252.

Sathiamma, B. Chandrika Mohan and Murali Gopal (2001). "Biological potential and its exploitation in coconut pest management" In." Biocontrol potential and its exploitation in sustainable agriculture" Vol 2. Insect pests R.K. Upadhay, K.G. Mukerje and B.P. Chamola (Eds.). Kluwer Academic/Plenum Publishers, New York. pp: 261-283.

Singh, H. P. and Rethinam, P. (2003). Coconut eriophyid mite-issues and strategies. In: Coconut Eriophyid Mite- Issues and Strategies. (Proceedings of the International Workshop on Coconut Mite held at Bangalore), H. P. Singh and P. Rethinam (Eds.), Coconut Development Board, Kochi (India), 146 pp.

Chapter 11

Organic Farming Practices in Palms and Cocoa: Field Level Scenario and Future Strategies

✩ *D. Jaganathan, C. Thamban, S. Jayasekhar, C.T. Jose,*
V. Krishnakumar, P. Anithakumari and K.P. Chandran

1. Introduction

Plantation crops like coconut, arecanut, cocoa, coffee, tea and rubber play an important role in social, cultural and economic life of people in India. Coconut, arecanut and cocoa are the important perennial crops which are largely grown in south India especially in the states *viz.,* Kerala, Karnataka, Tamil Nadu and Andhra Pradesh. India is one of the major producers of coconut in the world and about 12 million people are dependent on coconut farming and its allied activities. India has produced 21,665 million nuts in the year 2014 from an area of 2.14 million ha with a productivity of 10,122 nuts per hectare. Kerala, Karnataka, Tamil Nadu and Andhra Pradesh are the four major states which account for 88.8 per cent of area and 91.2 per cent of India's production (CDB, 2015). Arecanut is grown in parts of Karnataka, Kerala, Assam, Meghalaya, West Bengal, Tamil Nadu and Andaman and Nicobar Islands. India is the largest producing country with a production of 7.46 lakh tonnes from an area of 4.50 lakh ha (DASD, 2014). Karnataka, Kerala and Tamil Nadu are the major southern states where arecanut is cultivated in an area of 3.20 lakh ha with a production of 5.97 lakh tonnes. (DASD,2014). Cocoa is considered to be the food of God and is native to Amazon region of South America. Cocoa was introduced in India as a mixed crop/intercrop in coconut and arecanut plantations during 1970s. At present, it is cultivated in 65,500 hectares in India as component crop in coconut, arecanut and oil palm plantations with a production of 13,400 tonnes (DCCD, 2014). It is mainly cultivated in four southern states *viz.,* Karnataka, Kerala, Tamil Nadu and Andhra Pradesh. There is a tremendous scope for area expansion in cocoa because of heavy demand in Indian chocolate industry and confectionaries which is portrayed as 60,000 mt for the year 2025.

Research on coconut, arecanut and cocoa started in India during 1916, 1956 and 1970, respectively which resulted in generation of large number of viable technologies *viz.,* improved varieties/hybrids, agro techniques, cropping/farming system, pest and disease management, processing and value addition. Diverse production systems had been practiced by farmers due to diverse agro-climatic and socio economic conditions across different parts of the country. Organic farming is getting prominence in various parts of India since 1990s, but no systematic and institutional work had happened till 2000. National Programme for Organic Production (NPOP) was launched in May, 2000 by the government with the objective of promoting organic farming in India. Since then, India is showing rapid progress in organic sector. In 2010-11, 6.0 lakh hectares of cultivated land were under certified organic and another 1.75 lakh hectares were under conversion (NCOF, 2011). In India, National Steering Committee comprising of Ministry of Commerce, Ministry of Agriculture, Agricultural and Processed Food Export Development Authority (APEDA), commodity boards and various other government and private organizations associated with the organic movement is monitoring the overall organic activities under the National Programme for Organic Production. APEDA is the coordinating agency for organic food production and export under the brand name "India Organic". The steps involved in certification are, registration of the producers and processing industries, provision of basic information on the crop and farm, inspection and verification of the field and processing unit, inspection of production methods and practices by the inspector of the certifying agency.

Realizing the potential of organic farming, National Project on Organic Farming was started in the year 2004. In India, the demand for organic produce increases year after year particularly in international trade market. Organic products produced in India are tea, spices, fruits and vegetables, rice, coffee, cashew nuts, oilseeds, wheat, pulses, cotton and herbal extracts. Products with potential in the domestic market are fruits, vegetables, rice and wheat, while those in the export market are tea, rice, fruits and vegetables, cotton, wheat and spices. Organic farming is a unique production management system which promotes and enhances agro eco-system health, including bio-diversity, biological cycles and soil biological activity and this is accomplished by using on-farm agronomic, biological and mechanical methods in exclusion of all synthetic off-farm inputs (FAO, 1993). Organic farming practices are gaining importance among farmers, trainers, entrepreneurs, policy makers, agricultural scientists, processors and extension personnel for varied reasons such as it minimizes the dependence of external inputs, thus, not only reduces the cost of cultivation but also safeguard as well as preserve quality of resources and environment. ICAR institutes, State Agricultural Universities (SAUs), Krishi Vigyan Kendras(KVKs), Non Governmental Organizations (NGOs), State Department of Horticulture/Agriculture and other government and private agencies have started advocating organic farming practices in horticultural crops including coconut, arecanut and cocoa.

In this context, systematic research for evolving organic package of practices for coconut, arecanut and cocoa has started recently at ICAR- CPCRI and different centres of the All India Coordinated Research Project on Palms keeping in view the sustainability aspects and other benefits. Many farmers cultivating coconut, arecanut and cocoa have started adopting organic farming practices due to various socio-economic and other related factors. In this chapter, field level scenario of organic farming practices in these crops adopted by such farmers is discussed in detail and future strategies for promotion of organic farming are outlined.

2. Socio-economic and Farming Details

An analysis of organic farming practices in coconut in south India conducted by Jaganathan *et al.* (2013) revealed that around 67 per cent of the farmers had less than two ha, with nearly 80 per cent having more than 15 years of experience in coconut cultivation. Around 63 per cent farmers maintained any livestock component in their farm, which is an integral part of organic farming. Only 12 per cent of the cultivators maintained any farm records, whereas, 23 per cent of farmers carried out soil testing as a basis for nutrient management and organic certification was adopted by less than 3 per cent of the farmers included in the study. In the case of arecanut (Jaganathan *et al.*, 2014), 58 per cent per cent of the farmers had less than two ha area cultivated with arecanut with 80 per cent having more than 15 years of experience in cultivation. Around 67 per cent maintained livestock component in their farms, while about 30 per cent of the farmers were found to maintain farm records. Soil fertility evaluation was done by 28 per cent of the farmers. Certification for organic farming was taken up only 3 per cent of the farmers surveyed. Majority of cocoa farmers (> 60 per cent) had less than two ha area under the crop and more than three fourth of farmers had livestock component in their farms. It was noticed that 30 per cent of farmers maintained farm records and soil testing was done by around same percentage of farmers. Organic certification was taken up by five per cent (Jaganathan *et al.*, 2015) of the farmers surveyed.

3. Adoption of Organic Farming Practices

In the conventional farming, the farmers depended heavily on external agencies for inputs like planting materials, fertilizers, pesticides *etc*. Moreover they had to wait for long time for getting the inputs for use in their farms. Farmers are becoming more aware about the ill effects of chemical farming on the health of human beings. The indiscriminate use of chemicals not only polluted the soil, water and air but also affected the health of human beings. This changed their mindset to go for organic farming practices and made them conscious of 'concern for human health'. The analysis of different reasons for adopting organic farming practices, in general, revealed soil and human health, economic as well as social issues as the major ones. Most of the farmers surveyed felt that 'maintenance of soil fertility' was very much necessary. Concern for human health, minimizing the environmental pollution, use of traditional farming practices, minimizing the use of external inputs, reducing the

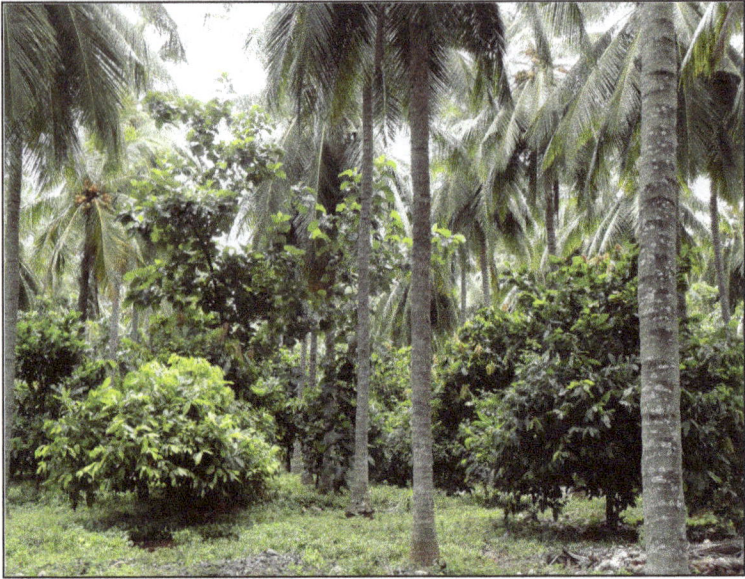

Figure 11.1: Mixed Cropping in Coconut Garden.

Figure 11.2: Growing Leguminous Crop in Coconut Garden.

production cost, efficient use of locally available resources, influence of institutions and other farmers were some of the other aspects considered by the farmers to change into organic farming. Growing of various intercrops for maintaining crop diversity and year round income generation, growing of green manure crops and,

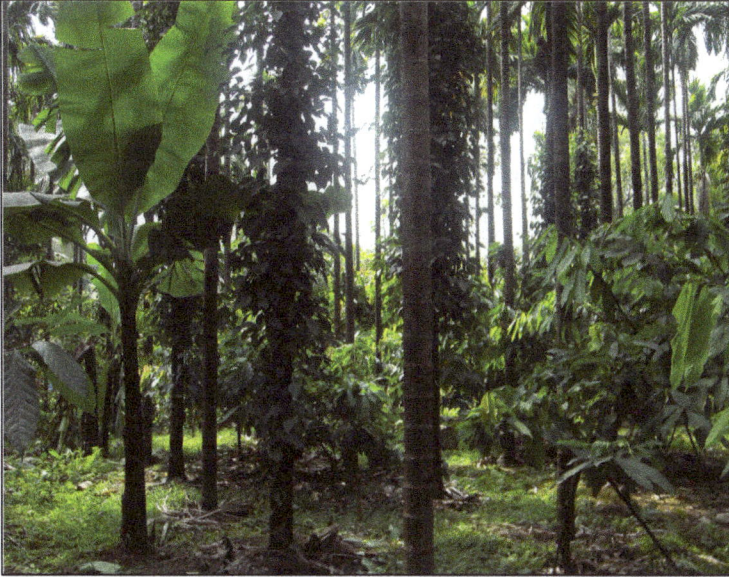

Figure 11.3: Mixed Cropping in Arecanut Garden.

Figure 11.4: Mulching using Cocoa Leaves.

use of green leaf manure for increasing soil fertility, mulching the plant basins and growing of cover crops for soil and moisture conservation *etc.* were the major agronomic practices adopted by the farmers who switched over to organic farming.

Intercrops like cocoa, nutmeg, banana and black pepper were mostly adopted by farmers of Coimbatore district (Tamil Nadu state) and Kozhikode district (Kerala state) because of suitable agro climatic and socio economic factors. Tuber crops and vegetables were mostly found in Thiruvananthapuram district (Kerala state).

Fodder crops, lemon *etc.* were found in Tumkur district (Karnataka state). Majority of farmers in Thanjavur district (Tamil Nadu state) did not cultivate any intercrop as they felt intercultural operations would be difficult. Mulching was practiced using residues of coconut, banana, cocoa *etc.* to conserve the soil and water. Green manure crops like sun hemp (*Crotalaria juncea*) and Kolinji (*Tephrosia purpurea*) were raised by the farmers for enriching the soil fertility. Green leaf manure crops like glyricidia, neem leaves, calotropis *etc.* were also used by farmers. Leguminous crops like *Stylosanthes gracilis*, *Calopogonium mucunoides*, *Vigna unguiculata etc.* were used as cover crops to prevent soil erosion and for enriching the soil fertility.

In order to maintain or improve soil fertility, use of organic inputs produced / prepared at the farm itself was resorted to by the farmers. Various crop residues, farm yard manure, cow dung slurry, vermicompost, ash and poultry manure were found to be the major organic inputs in organic cultivation. Crop residues of coconut, banana, cocoa and other weeds were used as organic matter for enriching the soil fertility. Organic inputs *viz.*, neem cake, neem based insecticides, biofertilizers, poultry manure, sheep manure *etc.* were the major inputs which were purchased from external sources for use in crop production in the organic farms.

4. Knowledge on Organic Farming Practices

Knowledge refers to the extent of information possessed by the farmers about organic farming. It is an important variable which will influence the adoption of organic farming practices. Majority of farmers cultivating coconut (68 per cent), arecanut (78 per cent) and cocoa (68 per cent) had medium level of knowledge about organic farming practices. The knowledge level of farmers on green manures, oil cakes, intercropping, mulching and vermicomposting was fairly

Figure 11.5: Pruning of Cocoa Plants.

high when compared to the knowledge on bio-control agents, botanical pesticides and bio-fertilizers. This warrants for conducting capacity building programmes *viz.,* trainings, exposure visits, method demonstration with respect to botanical pesticides, bio-fertilizers and bio-control agents in order to improve the knowhow of farmers which will result in better adoption.

5. Constraints in Organic Farming

Non availability of labour, non availability of quality organic inputs, lack of knowledge about organic farming practices, high cost for transporting organic inputs from outside the farm, lack of specialized markets for organic produces, high labour wages, low yield and profit during conversion period, lack of farmers' cooperatives for marketing, lack of standard package of practices for organic farming, lack of local certification agencies, and inadequate subsidies are the major constraints expressed by the farmers in adopting organic farming. Conventionally, prophylactic methods rather than curative methods were adopted for management of pests and diseases. Organic farming demands high technical know-how especially for pests and diseases management. Among the technical/extension constraints, difficulty in controlling pests and diseases by organic methods was expressed by majority of farmers. High cost for transporting organic inputs was the major one under economic constraints. For transporting inputs like farm yard and poultry manures, *etc.* vehicles are to be hired and the cost to be incurred for their application was also high. Farmers also felt that labour wages for weeding, and other cultural operations, irrigation *etc.* was also high, but the economic return for the produces remained more or the less during different years. Low yield and profit during conversion period was felt by farmers.

6. SWOT Analysis of Organic Farming

An attempt was made to analyze the strength, weaknesses, opportunities and threats (SWOT) in terms of farmers' and institutional perspective for adopting organic farming in coconut, arecanut and cocoa.

6.1. Farmers' Perspective

Strengths	Weaknesses
☆ Easy to adopt since it is often close to existing practice	☆ Labour intensive especially in initial phase
☆ Closely related to risk averse strategies of farmers	☆ Inadequate extension services
☆ Helps in developing local and traditional knowledge systems	☆ Non availability of specialized markets
☆ Local resources can effectively be used	☆ No demand and no premium price for organic produce
☆ Recycling and utilization of farm wastes	
☆ Free from risk of pesticides residues	
☆ Less dependency for external inputs which reduces the debt of farmers	
☆ Participatory farming	
☆ Build up of predators, natural enemies in the perennial system	

Opportunities	Threats
☆ Possibility for sustainable agriculture	☆ Belief in modernization
☆ Creation of new on-farm income generating opportunities	☆ Peer group ridicule
	☆ Subsidy for inorganic farming
☆ Possibility for premium prices in future if certified	
☆ Opportunities for greater social contacts through meetings, training *etc.*	

6.2.Institutional Perspective

Strengths	Weaknesses
☆ Donor agencies attracted to organic farming projects	☆ Lack of package of practices for different agro ecological regions
☆ Organic farming is to reduce the poverty among small and marginal farmers	☆ Demands time
☆ Government is encouraging organic farming under Mission for Integrated Development of Horticulture (MIDH) by establishing model organic plots at farmers' gardens.	☆ Less published and peer reviewed data to support organic movements, claims that organic farming increases sustainability
	☆ Lack of supportive policy framework
	☆ Lack of indigenous certification agencies

Opportunities	Threats
☆ Organic farming can make substantial contributions to sustainable environmental resource use	☆ Organic farming poses threat to established agribusiness interests
☆ Possibility for group movement by trained farmers for practicing organic farming	☆ Lack of adequate extension services and research capacity
☆ High potential for women's participation in organic farming	
☆ Opportunity for tying organic farming with other goals notably with respect to biodiversity, gender inequalities and potentially global warming	

7. Strategies for Promotion of Organic Farming

It is well known fact that organic farming practices is suitable for small and marginal farmers especially for their livelihood security and free from debt trap. Sustainable resource base, sustained crop yields without over reliance on costly external inputs, protection of the environment and bio-diversity *etc.* are other major benefits. Sustained efforts from research institutes, developmental organizations, progressive farmers, inputs dealers, processors and other stakeholders are warranted for better adoption of organic farming in plantation crops. Based on the discussion with scientists, experts, farmers and other stakeholders, the following strategies have been formulated for promotion of organic farming in plantation crops like coconut, arecanut and cocoa:

1. Promotion of research on organic agronomic practices, bio-control of diseases and pests, bio-fertilizers *etc.*
2. Development of package of organic farming practices for coconut, arecanut and cocoa suited to different agro ecological regions of the country.

3. Organize seminars, workshops, symposium *etc.* for better interaction of the farmers with the scientists, extension workers, government officials for further upgradation of the technologies.

4. Organize organic farmers' network to exchange ideas, technologies *etc*

5. Supply of bio-fertilizers, bio-agents, bio-pesticides and other organic inputs to small and marginal farmers in sufficient quantities at reasonable price as well as financial support by state and central governments to promote organic farming.

6. Farmers' participatory research for refining organic technologies for better adoption.

7. Establishment of model organic farming plots in different agro ecological regions.

8. Maintenance of data base at block/district level on area, production and productivity of different crops as well as organic nutrient resources and other materials needed for organic farming.

9. Simplification of certification processes, which are acceptable to small resource poor farmers.

10. Establishment of public warehouse for storage of organic farm produces as well as special marketing zones for organic products to promote domestic sales/export.

8. Conclusion

It is apparent from the results that the farmers who have implemented the organic farming practices do reflect the concern for sustainable agriculture. It becomes much more evident from the apprehensions expressed by the farmers regarding soil fertility, environmental impact and health hazards. Strikingly farmers also revealed the flexibility and freedom they might enjoy by minimal dependence on external agencies as an incentive of practicing organic farming. It is noteworthy that most of the farmers do not have awareness and expertise on advanced organic farming practices such as application of bio pesticides and bio fertilizers, which could be a point of intervention from the researcher front.

As a matter of fact, the organic farming culture in our country is experiencing a transition regime and yet to be evolved as an organized practice, especially in the case of perennial crops like coconut, arecanut and cocoa. Lack of good quality organic inputs, timely availability of the inputs and higher transaction costs, *etc.* are the important matters of concern from the farmer front. Possibility of occurrence of nutrient deficiencies/pests and diseases due to organic farming practices and appropriate management approaches is another area of concern. While proposing shift to organic farming, it is imperative to recommend a comprehensive package. Therefore, the lack of organized set up and inadequate infrastructure with meagre policy support might hamper the interest of those who are seriously practicing organic farming.

Farmers are more confronted with market-related difficulties such as low and highly fluctuating prices and it is a challenge to find favourable market outlets for the products. To realize the higher prices for organic produce it is imperative to obtain organic labeling for the product. Organic certification is certainly a cumbersome procedure and to materialize this, organic producers should join together as an Organic Producer's Society. The certification may be attempted through Participatory Guarantee Systems (PGS) which are locally focused quality assurance systems. They certify producers based on active participation of stakeholders and are built on a foundation of trust, social networks and knowledge exchange. Moreover, the Government of India supports PGS through the National Centre of Organic Farming (NCOF).

From the policy front it is of paramount importance to set up separate marketing facilities for organic produce through the existing channels of marketing of agriculture products such as the Amul/Milma, Supplyco and Horti-corp. From the researcher's point of view the vital challenge is to bridge the information asymmetry existing between farmers and the policy makers. Nevertheless, the concerted efforts from stakeholders of organic farming would ensure the sustainable cultivation and would also attract economic benefit. Above all, Product labeling (organic) followed by branding can open up a possible niche- international market for the organic product in the long run.

Selected References

Jaganathan, D. (2009). A multidimensional analysis of organic farming in Tamil Nadu. *Ph.D., thesis,* Indian Agricultural Research Institute, New Delhi.

Jaganathan, D., Thamban, C., Jose, C.T., Jayasekhar, S. and Anithakumari, P. (2013). Analysis of organic farming practices in coconut in south India. *Journal of Plantation Crops.* 41(1): 71-89)

Jaganathan, D., Thamban, C., Jose, C.T., Jayasekhar, S. and Chandran, K. P. (2014). Analysis of organic farming practices in arecanut in south India. *Book of Abstracts.* National conference on Sustainability of coconut, arecanut and cocoa farming – Technological advances and way forward. 22-23rd August, 2014, CPCRI, Kasaragod, p. 112.

Jaganathan, D., Thamban, C., Jose, C.T., Jayasekhar, S., Chandran, K.P. and Muralidharan, K. (2015). Analysis of organic farming practices in cocoa in India. *Journal of Plantation Crops* 43(2): 131 -138.

John, F.(2000). Prospects for conventional farmers adopting organic production techniques. In: *Organic farming in New Zealand: An evaluation of the current and future prospects including an assessment of research needs*. Ministry of Agriculture and Forestry, New Zealand. www.maf.govt.nz.

NCOF. (2011). *National Project on Organic Farming. Annual report (2010-2011)*. National Centre of Organic Farming, Ghaziabad, 88 p. (www.dacnet.nic.in/ncof)

Thomas George, V. (2010). Technological advances in organic farming of plantation crops. In: *Organic horticulture- Principles, practices and technologies*. (Eds.) Singh,

H.P and Thomas George V., Westville publishing house, New Delhi, pp. 32-47.

Thomas George, V., Subramanian, P., Krishnakumar, V., Alka Gupta and Chandramohanan, R. (2010). Package of practices for organic farming in coconut. Technical bulletin No. 64, CPCRI, Kasaragod, Kerala, 28 p.

Veeresh, G.K. (1997). *Organic farming and its relevance to present condition.* Agriculture Man and Ecology, Bangalore, 115 p.

Chapter 12

Quality Control Standards and Organic Certification for Plantation Crops

☆ Mathew Sebastian and M.P. Sajitha

1. Introduction

With the growing awareness for safe and healthy food among consumers and growing concern for deteriorating soil health and fertility and depleting natural resources among policy planners strategies are being drawn to adopt technologies promising safe and healthy food with due respect to the natural resource availability without compromising the productivity and food security. Organic farming has emerged as one of the viable option which promises food production in quantity and quality with resource sustainability. Demand for organic food is increasing day by day. That is the main reason for organic farming gaining importance over conventional farming. This is happening more so in crops like fruits, vegetables, spices, coconuts, tea, coffee, cocoa, grains, pulses *etc.* which are consumed on daily basis. Many farmers in India are shifting to organic farming due to the domestic and international demand for organic food.

Organic farming is a method of farming system which primarily aimed at cultivating the land and raising crops in such a way, as to keep the soil alive and in good health by use of organic wastes (crop, animal and farm wastes, aquatic wastes) and other biological materials along with beneficial microbes (biofertilizers) to release nutrients to crops for increased sustainable production in an eco friendly pollution free environment. As per the definition of the United States Department of Agriculture (USDA) study team on organic farming "organic farming is a system which avoids or largely excludes the use of synthetic inputs (such as fertilizers, pesticides, hormones, feed additives *etc*) and to the maximum extent feasible rely upon crop rotations, crop residues, animal manures, off-farm organic waste, mineral grade rock additives and biological system of nutrient mobilization and plant protection".

2. Certified Organic Products

Certified organic products are those whose production, processing, handling and marketing have been verified by an accredited certification body as being in conformity with specified organic standards. Once a product is certified organic, it can be labeled as organic. The certification process includes inspection to verify that production and handling are carried out in accordance with the standards against which certification is to be done; and certification to confirm that production and handling conforms to those standards. Certification procedures for the certification of organic products should make it possible to track and control the flow of products from primary production at farm level through each stage of manufacturing right to the final consumer product. Producers and exporters will have to obtain certification against organic standards applicable in those markets, in which they intend to sell their products with an indication that they are organic. Certification is needed in order to ensure that products labeled as "organic" are produced and handled in accordance with specified organic standards. In other words, certification creates trust in organic labeling and promotes fair competition in the market place. Many countries have formulated their own national regulations or standards for organic production like India's National Programme for Organic Production (NPOP), USDA's National Organic Programme, and European Union's Regulation EC No. 834/2007, Japanese Agricultural Standards (JAS) *etc*.

Third-party organic certification is a complex and formal process and the market began to demand it for sales transactions, and now it is required by the regulations of many governments for any kind of an "organic" claim on a product label. The certification agencies need to be accredited according to ISO 17065 Guidelines for offering organic certification for various scopes like agricultural and animal production and processing of such products, wild harvest and export-import of certified organic products according to respective standards.

3. Organic Quality Control

The organic quality control system is based on standards, accreditation, inspection, and certification

3.1. Standards for Organic Agriculture

Minimum requirements for a farm or product to be certified as 'organic' are precisely defined by organic standards. There are organic standards on the national as well as international level. For certification, the standards of the target market or importing country are relevant. Certain private labels such as Naturland, Demeter or BIO SUISSE have additional requirements in addition to their national standards. Organic standards have long been used to create an agreement within organic agriculture about what an "organic" claim on a product means, and to some extent, to inform consumers. Regional groups of organic farmers and their supporters began developing organic standards as early as in the 1940's. Currently there are hundreds of private organic standards worldwide; and in addition, organic standards have been codified in the technical regulations of more than 60 governments.

Third-party organic certification was first instituted in the 1970's by the same regional organic farming groups that first developed organic standards. In the early years, the farmers inspected one another on a voluntary basis, according to quite a general set of standards. Today third-party certification is a much more complex and formal process. Although certification started as a voluntary activity, the market began to demand it for sales transactions, and now it is required by the regulations of many governments for any kind of an "organic" claim on a product label. There are sets of guidelines and norms to be complied with, followed by the inspection by the certification agency before certification. These certifications are basically carried out on the basis of statutory certification norms and the voluntary certification norms.

3.1.1. Major International Standards

3.1.1.1. *United States Department of Agriculture/National Organic Programme Standards (USDA/NOP)*

The production rules for organic production according to the U.S. Standards have been laid down in the National Organic Program (NOP) of the United States Department of Agriculture (USDA). This Regulation is in force within the U.S.A., but also sets criteria for importing organic products to the country. This program establishes national standards for the production and handling of organically produced products, including a National List of substances approved for and prohibited from use in organic production and handling. The Indian conformity assessment system for organic certification bodies used by National Accreditation Body, Govt. of India has been determined by USDA Agricultural Marketing Service (AMS) to be sufficient to ensure conformity with the technical standards of USDA's National Organic Program.

3.1.1.2. *EU Regulation EC No.834/2007*

The regulations governs organic agriculture for European Union Member States set out for a complete set of objectives, principles and basic rules for organic production. This regulation is most relevant for exports to Europe. Organic products imported into the European Community should be allowed to be placed on the Community market as organic, where they have been produced in accordance with production rules and subject to control arrangements that are in compliance with or equivalent to those laid down in Community legislation. In addition, the products imported under an equivalent system should be covered by a certificate issued by the competent authority, or recognized control authority or body of the third country concerned.

At present India's National Programme for Organic Production Standards is considered equivalent to council regulation EC no. 834/2007 (for Category A and F)

☆ Category A- Live or unprocessed agricultural products (Crop Production)

☆ Category F – Seed/Propagating material.

3.1.1.3. INDOCERT Organic Standards for Non EU Country Operators (Considered equivalent to EC No.834/2007)

INDOCERT Organic standard reflects the current status of production practices in India and other third world countries where INDOCERT offers certification. This document lays down detailed INDOCERT Organic standards for Non- EU country operators for the certification of organic production systems in third world countries. It provides the basis for sustainable development of organic production systems as mentioned below:

☆ all stages of production, preparation and distribution of organic products and their control;

☆ the use of indications referring to organic production in labelling and advertising.

☆ The products mentioned below are covered in the scope of this document:

   a) Live or unprocessed agricultural products

   b) Processed agricultural products for use as food

   c) Processed agricultural products for use as feed

3.1.1.4. Japanese Agricultural Standards (JAS)

Japanese Agricultural Standards are the standards for the agriculture industry maintained by the Japanese Government. They are comparable to Japanese Industrial Standards for food and agricultural products. The JAS rule admits the equivalence with the European organic certification for vegetable products (in European countries). For livestock products the equivalence with European certification is not admitted. The certification based on these standards is widely recognized across the three key organic markets of US, EU and Japan and is trusted by a large section of importers and consumers.

3.1.1.5. Canadian Organic Standards

The Canada Organic Regime is the Government of Canada's response to requests by the organic sector and consumers to develop a regulated system for organic agricultural products. The Organic Products Regulations define specific requirements for organic products to be labeled as organic or that bear the Canada Organic logo. The regulations came into effect on June 30, 2009. All organic products bearing the Canada Organic logo or represented as organic in interprovincial and international trade must comply with the Organic Products Regulations.

The Canadian Food Inspection Agency (CFIA) is responsible for the monitoring and enforcement of the Regulations. Under the Canada Organic Regime, Certification Bodies are accredited based on the recommendation of Conformity Verification Bodies that are designated by the CFIA. The Certification Bodies are responsible for verifying the application of the Canadian Organic Standards.

National Programme for Organic Production (NPOP)

To provide a focused and well directed development of organic agriculture in the country, the Ministry of Commerce and Industry, Govt. of India launched the National Programme on Organic Production (NPOP) in the year 2000, which was formally notified in October 2001 under the 'Foreign Trade and Development Act' (FTDA). The NPOP provides for an institutional mechanism for implementation of National Standards on Organic Production (NSOP) through a National Accreditation Policy and Programme.

The National Programme for Organic Production provides for Standards for organic production, systems, criteria and procedure for accreditation of Certification Bodies, the National (India Organic) Logo and the regulations governing its use. The standards and procedures have been formulated in harmony with other International Standards regulating import and export of organic products. The aims of the NPOP include the following:

(a) To provide the means of evaluation of certification programme for organic agriculture and products (including wild harvest, aquaculture, live stock products) as per the approved criteria.

(b To accredit certification programmes of Certification Bodies seeking accreditation under this programme.

(c) To facilitate certification of organic products in conformity with the NSOP.

(d) To facilitate certification of organic products in conformity with the importing countries organic standards as per equivalence agreement between the two countries or as per importing country requirements.

(f) To encourage the development of organic farming and organic processing

The NPOP standards for production and accreditation system have been recognized by European Commission and Switzerland for unprocessed plant products as equivalent to their country standards. Similarly, USDA has recognized NPOP conformity assessment procedures of accreditation as equivalent to that of US. With these recognitions, Indian organic products duly certified by the accredited certification bodies of India are accepted by the importing countries.

3.1.1.6. National Standards for Organic Production

In 2000, the Government of India released the National Standards for Organic Products (NSOP) under the National Programme for Organic Production (NPOP). Products sold or labeled as 'organic' thereafter need to be inspected and certified by a nationally accredited certification body. In India, Agricultural and Processed Food Export Development Authority (APEDA) regulates the certification of organic products as per National Standards for Organic Production. "The NPOP standards for production and accreditation system have been recognized by European Commission and Switzerland as equivalent to their country standards. Similarly, USDA has recognized NPOP conformity assessment procedures of accreditation as equivalent to that of US. With these recognitions, Indian organic products duly certified by the accredited certification bodies of India are accepted by the importing countries. Organic food products manufactured and exported from India are marked with the India Organic certification mark issued by the APEDA.

Organic crop production management should cover a diverse planting scheme. The producer seeking certification under the NSOP shall be required to develop an organic crop production plan for their plantation. This plan shall include:

1. Description of the crops in the production cycle (main crop and intercrop) as per the agro climatic seasons.
2. Description of practices and procedures to be performed and maintained.
3. List of inputs used in production along with their composition, frequency of usage, application rate and source of commercial availability.
4. Source of organic planting material (seeds and seedlings).
5. Descriptions of monitoring practices and procedures to be performed and maintained to verify that the plan is being implemented effectively.
6. Description of the management practices and physical barriers established to prevent commingling and contamination of organic production unit from conventional farms, split operations and parallel operations.
7. Description of the record keeping system implemented to comply with the requirements.

National Standards for Organic Production include the following:

1. Crop production
2. Livestock, Poultry and Products
3. Beekeeping/Apiculture
4. Aquaculture Production
5. Food Processing and Handling
6. Any other category of products that the National Accreditation Body (NAB) may include from time to time

3.2. Accreditation

Accreditation is a process by which certification body's competency, authority and credibility are guaranteed. The accreditation process ensures that the certification practices of certification bodies are acceptable, typically meaning that they are competent to inspect and certify third parties, behave ethically, and employ suitable quality assurance. The accreditation system prevalent nationally for offering organic certification services is accreditation by National Accreditation Body (NAB), Govt. of India. The NAB is responsible for drawing up procedures for the evaluation and accreditation of certification programmes, as well formulating procedures to evaluate agencies implementing these programmes. The accreditation of inspection and certifying agencies is also the responsibility of the NAB. The NAB shall be serviced by APEDA. The NAB shall consist of members representing Ministry of Commerce, Ministry of Agriculture, FSSAI, MPEDA and various Commodity Boards such as the Tea Board, Spices Board and Coffee Board. The Additional Secretary (Plantations) shall be the Chairman of the NAB.

3.3. Inspection

The Certification Bodies shall have laid down policy and procedure on inspection methods and frequency which shall be determined by, among others:

- ☆ Intensity of production
- ☆ Type of production
- ☆ Size of operation
- ☆ Outcome of previous inspections and the operator's record of compliance
- ☆ Any complaints received by the programme
- ☆ Whether the unit or operator is engaged only in certified production
- ☆ Contamination and drift risk
- ☆ Complexity of production

If an organic farm has to be certified, it has to undergo an inspection at least once a year. The inspector evaluates the performance of the farm activities with the help of the farmer's statements and records and by viewing the fields, animals and farm buildings. He or she checks whether the statements and records are correct and plausible. In case of doubt, the inspector can take samples for laboratory testing or later conduct unannounced inspections. However, laboratory testing is only one tool for inspection in cases of suspicion of application of or contamination with prohibited substances. Chemical analyses just reveal whether a certain sample contains a specific substance at a certain moment.

There are un-announced inspections for the operators conducted by certification body. The selection of operators for unannounced inspection shall be based on risk analysis carried out by the Certification Body annually. At the end of inspection process, the inspector shall sign the inspection report, which will have to be countersigned by the operator and a copy of the report is handed over to the operator.

3.3.1. Inspection and Certification Process in India

The following steps are to be followed by a person for organic inspection and certification process

- ☆ Select an accredited inspection and certification agency
- ☆ The selection of agency is based on the acceptability of the exporting country and the standards prescribed by that country, costs of the inspection and certification
- ☆ File application form detailing preliminary information about the size of the farm/units, location of units, company activities, *etc.* to determine the cost of inspection and certification and based on application cost of inspection and certification is determined and a contract regarding the same is signed between the producer and the agency.
- ☆ Based on the information provided by the producer, the agency carries out farm inspection on a mutually convenient date and time.

☆ The inspection includes:

- ❏ Interviews with persons responsible for production
- ❏ Physical inspection of fields, premises, processing equipment, storage area *etc.,* Inspection of paper work, book keeping, *etc.*
- ❏ Testing for residue analysis is carried out if the inspector feels the need for the same

☆ After conducting the inspection, the agency provides the report of inspection, based on which, the agency decides whether or not to grant certification.

☆ If certification is not granted, the agency provides the reasons for rejection. If only certain parts of the business can be certified, the agency does so, providing the producer with certain recommendations for the remaining parts of the business which could not be certified.

☆ Each year following the official certification, the agency performs inspections to determine whether the requirements for certification are still met and issue certification decisions accordingly

3.4. Certification

It is a procedure in which a certification body assesses a farm or company and assures in writing that it meets the requirements of the organic standards. The inspector transmits his findings to the certification body as a written report. The report will be reviewed by an assessor not involved in the inspection process. The certification body compares the results of the inspection with the requirements of the organic standards. A certification committee decides whether certification may be granted or not. The organic certification process begins when an operator seeks certification information from the certification body. There are many different types of operations that need to be inspected such as crop farms, live stock operations, processing operations, grower groups and other handling systems. The specific type of inspection conducted depends on the process or production system requested for certification. Understanding the type of operation helps the inspector to perform a comprehensive inspection. Upon completing the inspection, the certification decision is taken by the Certification Committee, which reviews the complete file and issues certification decision and a scope certificate is issued.

3.4.1. Smallholder Group Certification

Grower Groups are organized group of farmers/producers who intend to produce organic products/engage in organic processes in accordance with national/international Standard of Organic Production. An Internal Control System (ICS) is a documented quality assurance system that allows an external certification body to delegate the annual inspection of individual group members to an identified body/unit within the certified operator.

The grower group shall be based on the Internal Control System (ICS) and shall apply to grower groups, farmers' cooperatives, contract production and small scale processing units. The producers in the group must apply similar production systems and the farms should be in geographical proximity. Individual farms with land holding of 4 ha (10 acres) and above can also be a part of the group but will have to be inspected separately every year by the accredited Certification Body. The total area of such farms shall be less than 50 per cent of the total area of the group. The grower group shall consist of minimum 25 and maximum 500 farmers. Processors and exporters/traders can own/manage the Internal Control System (ICS) but will have to be inspected annually by the external Certification Body. Separate certificates (Scope and Transaction Certificates) are required to be issued for the ICS, processors and traders to maintain the traceability of the product flow. The Certification Body shall not certify if there is no ICS as per NPOP and 100 per cent internal inspections are not conducted. In case the farmer group does not maintain an Internal Quality System, it cannot be treated as an ICS.

Group certification is based on the concept of an internal Quality Management System comprising of the following:

☆ Implementation of the internal control system

☆ Internal standards

☆ Risk assessment

An accredited Certification Body should be identified for conducting annual inspection of the individual group/unit. The accredited Certification Body shall evaluate the ICS by verifying the location of the ICS, quality manual, documentation, and its implementation, related to internal inspections, training, warehousing and purchase and sale. Thereafter the accredited Certification Body will approve the ICS and then conduct the external inspections.

All the farmers shall maintain the farm diary for noting.

ICS is developed to ensure quality Organic farming, implementation, storage, processing and marketing. It also aims to reduce the certification cost as low as possible. The ICS aim to organize all organic farms under one roof for better production and marketing. The implementation of an internal control system can help to save costs for external inspection and certification. Furthermore it supports the farmers in production and record keeping according to the standard's requirements. On the other hand, the set up and maintenance of an ICS needs considerable manpower and therefore also creates costs for salaries. If during the external inspection some of the group members are found to be not complying with the standards, the whole group risks to lose the certification.

The following are minimum requirements for setting up an ICS for grower groups:

☆ Development of Internal Control System (ICS) manual containing policies and procedures

☆ Identification of farmers in the group

☆ Creation of awareness about Grower Group Certification

☆ Identification of qualified/experienced personnel for maintaining the Internal Control System

☆ Give necessary training in production and ICS development

☆ Implementation of the policies and procedures

☆ Review and improvement of the ICS document for maintaining a harmonized quality management system

4. Conversion Requirements

1. The establishment of an organic management system and building of soil fertility requires an interim period, known as the conversion period. While the conversion period may not always be of sufficient duration to improve soil fertility and for re-establishing the balance of the ecosystem, it is the period in which all the actions required to reach these goals are started.

2. A plantation may be converted through a clear plan of how to proceed with the conversion. This plan shall be updated by the producer, if necessary and shall cover all requirements to be met under these standards.

3. The requirements prescribed under these standards shall be met during the conversion period. All these requirements shall be applicable from the commencement of the conversion period till its conclusion.

4. The Certification Body may calculate the start of the conversion period from the date of first inspection of the operator.

5. A full conversion period shall not be required where de facto requirements prescribed under these standards have been met for several years and where the same can be verified on the basis of available documentation. In such cases inspection shall be carried out in reasonable time intervals, before the first harvest.

5. Duration of Conversion Period

1. In the case of annual and biennial crops, plant products produced can be certified organic when the requirements prescribed under these Standards have been met during the conversion period of at least two years of organic management before sowing (the start of the production cycle).

2. In the case of perennial plants, the first harvest may be certified as organic after at least thirty six months of organic management according to the requirements prescribed under these Standards.

3. The accredited Certification Bodies shall decide in certain cases, for extension or reduction of conversion period depending on the past status/ use of the land and environmental condition.

4. Twelve months reduction in conversion period could be considered for annuals as well as perennials provided, documentary proof has been available with the accredited Certification Body that the requirements prescribed under these Standards have been met for a period of minimum three (3) years or more. This could include the land that has been certified for minimum three (3) years under the 'Participatory Guarantee System'

implemented by the Ministry of Agriculture. The accredited Certification Bodies shall also consider such a reduction in conversion period, if it has satisfactory proof to demonstrate that for three (3) years or more, the land has been idle and/or it has been treated with the products approved for use in organic farming as listed Annex 1 and 2 of the Appendix.

5. Organic products in conversion shall be sold as "produce of organic agriculture in conversion" or of a similar description, when the requirements prescribed under these Standards have been met for at least twelve months.

6. Choice of Crops and Varieties

i) All seeds and plant material shall be certified organic. Species and varieties cultivated shall be adapted to the soil and climatic conditions and be resistant to pests and diseases. In the choice of varieties, genetic diversity shall be taken into consideration.

ii) When organic seed and plant materials are available, they shall be used.

iii) When certified organic seed and plant materials are not available, chemically untreated conventional seed and plant material shall be used.

iv) The use of genetically engineered seeds, transgenic plants or plant material is prohibited.

7. Diversity in Crop Production and Management Plan

i) The basis for crop production in organic farming shall take into consideration the structure and fertility of the soil and the surrounding ecosystem, with a view to minimizing nutrient losses.

ii) Where appropriate, the organic farms shall be required to maintain sufficient diversity in a manner that takes into account pressure from insects, weeds, diseases and other pests, while maintaining or increasing soil, organic matter, fertility, microbial activity and general soil health. For non perennial crops, this is normal, but not exclusive, achieved by means of crop rotation preferably by leguminous crops.

iii) Soil fertility shall be maintained through, among other things, the cultivation of legumes or deep rooted plants and the use of green manures, along with the establishment of a programme of crop rotation several times a year and fertilization with organic inputs.

8. Nutrient Management

i) Sufficient quantities of biodegradable material of microbial, plant or animal origin produced on organic farms shall form the basis of the nutrient management programme to increase or at least maintain its fertility and the biological activity within it.

ii) Fertilization management should minimize nutrient losses. Accumulation of heavy metals and other pollutants shall be prevented.

iii) Non synthetic mineral fertilisers and brought-in bio fertilisers (biological origin) shall be regarded as supplementary and not as a replacement for nutrient recycling.

iv) Desired pH levels shall be maintained in the soil by the producer.

v) The certification programme shall set limitations to the total amount of biodegradable material of microbial, plant or animal origin brought onto the farm unit, taking into account local conditions and the specific nature of the crops.

vi) The certification programme shall set procedures which prevent animal runs from becoming over manuring where there is a risk of pollution.

vii) Mineral fertilizers shall only be used in a supplementary role to carbon based materials. Only those organic or mineral fertilizers that are brought in to the farm (including potting compost) shall be used when, the circumstances demand in accordance with **Annex 1.**

viii) Permission for use shall only be given when other fertility management practices have been optimized

ix) Manures containing human excreta (faeces and urine) shall not permitted to prevent transmission of pests, parasites and infectious agents.

x) Mineral fertilisers shall be applied in their natural composition and shall not be rendered more soluble by chemical treatment. The certification programme may grant exceptions. These exceptions shall not include mineral fertilisers containing nitrogen.

xi) The certification programme shall lay down restrictions for the use of inputs such as mineral potassium, magnesium fertilisers, trace elements, manures and fertilisers with a relatively high heavy metal content and/ or other unwanted substances, *e.g.* basic slag, rock phosphate and sewage sludge. All synthetic nitrogenous fertilisers are prohibited.

9. Pest, Disease and Weed Management

i) Organic farming systems shall be carried out in a way which ensures that losses from pests, diseases and weeds are minimized. Emphasis is placed on the use of a balanced fertilizing programme, use of crops and varieties well-adapted to the environment, fertile soils of high biological activity, adapted rotations, intercropping, green manures, *etc.* Growth and development shall take place in a natural manner.

ii) Weeds, pests and diseases shall be controlled through a number of preventive cultural techniques which limit their development in a balanced nutrient management programme, *e.g.* suitable rotations, green manures, early and pre drilling seedbed preparations, mulching, mechanical control and the disturbance of pest development cycles.

Accredited certification programmes shall ensure that measures are in place to prevent transmission of pests, parasites and infectious agents.

iii) Pest management shall be regulated by understanding and disrupting the ecological needs of the pests. The natural enemies of pests and diseases shall be protected and encouraged through proper habitat management of hedges, nesting sites *etc.* An ecological equilibrium shall be created to bring about a balance in the pest predator cycle.

iv) Products used for pest, disease and weed management, prepared at the farm from local plants, animals and microorganisms, shall be allowed. If the ecosystem or the quality of organic products might be jeopardized, the certification programme shall judge if the product is acceptable as per the procedure given to evaluate additional inputs to organic agriculture.

v) Thermic weed control and physical methods for pest, disease and weed management shall be permitted.

vi) Thermic sterilization of soils to combat pests and diseases shall be restricted to circumstances where a proper rotation or renewal of soil cannot take place. The certification programme on a case-by-case basis may only give permission.

vii) All equipment from conventional farming systems shall be properly cleaned and free from residues before being used on organically managed areas.

viii) The use of synthetic herbicides, fungicides, growth regulators, synthetic dyes insecticides and other pesticides are prohibited. The producer shall keep documentary evidences of the need to use the product.

ix) Commercial products used as inputs shall always be evaluated as per the criteria before approval is given for use.

x) The use of genetically engineered organisms or products is prohibited.

10. Contamination Control

i) In case of reasonable suspicion of contamination, the certification programme shall make sure that an analysis of the relevant products and possible sources of pollution (soil and water) shall take place to determine the level of contamination.

ii) All relevant measures shall be taken to minimize contamination from outside and within the farm.

iii) Buffer zones shall be maintained to prevent contamination from conventional farms. The buffer zone should be sufficient in size to prevent the possibility of unintended contact of prohibited substances applied to adjacent conventional land areas/farms.

iv) Polyethylene and polypropylene or other polycarbonates coverings such as plastic mulches, fleeces, insect net and silage wrapping, only are allowed. These shall be removed from the soil after use and shall not be burnt on the farmland. The use of polychloride based products is prohibited.

11. Soil and Water Conservation

i) Soil and water resources shall be handled in a sustainable manner. Relevant measures shall be taken to prevent erosion, salination of soil, excessive and improper use of water and the pollution of ground and surface water.

ii) Clearing of land through the means of burning organic matter, *e.g.* slash-and-burn, straw burning shall be restricted to the minimum. The clearing of primary forest is prohibited.

iii) The certification programme shall require checking appropriate stocking rates which does not lead to land degradation and pollution of ground and surface water.

12. Labeling

Operators are given guidelines on how to use the certification body's organic seal, organic certificate, and transaction certificate *etc.* by the certification body. Operators must continue to follow the standards and maintain adequate records while certified.

13. Organic Certification Mark

13.1. India Organic Logo

A trademark – "India Organic" will be granted on the basis of compliance with the National Standards for Organic Production (NSOP). Communicating the genuineness as well as the origin of the product, this trademark will be owned by the Government of India. Only such exporters, manufacturers and processors whose products are duly certified by the accredited Certification Bodies, will be granted the license to use of the logo which would be governed by a set of regulations.

The Accredited Certification Body is satisfied that the applicant is fit to use the Certification Trade Mark, then the Certification Body shall grant a license to use the logo. The license will have a validity period. The Certification Trade Mark shall be

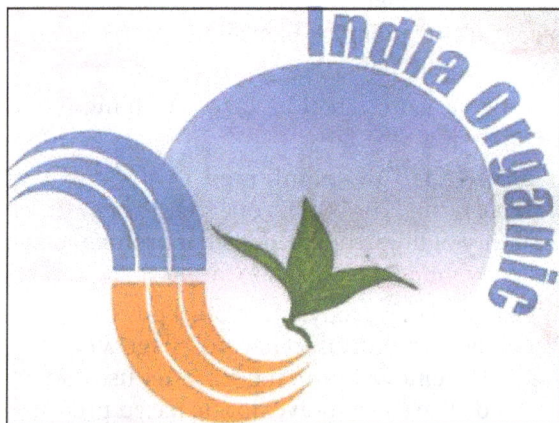

applied to only such types, grades, classes, varieties, sizes of the products for which the license has been granted. The manner, in which the licensee proposes to place or use the Certification Trade Mark, must be approved by the Certification Body. In the event of a withdrawal of the right to use the aforesaid Certification Trade Mark, the certificate or the License shall be returned to the Accredited Certification Body.

13.2. Other Organic Logos

Some of the other logos that can be used by the certified client with label approval include:

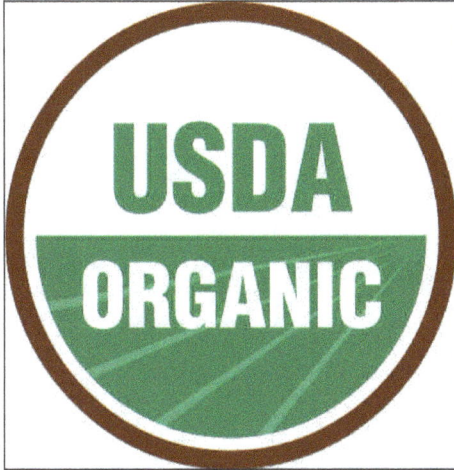

European Union Organic Logo

Chapter 13

Transition towards Organic Farming: Policies, Problems and Prospects

☆ *S. Jayasekhar, C. Thamban, K.P. Chandran*
and D. Jaganathan

1. Introduction

In the recent past we have witnessed an overwhelming popularity and scientific acceptance of organic farming in the western world, especially USA, Germany and the Scandinavian countries. Organic agriculture is now practiced in almost all countries of the world, and its share of agricultural land and farms is growing. The market for organic products is growing, not only in developed countries but also in many other countries, including developing countries. As a matter of fact, organic agriculture is penetrating the farmlands of India as well. Of late, organic farming has certainly found acceptance among diverse categories of farmers operating in different parts of the country under varied agro-ecological and financial conditions.

However, the farmers had to face several problems while converting from conventional to organic farming which include non-receipt of premium price for these products and lack of storage facility (Lanting, 2007). Sanghi (2007) has argued that organic farming is an intensive process, limited mostly to resource-rich farmers, and the export market and depends heavily on external support systems for price, market intelligence and certification of produce, among others. Hence, he has concluded that the scope of coverage and social relevance of the organic farming is also limited. Organic agriculture is often termed knowledge based rather than input based agriculture (Das, 2007). The scenario of organic farming practices of seasonal crops is entirely different from plantation crops like tea, rubber coffee and coconut. The standard duration of conversion for annual crops is 24 months, whereas, in the case of plantation crops it extends up to 42 months. There arise some pertinent questions in this context i) what exactly is the operating environment of organic farming in the world and India? ii) is there any convergence between various programmes/projects envisaged by the institutions regarding the implementation of organic farming programmes? iii) how efficient and effective are the forward and backward linkages with respect to market/price support and assured premium prices? We attempt to answer these questions by reviewing the present state of affairs in these regard both internationally and nationally and also by taking up a micro level case study at Kasaragod district of Kerala which is already declared as an organic district.

2. Organic Agriculture: The World Scenario

According to the latest FiBL-IFOAM survey, 2015 on certified organic agriculture worldwide, nearly 43.1 million ha land is being certified as organic in 170 countries. The countries with the most organic agricultural land are Australia (17.2 million ha), Argentina (3.2 million ha), and the United States (2.2 million ha) (Figure 13.1). The countries with the most producers are India (650'000), Uganda (189'610), and Mexico (169'703). About a quarter of the world's agricultural land (11.7 million ha) and more than 80 per cent (1.7 million) of the producers are in developing countries. Most of this category of land is used for cereals including rice (3.3 million hectares), followed by green fodder from arable land (2.4 million ha), oilseeds (0.8 million ha), vegetables (0.3 million ha), and protein crops (0.3 million ha). Permanent crops account for seven per cent of the organic agricultural land, amounting to 3.2 million ha. The most important permanent crops are coffee (with more than 0.7 million ha, constituting almost one quarter of the organic permanent cropland), followed by olives (0.6 million ha), nuts and grapes (0.3 million ha each), and cocoa (0.2 million ha).

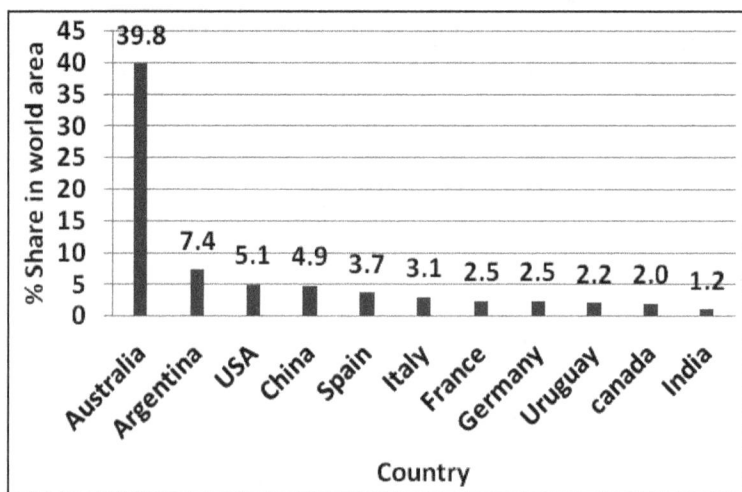

Figure 13.1: Countries with Largest Area of Organic Agricultural Land.

Global sales of organic food and drink reached 72 billion US dollars in 2013. Revenues have increased almost five-fold since 1999. Organic product sales have increased at a healthy rate over the last decade, and hopefully growth will continue in the coming years. Europe and North America generate over 90 per cent of global sales. Although Asia, Australasia, Latin America, and Africa have become important producers of organic agricultural crops, their markets for organic products remain small. In 2013, the countries with the largest organic markets were the United States (24.3 billion euros), Germany (7.6 billion euros), and France (4.4 billion euros). The largest single market was the United States (approximately 43 per cent of the global market), followed by the European Union (22.2 billion euros, 43 per cent) and China (2.4 billion euros) (Figure 13.2).

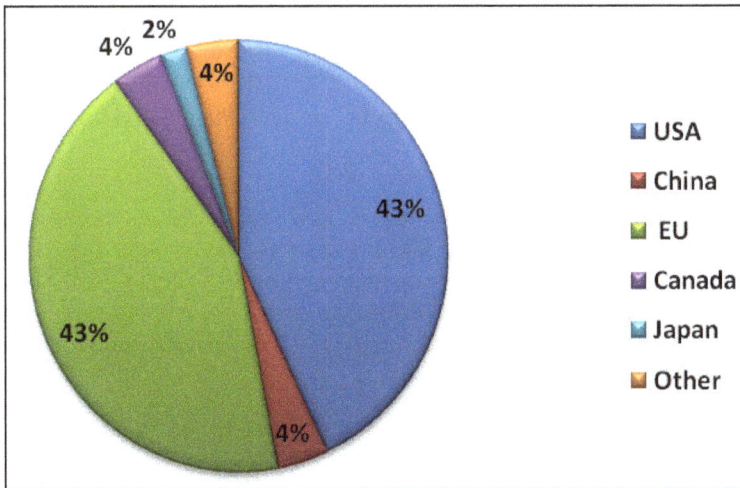

Figure 13.2: Global Market: Distribution of Sales Value of Organic Products by Countries.

The highest per-capita consumption was in Switzerland followed by Denmark, and Luxembourg (Figure 13.3). Liechtenstein, Austria, Sweden, Germany. US, Canada and Norway also made their entry into the top 10 consumers of organic products (per capita consumption).

3. Organic Agriculture: The Indian Scenario

Currently, India ranks 10th among the top ten countries in terms of cultivable land under organic certification. The certified area includes 15 per cent cultivable area with 0.72 million hectare and rest 85 per cent (3.99 million hectares) is forest and wild area for collection of minor forest produces. The total area under organic certification is 4.72 million Hectares (2013-14). India produced around 1.24 million tonnes of certified organic products which includes all varieties of food products namely Sugarcane, Cotton, Oil Seeds, Basmati rice, Pulses, Spices, Tea, Fruits, Dry fruits, Vegetables, Coffee and their value added products. The production is not limited to the edible sector but also produces organic cotton fiber, functional food products *etc*. Among all the states, Madhya Pradesh has covered largest area

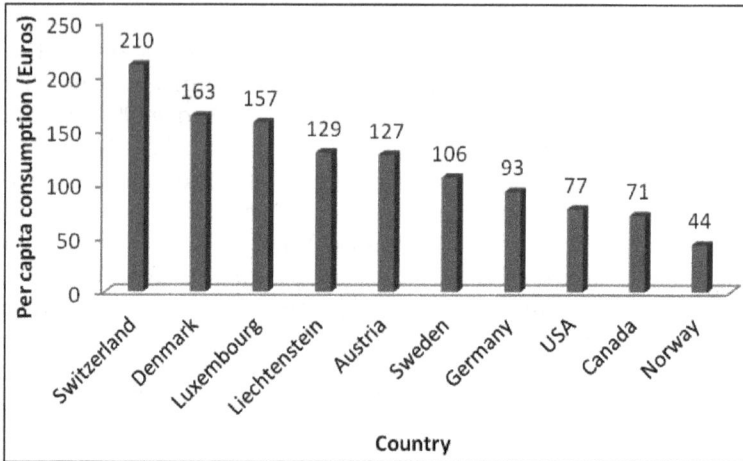

Figure 13.3: Top 10 Consumers of Organic Products (per capita consumption).

under organic certification followed by Himachal Pradesh and Rajasthan. The states of Uttarakhand and Sikkim have declared their states as 'organic states'. In Maharashtra, since 2003, about five lakh ha area has been under organic farming. Most of the area in the northeastern states is being used for organic farming. In Nagaland, 3000 ha area is under organic farming. States like Rajasthan, Tamil Nadu, Karnataka, Kerala, Madhya Pradesh, Himachal Pradesh and Gujarat are promoting organic farming vigorously.

Ministry of Commerce, Govt. of India launched the "National Programme on Organic Production" (NPOP) defining the National Standards for Organic Production (NSOP) and the procedure for accreditation and certification in 2000. India now has 30 accredited certification agencies for facilitating the certification to growers. For area expansion and technology transfer, Ministry of Agriculture launched a National Project on Promotion of Organic Farming (NPOF-DAC) and earmarked funds for setting up of organic and biological input production units, vermicompost production units and for organic adoption and certification under various schemes such as NHM (now MIDH), NMSA and RKVY. To empower farmers through participation in certification process and to make the certification affordable for domestic and local markets, Ministry of Agriculture has also launched a farmer group centric organic guarantee system under PGS-India programme.

India exported 135 products during 2013-14 with the total volume of 194088 MT including 16,322 MT organic textiles. Organic products are exported to US, European Union, Canada, Switzerland, Australia, New Zealand, South East Asian countries, Middle East, South Africa *etc*. Oil seeds - soybean (70 per cent) lead among the products exported followed by cereals and millets other than basmati (6 per cent), processed food products (5 per cent), basmati rice (4 per cent), sugar (3 per cent), tea (2 per cent), pulses and lentils (1 per cent), dry fruits (1 per cent), spices (1 per cent) and others. Domestic market is also growing at an annual growth rate of 15-25 per cent.

4. Organic Agriculture: The Kerala Scenario

There is rich potential for promoting organic farming in Kerala in the light that intensity of inorganic agriculture here is not that severe compared to that in other States in the country. This implies the positive side of agriculture in Kerala in terms of the already low levels of consumption of hazardous chemicals and, therefore, chances of redeeming farmers to organic agriculture are quite high. Realizing these facts, the State Department of Agriculture commenced organic farming promotional activities since 2002-03. In the following year, the Department set up a cell for Promotion of Sustainable Agriculture and Organic Farming. It has also launched two brands, namely 'Kerala Organic' and 'Kerala Naturals' to market organic farm produces (Balachandran, 2005).

Currently there are a number of certified organic farmers in the state, those cultivating cash crops such as spices, tea, and coffee, mainly targeting export market and also non-certified organic farmers who focus on food crops and biodiversity. All of them, whether certified or not, focus clearly on soil health improvement. Kerala also has an accredited organic certifying agency catering to the needs of the farmers. Some of the farming systems such as *Pokkali* and *Kaipad* cultivation, cultivation of *Jeerakasala* and *Gandhakasala* varieties of paddy in Wayanad and, homestead farming systems all over the state are default organic. Studies have established the economic viability and productivity of homestead farms in the State and elsewhere. Recently the Adat panchayath in Thrissur district has started organic cultivation of rice in an area of 2,500 acres, promoting integrated farming system, which is known as Adat model. Similarly Marappanmoola in Wayanad has another model organic farming system involving hundreds of farmers (GoK, 2008). Marketing of organic produce is also being experimented in many places like Organic Bazaar in Thiruvananthapuram, Eco-shops in Thrissur and Kozhikode and, Jaiva Krishi Sevana Kendram in Kannur. Self help groups of women are encouraged to undertake organic farming of vegetables in many panchayats of the state.

5. Looming Perplexity and Need for in-depth Studies

Government of Kerala envisages converting the state into an organic state in a phased manner through a gradual approach. A rapid, superficial shift could lead to failure jeopardizing the possibility of continuing with the vision on organic agriculture. Therefore, it is essential to generate knowledge on select areas to move forward to other regions with the organic farming agenda. The lessons from such exploratory studies are going to be crucial and can be generalized and applied to other regions where conditions are almost similar. Moreover, such studies would clarify the possibilities of organic farming in similar regions as well as generate information on specific problems and constraints that organic agriculture could face, and possible solutions. Considering the eco-climatic-cultural variations in Kerala's agriculture, so far there has not been any serious effort/study taken up to understand the regional dynamics of principles and practices of organic farming in the state. The micro-level in-depth study is of paramount importance, since it is almost impossible to derive a generic model of organic farming for the entire state. As of now indubitably there exists an asymmetric understanding and knowledge gap among the policymakers, farmers, institutional agencies and other stakeholders' regarding the implementation, adoption, economic viability and sustenance options of the organic farming. As a matter of fact if organic agriculture were to be recognized as a sustainable alternative, it would require enormous support from the state for its propagation and implementation.

6. Farmer's Perception and Field Level Reflections

Although majority of farmers were aware of the fact that Kasaragod is officially declared as an organic district they were not aware of the detailed concept and practice of organic farming with respect to the crops and enterprises they managed. The institutional void is evident from the fact that many of the farmers have not even heard about the government initiative in this regard. It assumes even more importance especially when government is gearing up to declare the entire state as organic. It was revealed that most of the farmers were aware about the extension programmes for popularizing organic farming implemented by the Department of Agriculture. However, only 56 per cent of the farmers have attended such awareness programmes. Further, about one third of the farmers (34 per cent) did not possess adequate knowledge about the organic farming practices recommended for the crops they cultivated. Majority (74 per cent) of the farmers were not aware about organic certification procedures and none of them obtained organic farming certification yet though 5 per cent of the farmers made some efforts for securing the same.

According to the farmers, major constraints in adopting organic farming included lack of knowledge about the organic farming practices recommended for the crops they cultivated, extension support to local market facility, non availability of quality organic inputs, high cost of inputs, lack of availability and high wage rate of labor, *etc.* the major hindrance is the underdeveloped or rather undeveloped market for organic products. Infact in the absence of a well evolved organic market and pegged up prices it would be very difficult to sustain the organic farming even though we attempt forced institutional measures. The empirical evidence reflects that as of now around 43 per cent farmers in the district apply organic nutrients (exclusively) and in the case of coconut 82 per cent farmers apply organic manures (Table 13.1). These figures indicate that the state intervention in this regard is in the right direction and in near future.

Major constraints as perceived by farmers in adopting the eco friendly organic methods of pest and disease management in crops were lack of knowledge about recommended practices, lack of availability of eco friendly organic pesticides and bio inputs, high cost of bio inputs and lack of extension support for adopting appropriate methods of pest and disease management. It is interesting to know that even though plant protection chemicals are banned in Kasaragod district, many farmers are able to get the chemicals from adjoining localities in the neighbouring district of Kannur and Dakshina Kannada districts of Karnataka.

After the commencement of the organic farming initiates by the government there is an increased awareness among the farmers regarding the ill effects of indiscriminate and unscientific use of chemical pesticides and the need to shift towards eco-friendly organic practices. However, majority of the farmers are not

Table 13.1: Crop-wise details of Nutrient Management Practices Adopted by Farmers

Sl.No.	Crop	Total No. of Farmers	Farmers Applying only Organic Manure		Farmers Applying only Chemical Fertilizers		Farmers Applying both Organic Manure and Chemical Fertilizers	
			No.	Per cent	No.	Per cent	No.	Per cent
1.	Coconut	50	41	82	0	0	9	18
2.	Arecanut	35	27	77	0	0	8	23
3.	Rubber	30	6	20	3	10	21	70
4.	Banana	30	2	7	0	0	28	93
5.	Rice	30	3	10	2	7	25	83
6.	Cucurbits	15	7	47	0	0	8	53
7.	Bhindi	10	2	20	0	0	8	80
8.	Cowpea	15	5	33	0	0	10	67
	Total	215	93	43	5	2	117	55

properly educated about the appropriate organic farming practices to be adopted for raising different crops. Further, the inputs recommended for crop and soil health management under organic system are not available in the required quality and quantity. This situation has given opportunities for unscrupulous elements to flood the market with products of questionable properties. Many a times the organic inputs supplied by various agencies are found to be inferior in quality. Farmers are of the opinion that a continuous and intensive quality control mechanism is required to prevent the exploitation by such elements.

Some farmers from the district procure the chemical pesticides from the neighbouring district i.e Kannur in the south and Mangalore town in the north when their crops are threatened with pests and diseases. According to these farmers they are forced to resort to this practice as there are no viable substitutes to manage pests and diseases to save the crops in the organic way, especially when the incidence is severe. The rationality/possibility of imposing restrictions on use of chemical inputs restricted to the farming community of a particular district while the same are freely available in the nearby localities of the adjacent district ie Kannur in the south and beyond Thalappady in the North in Dakshina Kannada district of Karnataka state is also quite debatable.

7. Conclusion

The shift to an organic region/state/district can be viewed as a complex process of transition which demands radical change in knowledge, skill and organizational pattern. It is more or less like participation in an alternative food chain and bears very high switch over cost. The major worry is the response time of the farmers and the effective formulation of alternatives by the institutional agencies. Moreover, farmers as of now are perplexed because of lack of concrete recommendations, and there exists an obvious apprehension on possible exploitation by intermediaries who supply the prohibited items at exorbitant prices. Although multitude of governmental programmes on organic farming is in pipeline, the air of ambiguity is very much prevalent regarding the implementation of various programmes. However, at the same time increasing health consciousness and increasing disposable income among Indians is ceaselessly increasing the demand for organic food. Therefore, strong national organic policy is the main need of the current position which will give an important place to organic farming addressing the current issues and obstacles. Government needs to do a meticulous and in-depth evaluation of the general picture of the organic sector policies, programmes and plans to understand how they affect the current organic sector. An action plan for the organic sector should be developed based on the analysis of the state of the sector, participatory consultations, a need evaluation and proper sequencing of the actions.

Selected References

Balachandran, V. (2005). Future in the Past: A study on the status of organic farming in Kerala. *Discussion Paper 82*. Kerala Research Programme on Local Level Development, Central for Development Studies, Trivandrum 111p.

Das Kasturi. (2007). Towards a smoother transition to organic farming, *Economic and Political Weekly*, **42**(24): 2243-2245.

FiBL- IFOAM Survey, (2015): Organic Agriculture Worldwide: Current Statistics, 2014 Helga Willer, Research Institute of Organic Agriculture (FiBL), Frick, Switzerland.

Government of Kerala. (2008). *Kerala State Organic Farming Policy: Strategy and Action Plans*. 24p.

Sanghi, N.K. (2007). Beyond certified organic farming: An emerging paradigm for rainfed agriculture, *Proceedings of the National Workshop on New Paradigm for Rainfed Farming: Redesigning Support Systems and Incentives*, 27-29 September, IARI, New Delhi.

Lanting, H. (2007). Building a farmers owned company (Chetana) producing and trading fair trade–Organic products, Proceedings of the National Workshop on New Paradigm for Rainfed Farming, published by WASSAN, New Delhi.

Index

Figure 2.1: Coconut Nursery Seedlings.
p.31

Figure 2.2: Coconut Poly Bag Seedlings.
p. 31

Figure 2.3: Coconut Leaves as Mulch.
p. 35

Figure 2.4: Mulching Coconut Basin
using Husk. p. 35

Figure 2.5: Mulching Coconut Basin
using Coir Pith. p. 36

Figure 2.6: Catch Pit with Pineapple on
Bund. p. 37

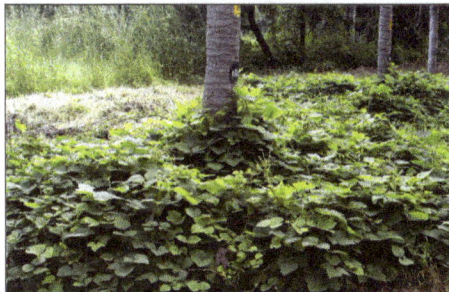

Figure 2.7: Green Manure Crop in
Coconut Basin. p. 39

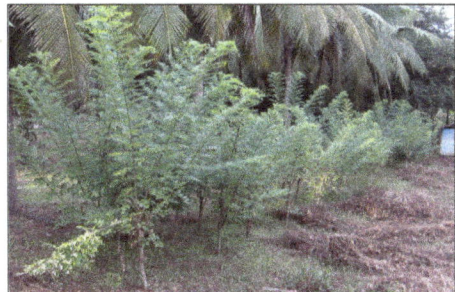

Figure 2.8: Growing Glyricidia as Green
Manure Crop. p. 40

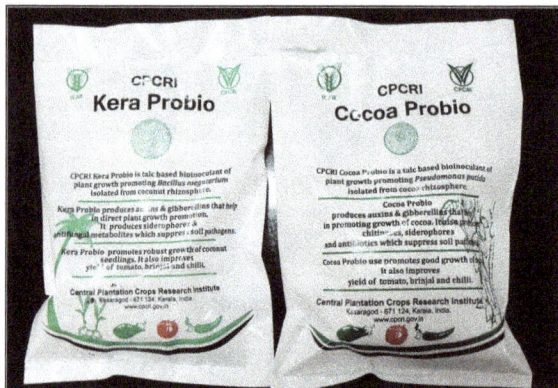

Figure 2.9: Kera Probio and Cocoa Probio. p. 43

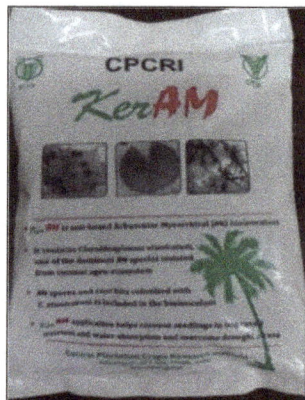

Figure 2.10: CPCRI 'Ker AM'. p. 43

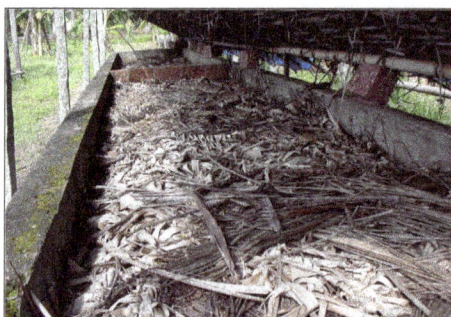

Figure 2.11: Vermicomposting in Tanks.
p. 48

Figure 2.12: Kalpa Organic Gold.
p. 48

Figure 2.13: Multiplication of Earthworm.
p. 49

Figure 2.15: Kalpa Soil Care. p. 51

Figure 2.14: Collection of Vermiwash. p.
50

Figure 2.16: Biochar from Coir Pith. p. 52

Figure 2.17: Biochar from Tender Nut Waste. p. 52

Figure 2.18: Cocoa as Mixed Crop with Coconut. p. 54

Figure 2.19: Coconut-based High Density Multispecies Cropping System. p. 55

Figure 2.20: Mixed Farming in Coconut Garden with Dairy Cows. p. 56

Madhuramangala

Mangala

Mohitnagar

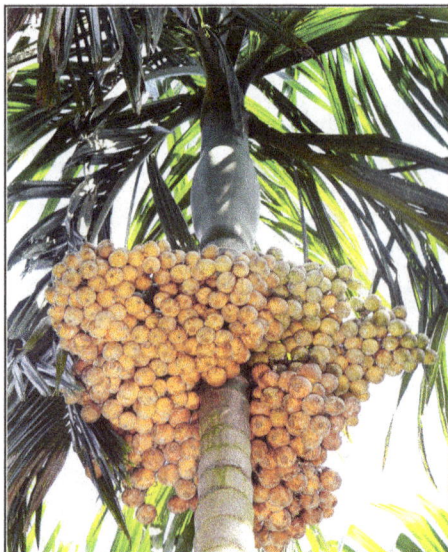

Nalbari

Figure 3.1a: Varieties of Arecanut. p. 71

Sreemangala

Sumangala

VTLAH 1

VTLAH 2

Swarnamangala

Figure 3.1b: Varieties of Arecanut. p. 72

Figure 3.2: Hybrids of Arecanut. p. 73

Figure 3.3: Arecanut Nursery Seedlings. p. 75

Figure 3.4: Cocoa as Mixed Crop in Arecanut Field. p. 77

Figure 3.5: Arecanaut Leaves for Recycling. p. 79

Figure 3.6: Harvested Mature Nuts. p. 82

Figure 3.7: Spreading Harvested Nuts for Drying. p. 85

Figure 4.2: Cocoa Nursery Seedlings. p. 102

Figure 4.3: Cocoa as Mixed Crop in Coconut Garden. p. 103

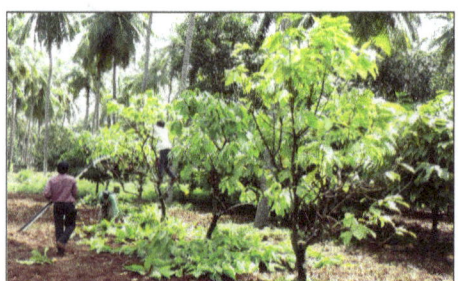

Figure 4.4: Pruning of Cocoa Plants. p. 109

VTLCC 1

VTLCH 1

VTLCH 2

VTLCH 3

Figure 4.1a: Cocoa Varieties. p. 97

VTLCH 4

VTLCS 2

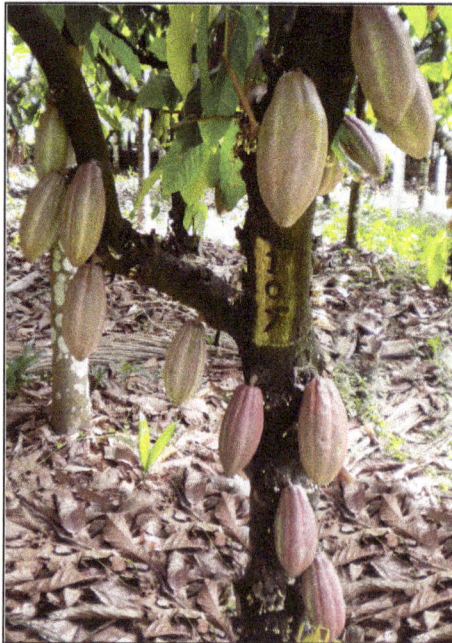

VTLCS 1

Figure 4.1b: Cocoa Varieties. p. 98

Figure 4.5: Harvested Mature Pods. p. 117

Figure 4.6: Fermentation of Beans. p. 119

Figure 4.7: Drying of Beans. p. 120

Figure 5.1: Cashew Nut in different Maturity Stages. p. 125

Amrutha

Dhana

K22-1

Kanaka

Figure 5.2: Varieties of Cashew. p. 131

Figure 5.3: Soft Wood Grafts. p. 133

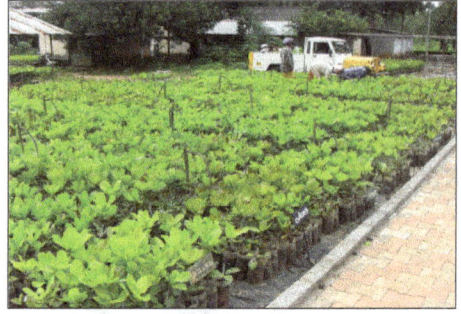

Figure 5.4: Cashew Nursery Plants
Ready for Planting. p. 133

Figure 5.5: Cashew Plantation. p. 136

Figure 5.6: Turmeric as Intercrop in
Cashew Plantation. p. 139

Figure 5.7: Pineapple as Intercrop in
Cashew Plantation. p. 139

Figure 5.8: Rejuvenation after Top
Working. p. 141

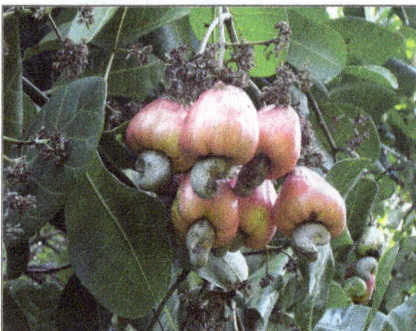

Figure 5.9: Mature Nuts Ready for
Harvest. p. 154

Figure 5.10: Cashewnuts Ready for
Peeling. p. 155

Figure 5.11: Cashew Apple. p. 156

Figure 5.12: Cashew Apple Pickle. p. 156

Figure 6.1: Robusta Coffee. p. 164

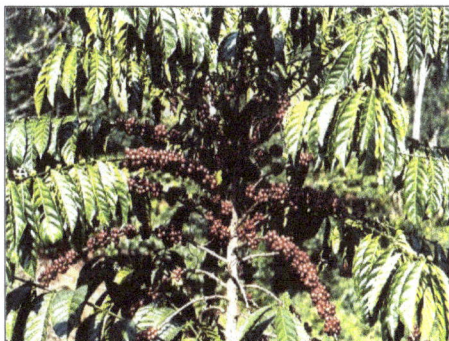

Figure 6.2: Arabica Coffee. p. 164

Figure 6.3: Coffee grown under Mixed Shade. p. 168

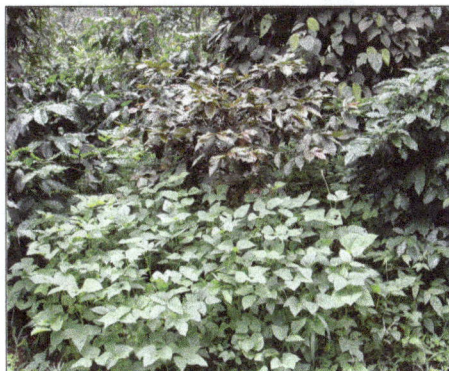

Figure 6.4: Cow Pea as Green Manure Crop. p. 169

Figure 6.5: Composting in the Coffee Plantation. p. 173

Figure 6.6: Composting Outside the Coffee Plantation. p. 173

Figure 7.1: General View of Tea Plantation. p. 193

Figure 7.2: Vegetative Propagation in Tea. p. 195

p. 201

Figure 7.4: Tea Field with Shade Trees. p. 202

Figure 7.5: A Tea Plantation Ready for Plucking. p. 211

Figure 8.1: Black Pepper Garden. p. 218

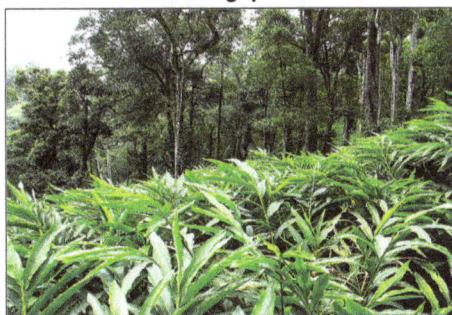

Figure 8.2: Cardamom Plantation. p. 225

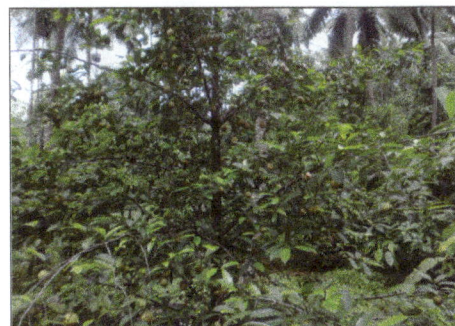

Figure 8.3: Nutmeg as Mixed Crop in Coconut. p. 233

Figure 8.4: Yielding Clove Plant. p. 239

Figure 8.5: Garcinia Yielding Plant. p. 241

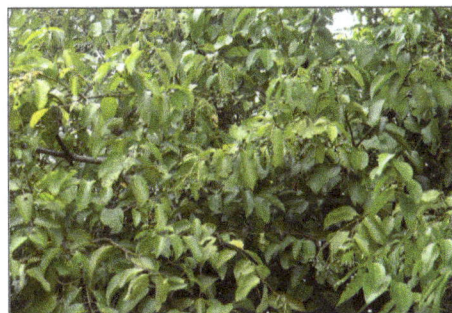

Figure 8.6: Cinnamon Plant. p. 244

Figure 9.1: Young Oil Palm Plantation
p.255

Figure 9.2: Yielding Oil Palm Tree. p. 256

Figure 9.3: Raising Seedlings in Primary
Nursery. p. 258

Figure 9.4: Raising Seedlings in
Secondary Nursery. p. 258

Figure 9.5: Growing Green Manure Crop
in Oil Palm Plantation. p. 262

Figure 9.6: Growing Coffee in Oil Palm
Plantation. p. 263

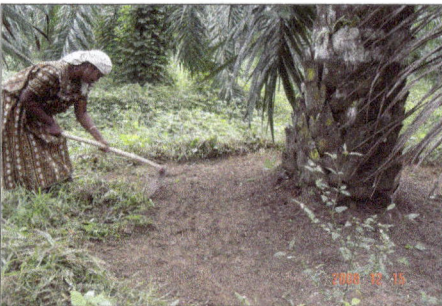

Figure 9.7: Weeding in Plant Basin.
p. 265

Figure 9.8: Ablation. p. 267

Figure 9.9: Harvested FFB for Processing. p. 278

Figure 10.1: *Goniozus nephantidis*, Larval Parasitoid of Coconut Black Headed Caterpillar. p. 285

Figure 10.3: Eriophyid Mite Infested Coconuts. p. 290

Figure 10.4: *Metarhizium anisopliae* Infected Grub of Rhinoceros Beetle. p. 293

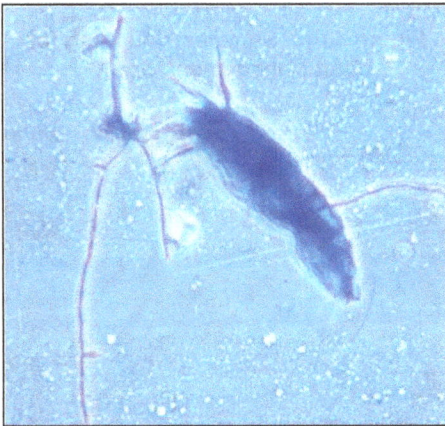

Figure 10.5: Coconut Eriophyid Mite Infected with *Hirsutella thompsonii*. p. 293

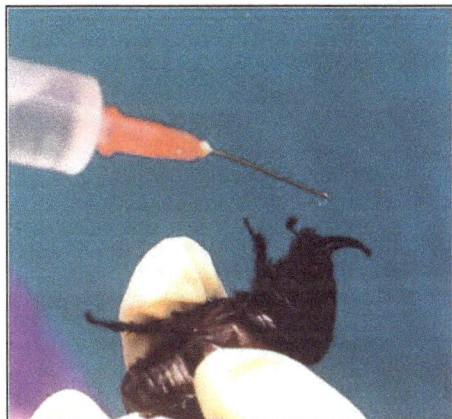

Figure 10.6: Inoculating Rhinoceros Beetle with Oryctes Rhinoceros Nudi Virus. p. 294

Figure 10.7: *Clerodendron infortunatum*
Plant. p. 298

Figure 10.8: *Clerodendron infortunatum*
Induced Malformation in Rhinoceros
Beetle. p. 298

Figure 10.9: Pheromone
Trap for Rhinoceros
Beetle. p. 299

Figure 10.10: Summer Ploughing for
Pest Management. p. 300

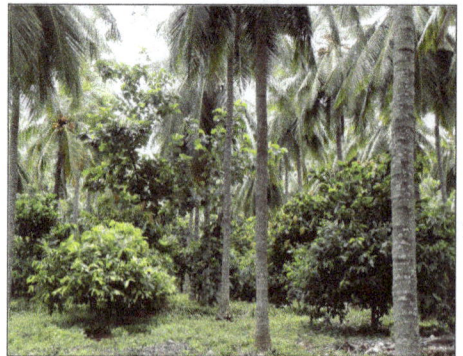

Figure 11.1: Mixed Cropping in Coconut
Garden. p. 308

Figure 11.2: Growing Leguminous Crop in Coconut Garden. p. 308

Figure 11.3: Mixed Cropping in Arecanut Garden. p. 309

Figure 11.4: Mulching using Cocoa Leaves. p. 309

Figure 11.5: Pruning of Cocoa Plants. p.310

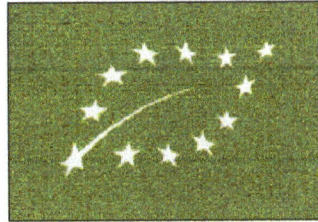

European Union Organic Logo p. 330-31

Figure 13.2: Global Market: Distribution of Sales Value of Organic Products by Countries. p. 335

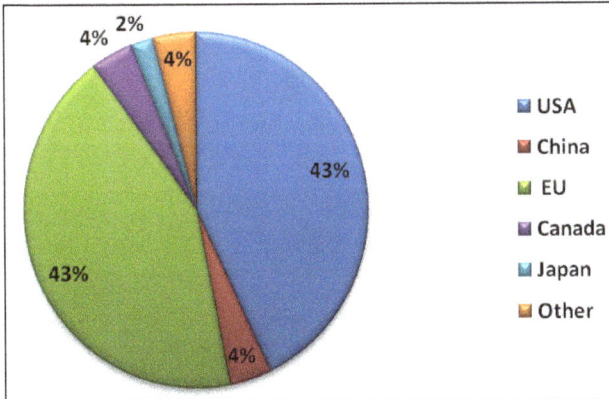

www.ingramcontent.com/pod-product-compliance
Lightning Source LLC
Chambersburg PA
CBHW050507190326
41458CB00005B/1462